NORTH CAROLINA
STAT
DEPT
ES

The Trans\vision Book of Health

The Trans\vision Book of Health

Encyclopædia Britannica, Inc.

Helen Hemingway Benton, *Publisher*

CHICAGO LONDON TORONTO GENEVA SYDNEY TOKYO MANILA

Library of Congress Catalog Card Number: 73-92150

International Standard Book Number: 0-85229-295-3

Printed in the United States of America

Preface

The major emphasis in this book is on the functioning of the human body—how it works and, sometimes, fails to work. The major aim of the book is to make better informed patients, and thereby to improve the relation between the patient and his doctor.

In line with that major emphasis, function rather than structure is emphasized throughout. *The Trans-Vision Book of Health* is not an anatomy book, strictly speaking. Considerations of structure are always secondary to considerations of function. Often function is better understood when structure is understood. But the main point is to know how things work when they are working right, and what happens when they start to work wrong.

In line with the major aim of the book—to make better informed patients—the authors of the various chapters have always kept in mind the question: "What would I really like my patients to know about themselves and their bodies when they come to me for help?" As is clear from the table of contents, a systemic approach has been taken in creating the book, with a chapter on each of the main systems of the body. Thus this question has applied, for each author, to the system that he was asked to cover. The book is successful if each chapter acquaints the reader with the normal function of the system with which it deals, and informs him about what can go wrong with it.

This does not mean that either the authors of the various chapters or the editors wish to encourage irresponsible self-diagnosis and self-treatment. The human body as a whole, and the various systems that make it up, is an enormously complicated—even miraculous—mechanism. A good physician will always know more than even the best-informed patient. Thus, for example, rather than providing information that would tempt a reader to a diagnosis of a specific skin disease, the author of Chapter 2 was asked to point out that an increase in redness of the skin is usually the result of enlargement of the surface blood vessels. There can be many reasons for this happening, most of them not at all serious. Similarly, there are perhaps a hundred possible causes of headache besides a brain tumor— each of them more likely than a brain tumor.

Nevertheless, authors were not asked to avoid mentioning common diseases. While the book emphasizes the normal and employs illustrations that emphasize health rather than disease, it is recognized that understanding the abnormal often helps us to understand the normal, and describing dysfunction helps to make function clear. Thus, authors were asked not only to give passing mention to typical diseases of the system they were covering but also to go into detail on a few of the most common ones. By *common* is meant not only statistically common—disease that many people have, like the common cold—but also ailments that, partly owing to public relations efforts, are much in the public consciousness, even if statistically quite rare.

Authors were asked specifically to do the following things and cover the following areas in their various chapters:

1. Develop a colorful introduction that gives an overall view of the subject matter and creates in the reader a desire to proceed further for more information.
2. Cover the assigned system completely. Limitations of space naturally forced decisions regarding priorities. It was felt that the ideal chapter would combine accuracy, brevity, and clarity. The authors achieved this combination to an extent very gratifying to the editors.
3. Emphasize the wonder of the human body. The goal was to win the reader's attention, to whet his curiosity, and to drive home a lasting appreciation of the functioning of his own physical equipment. In short, always try to be interesting—an attempt in which, the editors feel, the authors were eminently successful.
4. Remember that it is normal and to be expected for a patient to regress emotionally when he is faced with the prospect of disease, or the fear thereof. In the past, many physicians have tended to reinforce this kind of regression on the assumption—mistaken, the editors feel—that a childlike patient is easier to treat than an adult, aware patient. It may be easier to diagnose a childlike or unaware patient, but it is harder to treat him because successful treatment is partly an accommodation to disease or dysfunction.

We, the editors, do not want to give the impression that we think of all readers as patients. Just the opposite; most readers will not be patients, and we wish all readers the best of health, now and in the future. If nothing ever happens to you and you never get sick, we think you will enjoy this book because it can tell you a lot about how your body works. An if, unfortunately, you should suffer an accident or become ill, we hope it will help you to endure what must be endured and to get well again very soon and with as little pain and anguish as possible.

The Contributors

Samuel Bluefarb, M.D., contributor of chapter 2, "Skin and Appendages," is Professor and Chairman of the Department of Dermatology, Northwestern University Medical School. He attended college and medical school at the University of Illinois, from which he received an M.D. degree in 1937. Dr. Bluefarb is active in a number of professional groups. He is also the author of several books and more than a hundred scientific papers.

William J. Kane, M.D., who wrote chapter 3, "Musculoskeletal System," is Chairman of the Department of Orthopaedic Surgery at Northwestern Memorial Hospitals and Ryerson Professor of Orthopaedic Surgery at Northwestern University Medical School. After attending Holy Cross College, Dr. Kane received his M.D. degree from Columbia University's College of Physicians and Surgeons in 1958. In addition, he received a Ph.D. degree from the University of Minnesota in 1965 for his research into blood flow to bone.

Norman H. Olsen, D.D.S., co-author of chapter 4, "Dental System," is Dean of the Northwestern University Dental School, where he also holds the position of Professor of Pedodontics. Dr. Olsen received his undergraduate education from the University of Idaho and Creighton University, from which he acquired a D.D.S. degree in 1951. A fellow in the American College of Dentists, he has long been active in dental education and in community activities.

Charles R. Martinez, D.D.S., co-author of chapter 4, "Dental System," is Assistant Professor of Pedodontics at Northwestern University Dental School, the institution from which he received his D.D.S. degree in 1969. Dr. Martinez, who specializes in dentistry for children, has also performed dental research, for which he earned an M.S. degree in 1973. Prior to that time, he served in the U.S. Public Health Service's Indian Health Division.

David W. Cugell, M.D., who wrote chapter 5, "Naso-Respiratory System," is Bazley Professor of Pulmonary Diseases at the Northwestern University Medical School. A native of Connecticut, Dr. Cugell studied at Yale University and received his M.D. degree from the Long Island College of Medicine in 1945. Among the professional groups of which he is a member is the American Thoracic Society. Dr. Cugell has written at length on pulmonary conditions.

Sumner C. Kraft, M.D., author of chapter 6, "Digestion," is Professor of Medicine at the University of Chicago. After study at Tufts and Boston universities, he pursued his medical education at the University of Chicago, from which he received his M.D. degree in 1955. Dr. Kraft has written extensively in the field of gastroenterology. He is also a Fellow of the American College of Physicians.

Morton D. Bogdonoff, M.D., who contributed chapter 7, "Circulation," is Professor and Chairman of the Department of Medicine, the Abraham Lincoln School of Medicine of the University of Illinois. He received his M.D. degree in 1948 from the Cornell University Medical College. In addition to his academic duties, Dr. Bogdonoff is chief editor of the *Archives of Internal Medicine* and has served on the Presidential Advisory Panel on Heart Disease.

Hau C. Kwaan, M.D., author of chapter 8, "Blood and Lymph," is Director of the Hematology, Oncology, and Thrombosis Research Laboratories of the Veterans Administration Research Hospital in Chicago. A past resident of Hong Kong, Dr. Kwaan received the equivalent of an M.D. degree from the University of Hong Kong in 1952 and the equivalent of a Ph.D. degree from the same institution in 1958. He has written extensively on blood problems and is a charter member of the International Society on Thrombosis and Haemostasis.

Murray L. Levin, M.D., contributor of chapter 9, "Renal System," is Chief of the Renal Section of the Veterans Administration Research Hospital in Chicago. He is also Associate Professor of Medicine at Northwestern University. Dr. Levin did his undergraduate work at Harvard College and received his M.D. *cum laude* from Tufts University School of Medicine. He is also an executive officer of the American Federation of Clinical Research.

Michael Newton, M.D., author of chapter 10, "Sex and Pregnancy," is Professor of Obstetrics and Gynecology at the Pritzker School of Medicine, University of Chicago. Born in Great Britain, he received academic degrees from Cambridge University and an M.D. degree from the University of Pennsylvania in 1943. Coeditor of a medical textbook, Dr. Newton has also written many scientific articles. Among the many professional societies of which he is a member is the American College of Obstetricians and Gynecologists.

Will G. Ryan, M.D., contributor of chapter 11, "Endocrine System," is Associate Director of the Section of Endocrinology and Metabolism, Presbyterian-St. Luke's Hospital in Chicago and Associate Professor of Medicine, Rush Medical College. He received his M.D. degree from the Baylor University College of Medicine in 1956. After medical service in the U.S. Air Force, Dr. Ryan became a board-certified specialist in internal medicine.

Nicholas A. Vick, M.D., contributor of chapter 12, "Nervous System," is Assistant Professor of Neurology at the Pritzker School of Medicine. He also serves as Director of the Muscular Dystrophy Clinic, University of Chicago Hospitals and Clinics. Dr. Vick studied at the University of Michigan and received his M.D. degree from the University of Chicago School of Medicine in 1965. One of his research interests has been virus-caused brain tumors.

Roy R. Grinker, Sr., M.D., author of chapter 13, "The Integrative System: The Psyche," is Director of the Institute for Psychosomatic and Psychiatric Research and Training at Michael Reese Hospital in Chicago. After receiving his M.D. degree from Rush Medical College in 1921, Dr. Grinker did postdoctoral work at several European schools, including the University of Vienna. He has performed editorial work for psychiatric journals, served on government committees for mental health, and written several hundred scientific papers or books.

Table of Contents

At the end of each of the chapters on the various systems of the body (that is, Ch. 2-13), the reader will find references to specific articles in the Macropædia of *Encyclopædia Britannica* that discuss in depth various aspects of the subject matter covered in the chapter, as well as references to topics in the Outline of Knowledge (the Propædia) where other material in the Macropædia is cited.

The following general articles in the Macropædia are relevant to the overall theme and aims of this book and are therefore listed here rather than at the end of particular chapters.

AGING, HUMAN
ALLERGY AND ANAPHYLACTIC SHOCK
ATROPHY
BIRTH DEFECTS AND CONGENITAL DISORDERS
BODY CAVITIES AND MEMBRANES, HUMAN
BURNS
CANCER
CHILDHOOD DISEASES
COMPRESSION AND DECOMPRESSION INJURIES
DEHYDRATION
DIAGNOSIS
DISEASE, HUMAN
EXERCISE AND PHYSICAL CONDITIONING
FROSTBITE
GENETICS, HUMAN
HEALTH, HUMAN
IMMUNITY
INFECTIOUS DISEASES
INFLAMMATION
MOTION SICKNESS
NUTRITION AND DIET, HUMAN
QUARANTINE AND ISOLATION
RADIATION INJURY
SHOCK, PHYSIOLOGICAL
THERAPEUTICS
TRANSPLANTS, ORGAN AND TISSUE
WOUND

Under polarized light and a microscope, urate crystals—salts of uric acid—display vivid colors. Uric acid is one of the products of the body's use of biochemicals in its daily activities. Photo, Carroll H. Weiss, RBP

Mr. Charles Van Doren (left) and Dr. Richard H. Kessler (right) participated in the frank dialogue maintained in the opening chapter.

1

A Dialogue Between a Layman and a Physician

The editors of this book considered many different candidates for its first chapter. They thought of asking a prominent physician to write a statement about the book's meaning, purpose, and importance to its readers. They also thought about asking a layman, a potential patient, to write down his ideas about the physician-patient relationship—a relationship that, more than any other goal, the editors hoped to improve by publishing the volume. In the course of these deliberations, two of the editors, Dr. Richard H. Kessler and Mr. Charles Van Doren, met several times to talk about the problem. The more they talked, the more it became apparent to them and to their associates that the best way to make the points that needed to be made was to transcribe, more or less verbatim, the actual conversation they were engaging in. Hence this *Dialogue*—which serves not only to begin this book but also, it is hoped, to throw light on some problems and difficulties that concern all patients as well as physicians when they meet to discuss illness or injury, the treatment of it, and the prospects (or lack of them) for cure.

Dr. Kessler is Associate Dean of the Medical School of Northwestern University in Chicago. He has had a very wide experience in medicine—as practicing physician, medical researcher, and hospital and medical school administrator. He is above all what might be called a philosophical doctor, one who is deeply concerned not only with the mechanics of medicine but also with its human—its emotional and psychological—aspects. Mr. Van Doren is a writer and editor and an executive of Encyclopædia Britannica. But it is in his capacity as a layman—a non-physician—that he joins in the conversation. As such, he is like all other laymen—all other persons who are not trained in medicine and do not possess a medical degree. He is thus the representative of the reader.

Van Doren. *I hate it when I have to see a doctor. Is that a feeling that you often find in patients?*

Kessler. *In all patients.*

Van Doren. *Even so-called hypochondriacs—people who are supposed to like going to see their doctor?*

Kessler. *Yes, even in them. In their case it's hidden, of course. On the surface they seem to be almost glad that they're sick, so that they can describe their symptoms to their physician. But underneath this apparent gladness is the same dislike—and probably fear—that I assume you experience yourself, that leads you to say you hate to visit your doctor.*

"My physician can't feel what I feel. He can perhaps understand my pain, but he can't feel it himself . . . this privacy of pain seems to put me at a disadvantage . . ."

Van Doren. *What we're dealing with—the doctor and I—when I go to see him about some medical problem is my own body, not his. My body is enormously important to me—as everybody's is to him or her. And I suppose I feel that my body can't be of similarly great importance to my physician. My physician can't feel what I feel. He can perhaps understand my pain, but he can't feel it himself. I'm the one who is suffering, not he. And this privacy of pain seems to put me at a disadvantage and makes me awfully uncomfortable—on top of the actual suffering!*

Kessler. *I understand what you mean, for of course I—like every physician—am sometimes a patient myself. But I want to correct what may be a misapprehension on your part. The patient's body is terribly important to the physician, too. For example, if my primary self-description—what I mean to myself—is as a healer, then I have a primary and deep concern about the pain and anguish that you, the patient, are suffering. And not only the pain from a disease or a wound. I mean also the anguish that I realize you are suffering because you had to seek my help.*

Van Doren. *Yes, when I go to a doctor I'm admitting that I'm helpless. There's something I can't do for myself.*

Kessler. *Physicians understand that perhaps better than you realize. But there's something else that's very important. All physicians identify more or less with their patients—vicariously share their problems, fears, and pains. They are to varying degrees sympathetic, which is an emotional response, or empathetic, which is an intellectual response. However, to preserve his own self-description—as a healer—the doctor has to limit his emotional involvement.*

Van Doren. *You mean the doctor feels the patient's pain more than he admits?*

Kessler. *I'm sure good doctors do. It's necessary to do so. This is an ancient problem, which was perhaps best articulated by Hippocrates, the Greek "father of medicine" who lived in the 5th century* B.C.

Van Doren. *The author of the Hippocratic Oath.*

Kessler. *Yes, but I have something else in mind. Hippocrates, who was an excellent doctor and a wise man, was probably the first to see that an emotional tie between a patient and his physician can be an obstacle, rather than an aid, to rational* *and effective therapy. He therefore proposed the system that we now call "professional courtesy"—an agreement among physicians such that they care for one another's families, and never their own.*

Van Doren. *That's still done, isn't it? I suppose I always thought it was to avoid paying fees!*

Kessler. *You're too cynical. Fees are avoided—but that's not the purpose of professional courtesy. Strong emotions—the kind that a physician would have about his wife or his child—interfere with intellectual function. Although we all—physicians as much as laymen—want sympathetic care when we're sick or hurt, we also want a healer who is able to maintain command of his intellect.*

Van Doren. *As with so many fundamental questions, then, there are two sides to this. In the physician-patient relationship the key is to combine empathy with intellect—to feel and demonstrate appropriate concern without the excess of emotional involvement that impedes performance.*

Kessler. *Exactly. If I were your doctor, and you were ill, I think you would prefer me to be cool and collected, to act as if I knew precisely what had to be done for you, and as if I were able to do it. This might to some extent be an illusion; I might not be as certain as I seemed to be. But you would want me to maintain that illusion. And maintaining it would be good for you. It would, to some extent at least, help you to get well.*

Van Doren. *Because it would give me confidence, and confidence is a part of successful therapy?*

Kessler. *Yes. And in saying that, you're touching on a terribly important fact about medicine. We should never forget that medicine is a very special art. Most arts deal with inanimate matter—the paint and canvas of the painter, the stone of the sculptor, the brick and mortar of the architect. But the physician—he shares this with the farmer and the teacher—deals with a living thing. The farmer helps plants and animals to grow, he doesn't make them grow. The teacher helps minds to learn, he doesn't make them learn. It's the student who does the learning. Similarly with the physician. He can only help the patient to get well—and sometimes he doesn't help much. But it's the patient who gets well—or his body.*

Van Doren. *Sometimes, of course, there's no hope for cure. And that brings up a subject that I hesitate to mention, but that I very much want to ask you about nevertheless. I'm not prepared to say—in fact I don't really know—the exact nature of the connection between the mind and the body. I certainly know very little about the connection between my mind and my body. Yet I'm certain there is a strong connection. For example, I'm pretty sure that if I didn't possess a body—if my body died—I wouldn't have a mind, either. Or at least the same kind of mind. My body, therefore, in some sense is me; I wouldn't be me if I didn't possess my body—or, again, I would be a very different me. And yet I know that sometime, sooner or later and in some place not yet known to me, some physician will tell me that my body is dying or is soon to die. When that happens, another human being who is very different and distant from me because he is a physician will tell me, in effect, that I will shortly cease to exist. My doctor will impose a sentence of death upon me. Is there any wonder that I fear him?*

Kessler. *You are correct in thinking that the relationship between patient and physician is highly emotional—more so for the patient than for the physician. There are several elements in this emotional charge and you touch on one when you speak of your feelings about your own body—particularly when you identify the physician as the bearer of the tidings of your death. Even the most trivial illnesses remind us of our fragility, of our mortality, which few of us—except, strangely*

"If I were your doctor, and you were ill, I think you would prefer me to be cool and collected, to act as if I knew precisely what had to be done for you, . . ."

Gross Anatomy of the Male

1. Aorta, 5, 6, 11, 12, 15
2. Appendix vermiformis, 4, 13, 14
3. A. brachialis, 5, 6, 15
4. A. carotis communis, 5, 15
5. A. femoralis, 5, 6, 15
6. A. iliaca communis, 5, 6, 15
7. A. iliaca externa, 5, 6, 15
8. A. iliaca interna, 5, 6, 15
9. A. lienalis, 5
10. A. mesenterica inferior, 5, 6, 15
11. A. mesenterica superior, 15
12. A. pulmonalis (and branches), 5, 12, 15
13. A. renalis, 6, 15
14. A. subclavia, 5, 6, 15
15. Atrium sinistrum, 12
16. Bronchus principalis, 5, 15
17. Cartilago thyreoidea, 3
18. Cerebellum, 12, 15
19. Cerebrum, 12, 15
20. Clavicula, 1, 2, 7, 8, 9, 10, 15
21. Colon ascendens, 3, 4, 13, 14
22. Colon descendens, 3, 4, 13, 14
23. Colon sigmoideum, 3, 4, 13, 14
24. Colon transversum, 3, 4, 12
25. Concha nasalis inferior, 4, 7, 12, 15
26. Concha nasalis media, 4, 7, 12, 15
27. Concha nasalis superior, 4, 7, 12, 15
28. Corpus callosum, 12, 15
29. Cranium, 2
30. Diaphragma, 2, 3, 4, 5, 6, 10, 11, 12, 15
31. Ductus choledochus (common bile duct), 4, 5, 11, 12, 14
32. Duodenum, 5, 13, 14
33. Falx cerebri, 13, 14
34. Femur, 7, 8
35. Fossa ovalis, 9
36. Glandula lacrimalis, 2
37. Glandula parotis, 1, 2
38. Glandula submandibularis, 1, 2
39. Glandula suprarenalis (adrenal gland), 5, 6, 15
40. Glandula thyreoidea (thyroid gland), 3, 4, 11, 12
41. Glans penis, 1
42. Hepar (liver), 3, 4, 10, 11, 12
43. Humerus, 7, 8
44. Intestinum tenue, 3, 4, 11, 12
45. Larynx, 4, 7, 12, 15
46. Lien (spleen), 5, 13, 14
47. L. inguinale, 1, 2, 3, 5, 6, 9, 14, 15
48. Lingua (tongue), 3, 4, 5, 6, 7, 12, 15
49. Mandibula, 1, 2, 3, 4, 5, 6, 7, 8, 12, 15
50. Maxilla, 2, 3, 4, 5, 6, 7, 8, 12, 13, 14, 15
51. Medulla oblongata, 12, 15
52. Medulla spinalis, 7, 15
53. M. biceps brachii, 5, 6, 11, 12
54. M. deltoideus, 5, 6, 9, 10, 15, 16
55. M. latissimus dorsi, 16
56. M. masseter, 1
57. M. obliquus externus abdominis, 1, 9, 16
58. M. pectoralis major, 1, 2, 5, 9, 10, 15
59. M. psoas major, 7, 8, 15
60. M. quadriceps femoris, 3, 4, 5, 6, 15
61. M. rectus abdominis, 1, 2
62. M. serratus anterior, 1, 9
63. M. sternocleidomastoideus, 1, 2, 9, 10, 11, 16

enough, many of the elderly—are able to contemplate without fear and anxiety. Patients almost always approach their doctor afraid that he will find something terribly wrong with them—that he will discover cancer or some other dread disease.

Van Doren. *And therefore I am not only fearful but also angry at you, my doctor, my executioner.*

Kessler. *You're exaggerating when you say that, although I understand what you mean—because, remember, I have been a patient myself and will be again. What you may be forgetting, however, is that when you have such feelings you're probably regressing emotionally.*

Van Doren. *In recognizing my dependency on you, the doctor, I'm recalling a condition of real dependence—my childhood?*

Kessler. *Yes. And then you may take the next step—many patients do—and act like a child, making me your temporary father. These are hardly circumstances for a satisfactory doctor-patient relationship!*

Van Doren. *But more easily described than avoided.*

". . . my ability to diagnose your condition — and therefore to cure it — depends largely on the quality of the (medical) history you provide."

Kessler. *I should add that the fault doesn't lie exclusively with the patient. Some physicians delight in the role of pseudo parent and even encourage—directly or, more likely, indirectly and unintentionally—a kind of childishness in their patients.*

Van Doren. *Doctor-parents are bigger, stronger, and wiser than patient-children.*

Kessler. *Even if only in fantasy.*

Van Doren. *But that, too, might conceivably be an effective—ritual, shall we call it?—to go through. Getting back to the confidence that the patient needs to feel, I suppose any way he can come by it is, in the end, all to the good.*

Kessler. *Not always. Your anger at being sick (Why me?) may be expressed in ways as irrational as the anger itself, and in ways that are ultimately self-destructive. The size of my fee, the time you have to spend in my waiting room, your disagreement with my diagnosis or my program of therapy —all these may form a barrier between you and me that may adversely affect your symptoms.*

Van Doren. *Make me feel worse instead of better, which is just the opposite effect from the one I wanted.*

Kessler. *It is sometimes so. Look at the way in which emotions affect pain. How many times have you failed to notice, if you were engaged in a particularly invigorating and enjoyable activity, that you had cut your hand? Your hand is one of the most pain-sensitive structures in your body, yet you were so distracted by the occupation that you weren't even aware of the injury. Now recall tiny wounds—for example, the pinprick by the nurse in my office to obtain a blood sample—that hurt like the devil. The emotional setting is a determinant of the severity of pain.*

Van Doren. *You're right. I cut my hand while I was playing tennis the other day and didn't even notice it until I saw the blood on the racket. Even then it didn't hurt.*

Kessler. *You were winning.*

Van Doren. *How did you know?*

Kessler. *If you were losing it would have hurt.*

Van Doren. *Am I so—childish, I guess is the word?*

Kessler. *It's not childish, it's only human. Your fear and anger, too, affect your relationship to me, as your doctor. For one thing, you're liable to forget or minimize important symptoms. You're not deliberately hiding the facts from me, but involuntarily suppressing aspects of your condition that you consider—perhaps wrongly—to be life-threatening. You see how important this is when you realize that my ability to diagnose your con-*

dition—and therefore to cure it—depends largely on the quality of the history you provide. If you can't tell me all of the details of how you feel, you limit the information I require to give you optimum care.

Van Doren. *I presume that good doctors are aware of this tendency.*

Kessler. *Of course. A good part of my task, as your physician, is to assess your fears and help you to overcome these and other strong emotions that can cause you to hide facts about yourself from yourself and, therefore, from me. Your task in this relationship, stated in broadest terms, is to know your own attitudes, your anxieties about and your anger toward disease.*

Van Doren. *This doesn't mean, does it, that I should cease to have emotions? I couldn't, even if I should.*

Kessler. *Of course not. The emotional set that you have toward your body is a part of you that will probably always be present. And it should be. Emotions give color to life. What I'm trying to stress is that your excessive or inappropriate attitudes toward disease may interfere with our relationship. They will accentuate and distort how you perceive and describe your symptoms. They will stimulate you to find reasons in* me *for your anger and will encourage you to seek a surrogate father in me.*

Van Doren. *But you mentioned before that some physicians encourage that in their patients.*

Kessler. *Naturally, I'm assuming that I and other physicians are emotionally wise. In fact, if I am deficient in emotional maturity or self-knowledge, then to some extent I will be inadequate as a physician.*

Van Doren. *I admire you for your frankness, which makes me all the more certain that you are not deficient in that way.*

Kessler. *I may be, nevertheless. I do know that I believe very deeply that it is an important part of my doctoring role to educate you, as a patient, where I think knowledge of yourself, both your body and your mind, are to the benefit of your health.*

Van Doren. *Unfortunately, not all doctors agree with you about that.*

Kessler. *No, and I can only guess at their reasons. Some physicians may prefer to have unquestioning, childlike patients to satisfy their own personal needs. Many believe—and to some extent truly, I think—that a little bit of knowledge can be a dangerous thing, leading not so much to self-diagnosis as to attempts at self-treatment. Others are so busy attending to the needs of their patients that they haven't time—or feel they don't—to offer explanations in any but the most cursory way.*

Van Doren. *I know you think that is a short-sighted view.*

"The educated patient . . . (who) understands something of the way his body works—is the better patient . . ."

Kessler. *Most emphatically so. The educated patient—one who, on the one hand, recognizes his susceptibility to unfounded fear and anger, and, on the other hand, understands something of the way his body works—is the better patient from several points of view. He presents himself for help sensitized to the way in which his body and mind function. He understands more about how doctors function. He knows the diagnostic value of the history of his symptoms. Given the opportunity to use his greatest gift, his intellect, he can assist me, the doctor, in helping him. He becomes more of a partner in solving the problems of his disease, responding to diagnostic and therapeutic suggestions in a rational, intelligent fashion rather than in fear, anger, and ignorance. That's the kind of patient I want and can help the most.*

Van Doren. *And that's the kind of patient I want to be. It's also the reason we've made this book.*

Kessler. *Exactly. And joined in this dialogue.*

Van Doren. *We might stop there. Yet I want to tell you something that may or may not surprise* *you. I studied biology, and chemistry, and even a little physiology, in school and college. I already know something about how my body works. I know, for example, that there are many bones in my hand, and many blood vessels through which my blood flows, and many nerves, and that the whole is covered with what I have learned to call a marvelous "organ," my skin.*

Kessler. *Right so far. But what are you leading up to?*

Van Doren. *Just this. Despite all this more or less superficial knowledge—I would admit that it's more rather than less superficial—when I hold my hand up before my face, and look at it, I don't understand it. No matter how long or hard I stare at it, it remains a mystery to me. And then take my heart. I think I have a heart—*

Kessler. *I can guarantee that you do.*

Van Doren. *It's funny to say it, but I'm serious, too. Of course I'm certain that I have a heart. Yet the fact remains that I've never seen my heart, or felt it, or smelled it, or tasted it. All my knowledge comes through my senses, yet the most important organ in my body is beyond my power to sense.*

Kessler. *You've heard it. You've felt your pulse.*

It's funny to say it, but . . . (although) I'm certain that I have a heart . . . I've never seen my heart, or felt it,. . . the most important organ in my body is beyond my power to sense."

Van Doren. *It may seem strange that I say this, but those are—to me—only secondary effects. They're derived conclusions, from books, not experience. Yes, I've heard and felt something thumping in my chest. I suppose it's my heart. Yet I don't really know it. Does what I'm saying make any sense?*

Kessler. *Yes and no. Yes, because I too have never seen my heart. I must take it on faith, just like you, that I have one—a healthy one. No, because I know—I really know—that I have a heart, and that you do, too. I know what my heart looks like, even though I've never seen it. I know how it functions, how long it's likely to function, and what can go wrong with it. If something does go wrong, I'll realize that pretty quickly.*

Van Doren. *Will I, a layman? Can I? I want to know what you know.*

Kessler. *You can't know everything that I know, just as I can't know everything that you know about literature, for example. But if you're willing to listen, I can tell you a lot.*

Van Doren. *Where shall we start?*

Kessler. *Let's begin with your body as a whole. I don't mean just your body, of course, but the human body—both male and female. Earlier in the chapter were two Trans-Visions, one of the adult male, the other of the adult female. I want to say something about them, not only for your benefit, Charles, but also—and more especially—for the benefit of the reader of this book.*

• • •

The reader is urged not to study the Trans-Visions after the fashion of a medical student. They will be most valuable when referred to within the context of material you read in this book. On many occasions a comprehension of the anatomical relationship between certain structures of the body will add importantly to your understanding of both normal and pathological, or diseased, function. For example, that the heart is separated from the stomach by a thin layer of diaphragmatic muscle certainly explains a person's confusion between the pain of an upset stomach and a heart attack.

Trans-Visions of the male figure stress the relationships of the major organs and organ systems. In these, the skeleton and other supportive structures of the body are emphasized. The illustrations of the female figure offer more detailed views of the chest (thorax), abdomen, and skull. Because the male and female Trans-Visions complement and supplement each other, you will find frequent cross-references to comparable views of particular value.

It is our purpose to facilitate your grounding in basic anatomical relationships by emphasizing the whole rather than regional structures. Using Trans-

"Trans-Visions enhance your ability to form a three-dimensional concept of how we are constructed . . . (but) the rest of the book tells you, and every reader, how the human body works."

Visions it is possible for you to contrast anterior with posterior views and to "dissect" visually by means of these sequential illustrations through the body. Much more information is available with these transparencies than is possible with the usual anatomical plates. Trans-Visions enhance your ability to form a three-dimensional concept of how we are constructed.

The numerical tables on the transparencies refer to the Trans-Vision index. The terminology employed is international; the language is Latin for the reason that it was the universal language of science when anatomy became a descriptive discipline. Not only are the Latin names recognized throughout the world but, often, they are descriptive of site, shape, or function. For example, Musculus (M.) biceps brachii translates as muscle (musculus) of the upper arm (brachii) having two heads or upper attachments (biceps).

Let me add a parenthetical word about muscle anatomy and function. The active, energy-requiring phase of muscle action is contraction, or shortening. By noting the attachments of the two ends of a muscle to bony prominences, you can easily visualize the effect of contraction. To extend the illustration, the biceps of the arm attaches to the top of the shoulder at two sites and below to the forearm (plates 5, 6, 11, 12 & 14). Its shortening forces flexion of the elbow.

As you read this book, you will note many references to key anatomical structures. In addition to using the illustrations that accompany the text, turn back to the Trans-Visions—this is when their value will be most apparent to you. In Dr. Newton's discussion of maturation in females as well as in the description of pregnancy and birth, knowledge of the bony components of the pelvis (pubic, iliac, sacral, and ischial bones) is particularly helpful. Get the three-dimensional concept by using these transparencies in the context of descriptions and explanations of human physiology.

• • •

Van Doren. *The Trans-Visions are wonderful. They tell me what I look like inside—as no one but a surgeon could ever see me, and even then I guess no surgeon would ever be able to see all of that.*

Kessler. *Not short of an autopsy.*

Van Doren. *Thanks a lot! In fact, though, these pictures, wonderful as they are, fail to tell me the most important things I want to know. They describe what I look like, but they don't tell me how I work.*

Kessler. *That's why this first chapter is only an introduction. The rest of the book tells you, and every reader, how the human body works.*

Van Doren. *But sometimes it doesn't work. Things go wrong.*

Kessler. *The book is about that, too.*

Van Doren. *"Let us go then, you and I." Strange, that line of poetry just occurred to me. It seems relevant to the business at hand.*

Kessler. *I suppose you're saying that some other time, in the course of some other dialogue, you'd like to talk about poetry. Then you could be the expert.*

Van Doren. *You're a very clever man, doctor.*

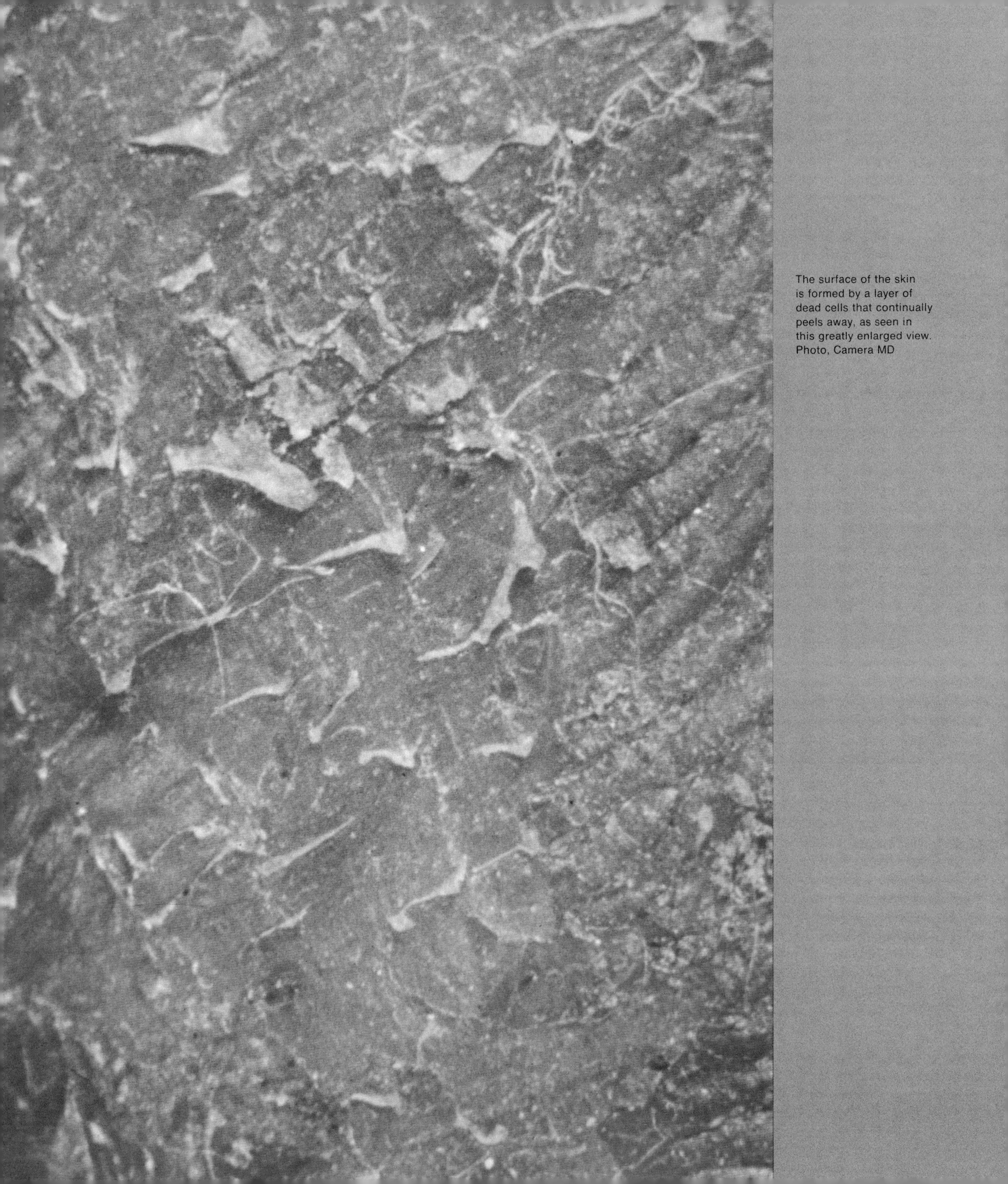

The surface of the skin is formed by a layer of dead cells that continually peels away, as seen in this greatly enlarged view. Photo, Camera MD

Van Doren. *Dick, I'm struck by this chapter's first sentence. Is the skin an organ?*

Kessler. *Sure it is. Any collection of cells with a single overall function can be called an organ. The skin definitely qualifies.*

Van Doren. *Apart from that, I was surprised and pleased to learn that the skin is waterproof!*

Kessler. *You're right, both as to the fact and as to its importance. The skin keeps the outside outside, and the inside inside. That's its most crucial function.*

2

Skin and Appendages

BY SAMUEL M. BLUEFARB, M.D.

The skin, or the integument, is an actively functioning and protective external organ. The largest organ of the human body, accounting for approximately one-sixth of the total body weight, it performs functions that are vital in maintaining bodily health. Human skin is rich with sensory nerves, as well as with blood vessels. These are linked with the other organs of the body. The skin also responds and adjusts to various hormonal signals that the body puts out. Sweating, for example, results from an interplay of sensory and hormonal signals. When the sensory nerves of skin send a "heat signal" to the sweat glands, sweat pours onto the skin. In addition, the thyroid gland also gets a "heat signal" and releases more thyroid hormone. This hormone acts as a vasodilator allowing the blood vessels on the skin to expand, or relax, so that they hold more blood, which is then cooled by the evaporation of perspiration. In response to cold, the blood vessels near and on the surface of the skin are made to contract, thereby conserving heat. Many of the skin's responses to stimuli occur without the individual being aware of them, and in that sense they might be considered instinctive. However, whether the body responds to signals consciously or unconsciously, or if it responds to signals that are readily recognized, such as sensations of touch or temperature, is not of great importance. It is important, however, to realize that the skin of man continuously performs functions vital to sound bodily health. Man's skin is indeed one of the most unique organs in nature and helps to make man the unique creature that he is.

Functions of the Skin

Dr. William Montagna, one of the outstanding authorities on the skin and hair in the United States, refers to the skin as "an effective shield against many forms of physical and chemical attack." Water and other fluids make up a large portion of the human body, some 70 percent of total body weight. The skin holds those fluids inside the body and helps to protect it by keeping out many foreign substances and microorganisms. The skin also wards off the harsh ultraviolet rays of the sun. Tanning is the skin's way of protecting the body against the sun's harmful burning rays. A prime function of the external skin layer is to provide a protective barrier of considerable toughness and resilience over the entire surface of the body. Furthermore, the skin helps regulate blood pressure and direct the flow of blood. The skin also provides the ability to record sensations such as touch and pain and heat and cold. The skin is able to distinguish between many, many kinds of texture: smooth, rough, soft, and hard. It is the skin, too, through a combination of many interrelated senses, that perceives whether something is liquid or solid. Not unimportantly the skin is one of the principal organs of human sexual attraction. The skin also helps to serve as identification for individuals. It shapes facial and body contours, and it provides distinctive and unduplicated fingerprints.

Of particular importance to man is the fact that the skin has the ability to repair and regenerate itself and thereby heal surface wounds. Scratches, minor cuts, and abrasions are so common that few persons could survive long if the skin could not repair itself. In a short time a mass of minor cuts, bruises, and scrapes would be highly susceptible to infection.

Structure of the Skin

To gain an understanding of the skin's basic structure, one should understand that the skin is divided into three distinct tissue layers. These are the epidermis, the dermis, and the subcutaneous fat.

The Epidermis

The epidermis forms the outer layer of the skin. It is made up of four or five sheets of cells tightly layered on top of one another. The topmost layer of the epidermis is made up of flattened dead cells and serves as the principal shield of the body. The cells are necessarily dead because living cells cannot survive exposure to air and water. This topmost layer of the epidermis is called the stratum corneum (*stratum* = cover and *corneum* = horn in Latin), or the horny layer.

The stratum corneum covers the entire body and is continuously shedding its cells through the wear and tear of daily activity, washing, the friction of clothing, and exposure to the environment.

The skin's property to regenerate and repair itself goes on continually and occurs at the basal, or base, cell layers of the epidermis. These deeper layers steadily replenish the horny layer by making more cells. As these cells are made they are pushed upward toward the surface of the skin. Inside each cell is an internal fiber system made up of tiny strands, called tonofibrils. It appears to serve a structural or supportive function for skin cells.

Other important structural components of the epidermis are desmosomes—numerous localized thickenings of skin cell membranes. As newly made cells move toward the surface of the skin they die by degrees as they produce keratin. This is the substance that eventually fills the space within the dead cell.

Keratin is a scleroprotein, a simple protein which is characterized by its insolubility, fibrous structure, and supportive and protective function. Keratin is tough and extremely resistant to changes in pH, (the acid-base balance of the skin), to heat and cold, and to digestion by the many enzymes the body produces. Keratin is the principal component of the body's dead surface structures, not only of

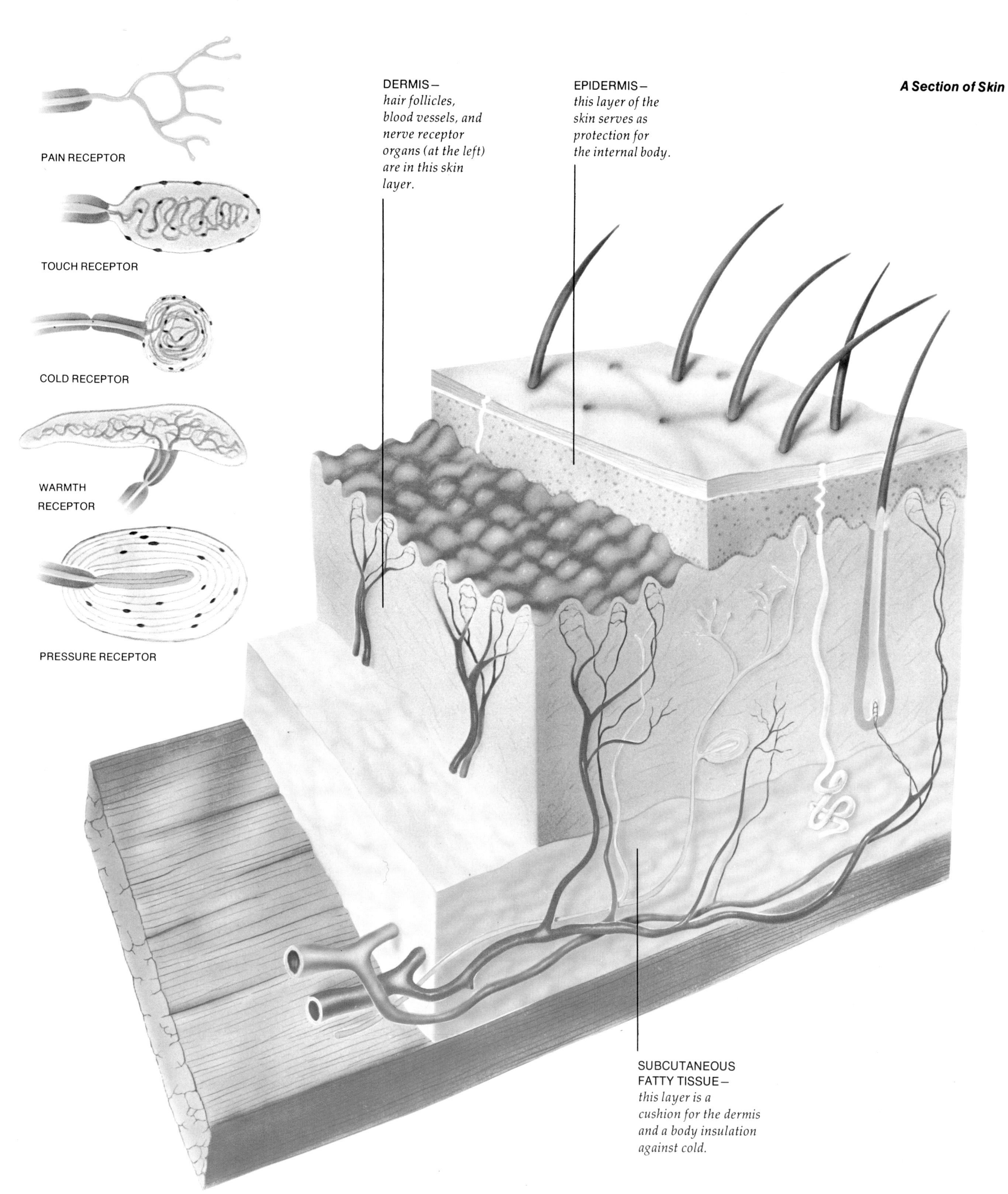
PAIN RECEPTOR
TOUCH RECEPTOR
COLD RECEPTOR
WARMTH RECEPTOR
PRESSURE RECEPTOR
DERMIS– hair follicles, blood vessels, and nerve receptor organs (at the left) are in this skin layer.
EPIDERMIS– this layer of the skin serves as protection for the internal body.
A Section of Skin
SUBCUTANEOUS FATTY TISSUE– this layer is a cushion for the dermis and a body insulation against cold.

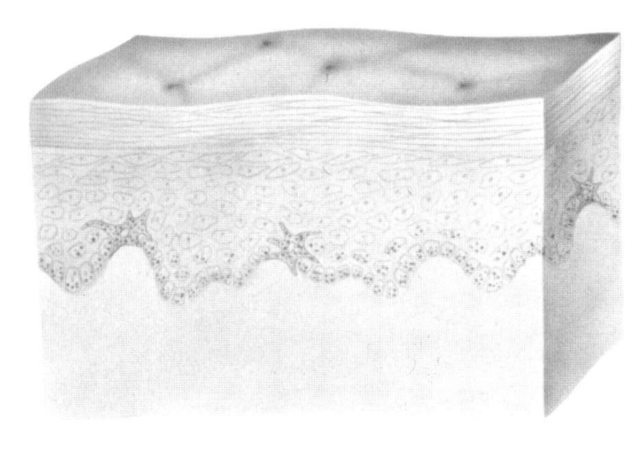

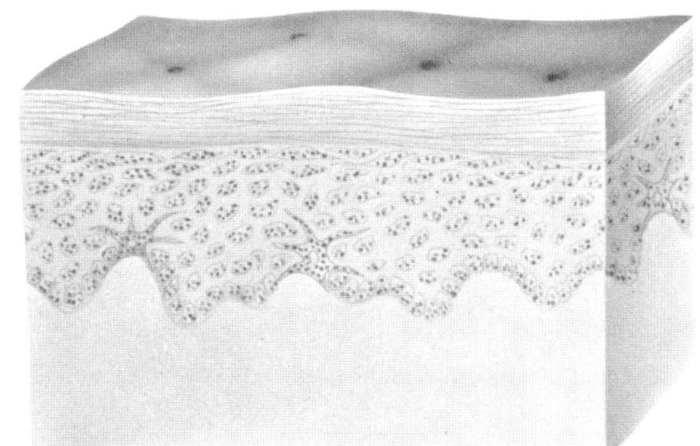

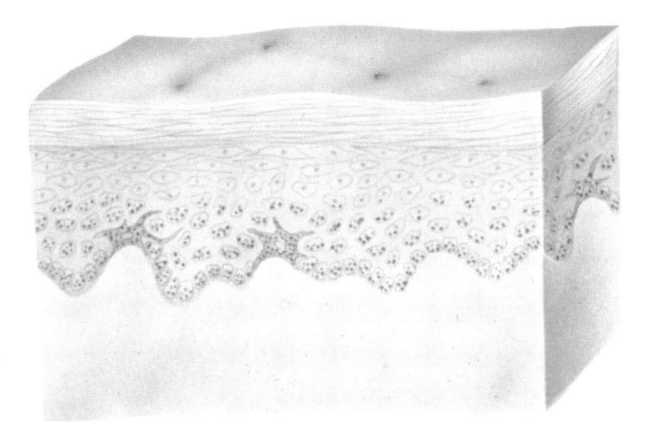

Skin color differences seen among people are the result of the amount of melanin, a dark pigment, interspersed throughout the epidermis. Cells called melanocytes at the base of the epidermis produce melanin.

the skin's horny layer but also of the hair, nails, and other skin appendages. Death is the fate of every epidermal cell and is achieved in an orderly manner.

Another conspicuous property of the epidermis is its pigmentation. Evenly scattered deep in the epidermis are the melanocyte cells that produce the dark pigment melanin. Spider-shaped with long tentaclelike processes, these cells inject granules of melanin into the surrounding epidermal cells. Here, the pigment forms a protective covering over each cell nucleus on the side toward the skin surface. The prime function of the pigment is to shield the cells by absorbing the ultraviolet rays of the sun. All human beings, regardless of race, have about the same number of melanocytes; the darker races are distinguished from the lighter ones only by the fact that their melanocytes manufacture more pigment. This protective capacity has evolved in the tropical peoples of the world as an asset with high survival value. In every race there are a few individuals who cannot manufacture melanin. This condition is called albinism, and the individuals affected with the condition are called albinos. In extreme cases they have skin of a milky white color, very light hair, and eyes with a deep red pupil and a pink or light blue iris. The lack of dark pigment in the eye admits light in excessive amounts and through other parts besides the pupil, and the vision of albinos is consequently poor.

The Dermis

The epidermis is supported by a second layer of skin, called the dermis. It is thicker than the epidermis and is composed chiefly of protein collagen fibers and the intertwined elastic fibers enmeshed in a gel-like matrix.

The dermis supports and binds the epidermis so that it conforms to the underlying bones and muscles. The collagen and elastic fibers of the dermis make the skin tough and elastic. Collagen is composed of bundles of banded fibers. It is the principal constituent of tendons and is one of the most abundant proteins in the animal kingdom.

Dense webs of blood vessels course through the skin all over the body. They transport through the skin a great deal more blood than is needed to nourish the skin itself. This immense circulation performs two functions. First, it acts as a cooling system. The sweat glands pour water onto the skin surface, and the evaporation of this water cools the blood circulating through the skin. When the outside environment is warm or the body is engaged in strenuous exercise, the blood vessels assist the cooling process by relaxing and allowing a maximum flow of blood through the skin, accounting for the flushed appearance of the skin at such times. On the other hand, when the environment is cold, the vessels contract rapidly and greatly reduce the blood flow in the skin, thus conserving the internal body heat. Second, the skin circulation serves to help regulate the blood pressure. The blood vessels in the skin have passages that can constrict like a rubber band to shut off the flow through the capillaries and thereby cause the blood to bypass them so that it flows directly from the small arteries into the veins. This mechanism for regulating blood flow acts like a safety valve and helps to control abnormal variations in blood pressure.

The extraordinary network of blood vessels in the skin is matched by an equally massive network of nerves. Much of this nervous system is concerned with controlling the glands, blood vessels, and other organs in the skin. Also, a vast complex of sensory nerve endings is present. These are particularly prominent in the most naked surfaces of the body—the fingers, palms, soles, lips, and even the transparent cornea of the eye, which has no blood vessels. These specialized nerves are sensitive to touch and heat. Without them the human body would be almost as out of contact with the outside world as it would be if it lacked the major sense organs.

Subcutaneous Fatty Tissue

The subcutaneous fatty tissue, merging with the deepest level of the dermis, is characterized by closely packed cells which contain considerable fat. The boundary between this layer and the dermis is not well marked, and the layer varies widely in thickness. The subcutaneous fat layer provides thermal insulation for conservation of body heat when the flow of blood to the skin is curtailed, and it also serves as a thermal barrier to protect underlying tissues from excessive environmental heat and retards loss of body heat in cold environments. Bundles of collagen (protein fibers) threading between the accumulation of fat cells provide a flexible linkage between the superficial skin layers and the underlying structures. Thus, the fat layer serves as a cushion for the dermis and epidermis and permits considerable lateral displacement in many regions of the body surface.

Hair

In man, hair is now largely an ornamental appendage. Apparently it survives mainly as a means of sexual attraction. A few special protective functions of hair remain important to the human body. The hairs inside the nostril slow incoming air currents, trap dust particles, keep out insects, and prevent the nasal mucus from pouring down over the lips. Aside from a few special functions such as the screening of the external passages, hair is not really essential to man.

The follicle that produces a hair is a tube that extends all the way through the skin and widens into a bulb—the hair root—at its deep end. When the hair has attained its characteristic length, it stops growing, except on the scalp, where the hair may grow very long if it is not cut. Having reached its limit, the hair forms a clublike base and puts out rootlets that anchor it to the surrounding follicle. The follicle shrinks and goes into a resting period. After a time, it forms the germ of a new hair, which in turn works its way toward the surface, loosening the old hair and causing it to be shed. Thus hair growth is a matter of alternate growing and resting periods. Human hair grows at the rate of about 0.35

Hair and Its Follicle

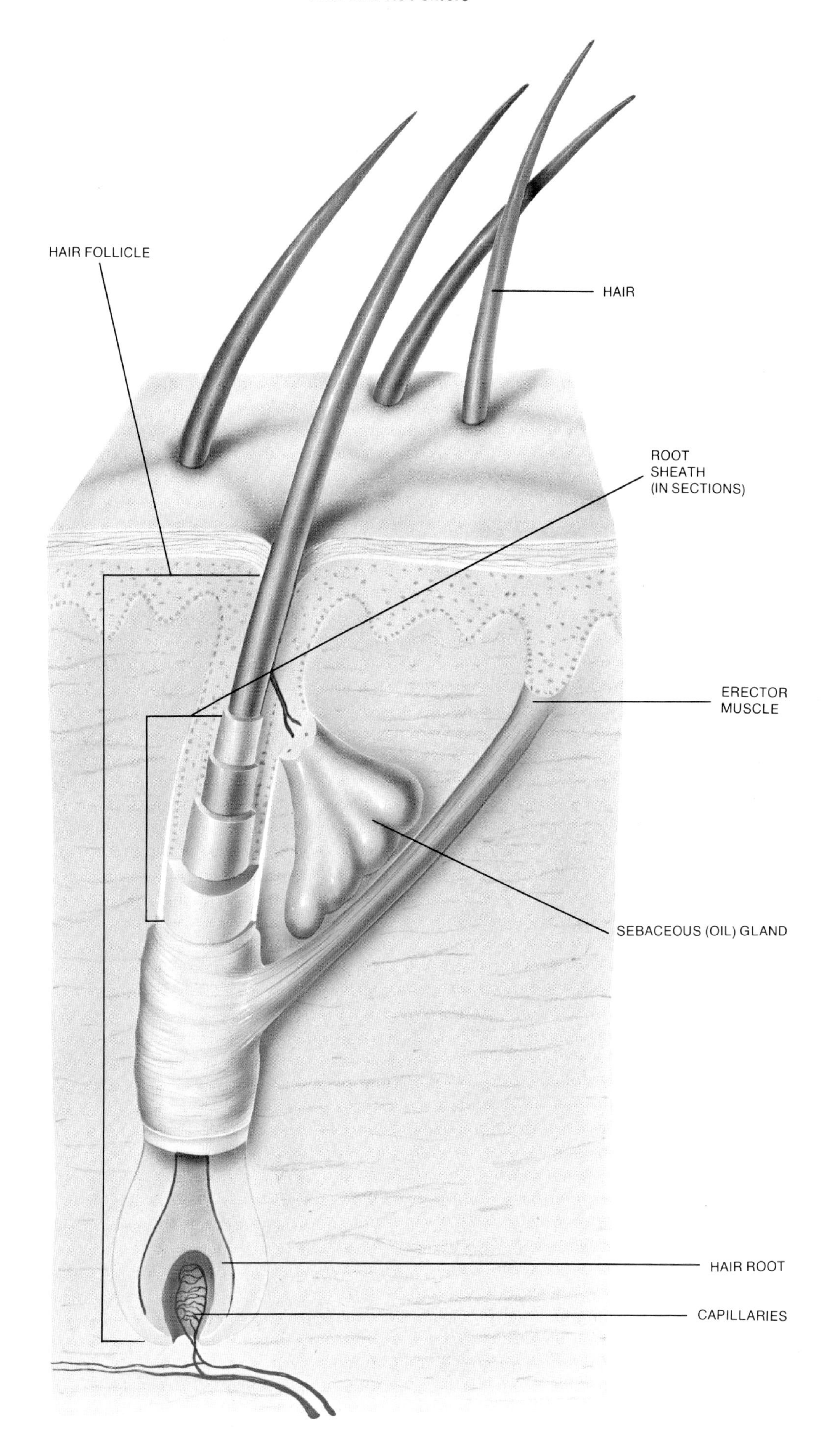

mm (millimeter) per day, or about an inch in every three months. The individual hairs of man may be round (which causes the hair to be straight), alternately round and oval (which makes the hair wavy), or ribbon-shaped (which makes hair kinky). The color of the hair depends on the amount and distribution of melanin in it and on the hair surface structure, which affects light reflection.

When the follicles on the scalp die, balding sets in. Some young men begin balding in their twenties. It is curious that this dramatic loss of hair should take place on the scalp, which, unlike other parts of the body, can grow as much as twelve feet of hair. The explanation lies in the action of androgenic, or male, hormones. Paradoxically, these hormones are responsible not only for the growth of all hair but also, under certain circumstances, for the reduction of scalp hair. Eunuchs—castrated males—have a deficiency of androgenic hormones and rarely become bald.

Eccrine sweat glands, under control of the nervous system, pour out the perspiration that through evaporation cools the body. More than two million eccrine sweat glands are distributed in the skin.

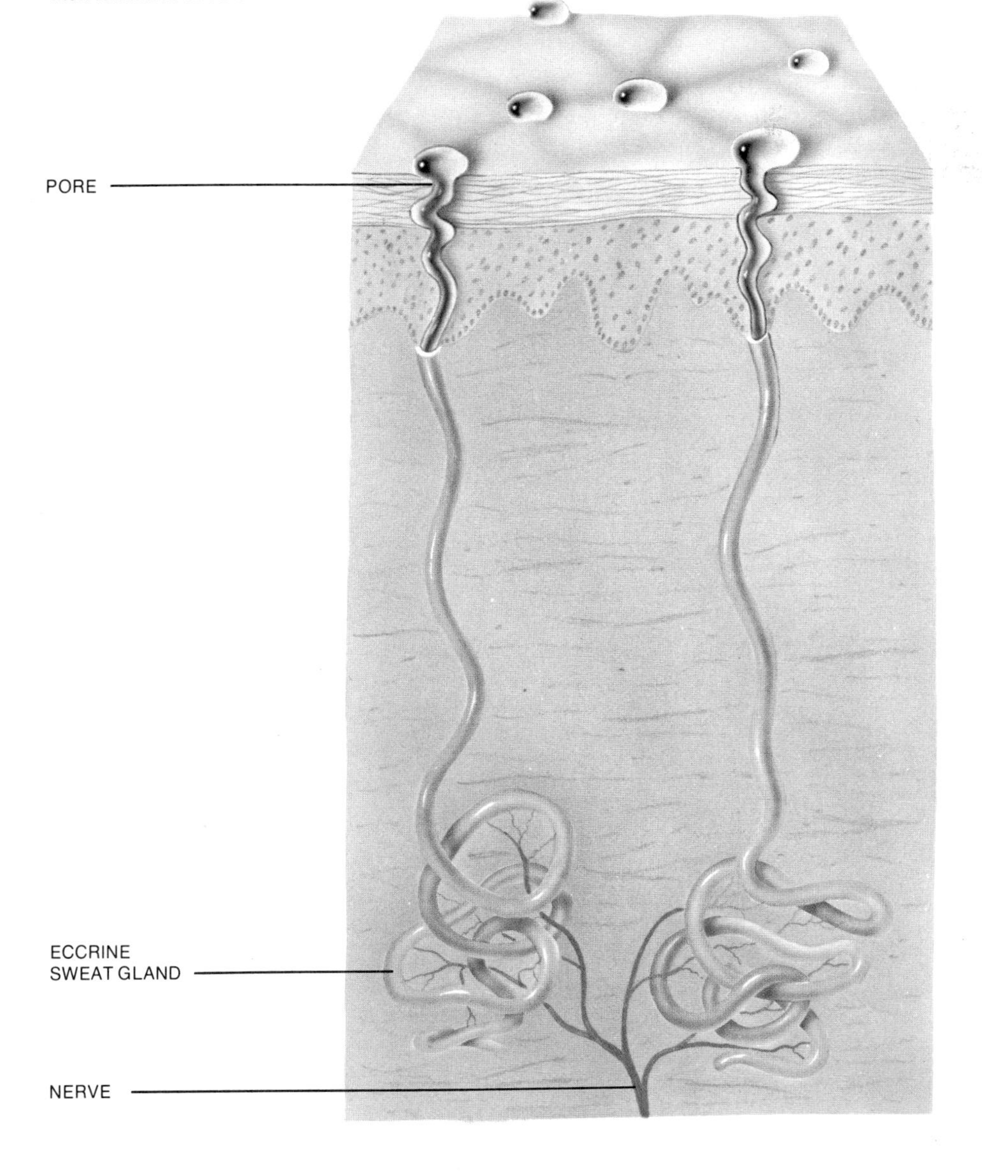

Skin Glands and Perspiration

The skin of man has two kinds of secretory glands—the sweat glands and the sebaceous glands. The latter produce a material called sebum, an oily mixture of fatty acids, triglycerides, cholesterol, and cellular debris. The production of sebum is somewhat like the production of the horny layer by the epidermis.

The human body skin carries more than two million sweat glands, which serve to control the body temperature and also to remove certain unwanted products from the body. In the healthy person, sweating is a normal, automatic function. More than a pint of perspiration, which is virtually all water, may pass from the body each day.

The body odor that most people are aware of is not caused by perspiration. Bacteria on the skin, however, can break down perspiration, and through the resulting formation of certain byproducts, an unpleasant body odor occurs.

Apocrine glands in the skin secrete the odorous component of sweat. They are primarily scent glands, and they produce their secretions in response to stress or sexual stimulation. The apocrine glands no doubt played an important role in human society before the deodorant and perfume industries usurped their function of creating a person's body odor. Aside from their odor-generating property, these glands are quite unnecessary to man.

The apocrine secretion is a milky, sticky fluid of varying color—pale gray, whitish, yellow, or reddish. Although the glands are large, the amount of their secretion is very small; most of the fluid in the copious sweat of the armpits is supplied by the eccrine sweat glands. These provide the vehicle for the spread of the odorous apocrine substances. The freshly secreted apocrine sweat is actually odorless. It becomes malodorous only when its substances are decomposed by the action of bacteria on the skin surface.

Sweat has two main functions—cooling the body by evaporation and moistening the friction surfaces. For example, in the palms and the soles moisturizing prevents flaking of the horny layer, improves the grip on surfaces, and assists sensitivity to touch.

The array of active sweat glands all over the body skin is a highly useful adaptation. It is an essential cooling system. Some people sweat more profusely than others, but this condition is not strictly related to the individual's total number of sweat glands. The difference lies, rather, in the relative activity of the glands.

The Nail

The nail consists of a matrix, the nail plate, or body, and the anterior portion. The matrix is the most important structure of the nail, for it is the living portion and generates nail growth. The outer boundary of the matrix is the lunula, or "moon," of the nail. In some persons the "moons" are very prominent; in others they are very small or may even be under the cuticle. The nail matrix contains nonfunctioning pigment cells, and therefore the nail is usually colorless. However, under certain circumstances these melanocytes are stimulated to produce melanin and pigmented areas may be found in the nails, especially in the nails of blacks.

In adults, the fingernails grow about 3 mm (a little more than 1/8th of an inch) per month. In infants and elderly people, the growth rate is slower. The nails grow slightly faster in warm climates and during summer months; slower during serious illness; faster if there is persistent injury or irritation such as biting or trauma. The nails of a right-handed person's right hand grow slightly faster than the nails of the left hand. The nail of the middle finger grows more rapidly, and that of the little finger more slowly, than the others. Toenails grow about one-fourth slower than fingernails. If a nail is completely removed, it will take about 120 to 150 days for it to regrow to normal size.

Abnormalities of the nails are of a wide range. For example, nails may be discolored or pigmented white, brown, black, yellow, or green; irregularly speckled or striped, crumbly, dull, softened, split, thinned, thickened, expanded, narrowed, ridged transversely or longitudinally, spoon-shaped, brittle, overgrown, roughed by spurs, or they may be shed completely.

In analyzing the abnormalities or diseases of the nail, remember that the nail plate is a dead structure and only reflects events in the past. Primary damage through infection or trauma must have occurred in the matrix, in the nail bed, or in the posterior nail fold. When the matrix is disturbed, affected, or diseased, the nail bed, the nail plate, or both, will be affected. When matrix function is inhibited or somehow becomes irregular, the nails will degenerate or waste away. Continuous or intermittent irritation of the matrix will increase function and produce hyperkeratosis, or thickening of the nail. In other words, any mechanical, organic, or physiologic disturbance—be it local, general, or congenital—can affect the matrix and the appearance of the nails.

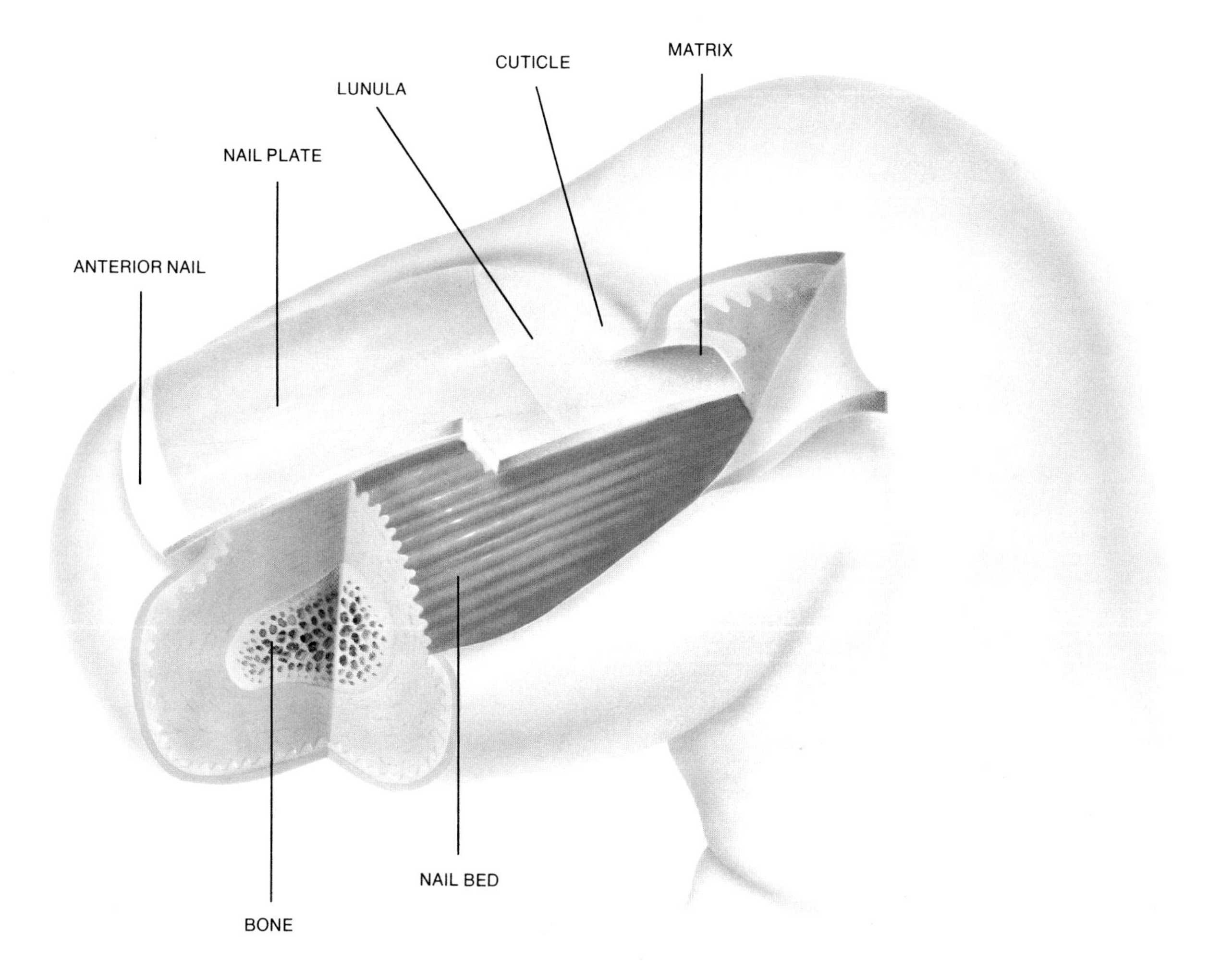

This illustration is a three-dimensional view of major parts of the thumb and its nail.

Common Skin Diseases (Dermatoses)

Skin disorders can range in seriousness from pimples to cancer. Dermatologists—physicians who are skin specialists—can diagnose a variety of disorders through direct examination of the skin.

Acne

Acne vulgaris, or pimples, in mild form is so common in adolescent males and females that it can almost be regarded as a physiological alteration associated with other pubertal changes. However, largely because of undue emphasis by modern society on an individual's personal appearance, the more serious forms of acne represent a disease, which, though seldom endangering health, is not to be lightly dismissed. This disease has disturbing psychological, social, and economic consequences.

Many adolescents are tormented by feelings of abnormality and inferiority and so shun all but essential social contacts. Girls are deprived of social engagements, young men are rejected for employment, and obstacles to marriage for either sex may be considerable. The degree of mental reaction is, of course, different with each individual, but frequently the reaction is of sufficient severity to produce serious symptoms of depression and withdrawal.

Severe acne is caused when keratin plugs skin pores and underlying sebaceous glands produce too much oil. Proper treatment can reduce the severity of the condition.

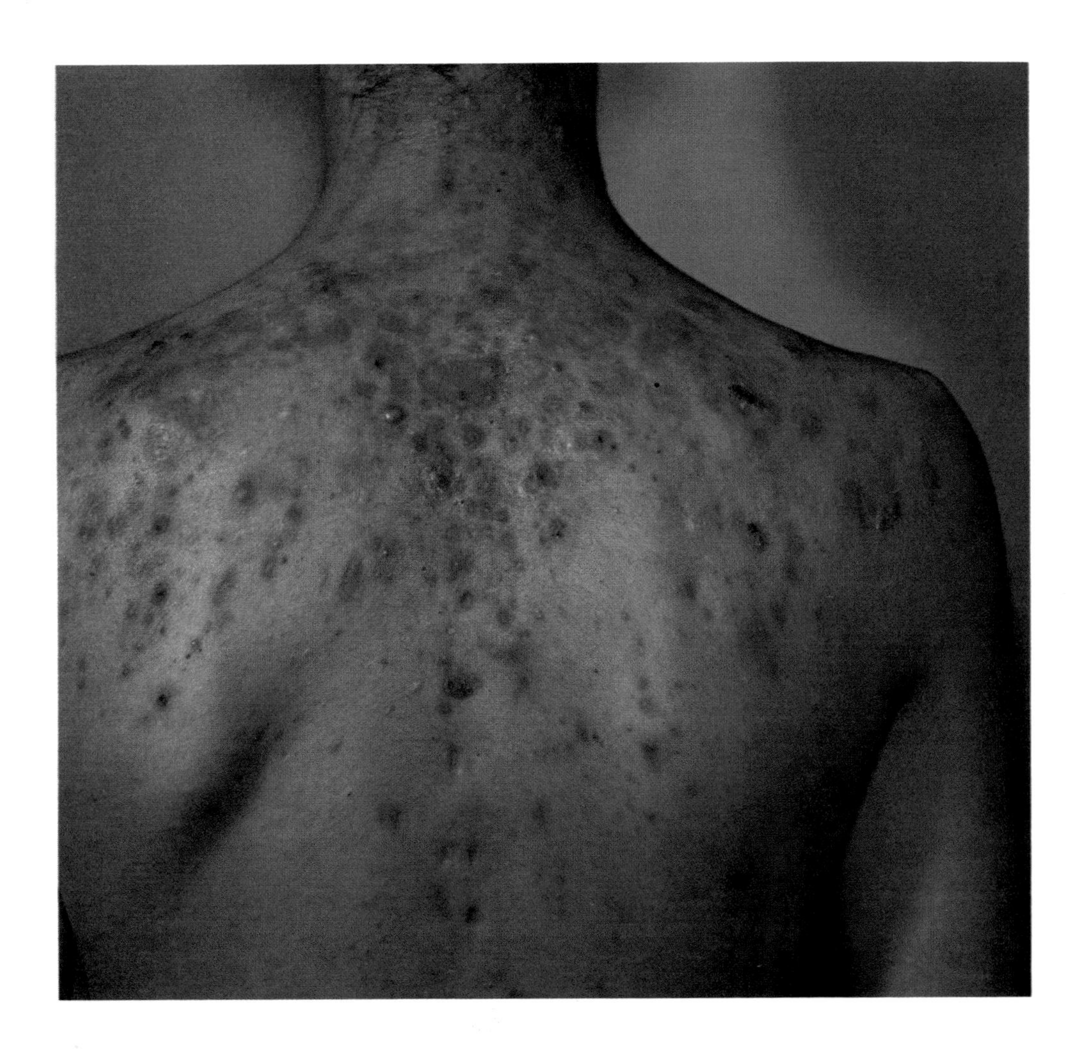

The association of acne lesions with the sebaceous glands was recognized more than 200 years ago. It is well known today that two main pathogenic factors are present—enlargement of sebaceous glands with subsequent seborrhea, or oversupply of oil, and plugging of the follicular pore through the production of too much keratin. The first development, and perhaps the second, is produced by the physiologically increased testosterone (a masculinizing sex hormone) levels in boys and increased progesterone (a feminizing sex hormone) levels in girls. The severity of the abnormal process is apparently determined by genetic factors. When an excess of keratin at the opening of the pore causes blockage of the sebum, the follicle enlarges. Stagnation and decomposition of sebum leads to inflammation, and secondary bacterial infection often follows. Pressure of the macroscopic or microscopic comedo, or blackhead, may result in fragmentation of the follicular wall and sebaceous gland. Surrounding cells may die. The progression of acne lesions through the stages of comedo, papule (elevated lesion), pustule (lesion containing pus), and cyst reflects the sequence of these pathologic changes. The occurrence of pitted scarring depends on the degree of destruction of the pilosebaceous apparatus—the hair follicle and its sebaceous gland—and surrounding tissue.

It is likely the sebum itself causes the early inflammation of acne. When the various components of sebum, a complex lipid mixture, are injected into human skin, the greatest amount of inflammation follows the injection of the free fatty acids, the response being similar to that produced by whole sebum. Therefore, the fatty acids are the most irritating compounds in sebum.

More direct evidence of a role for sebum in causing inflammatory acne is provided by the therapeutic effect of estrogens and antibiotics. Estrogens (feminizing hormones) suppress sebaceous gland secretion, and this decrease in sebum production brings concomitant improvement in acne. Some of the antibiotics are also of use in certain types of acne. By altering the composition of sebum and reducing the concentration of free fatty acids, which have been shown to be irritating to the skin, improvement follows in these cases.

The second question to be discussed briefly is whether acne is an endocrine, or hormonal, disorder. It is not difficult to cite evidence that might support the concept that acne is an endocrine disease. For instance, acne is usually absent in children but appears at puberty. Acne does not occur in castrated individuals. Yet when these people are given testosterone, they may develop acne lesions. In women, fluctuations of acne appear under two circumstances which indicate an endocrinologic basis for the disease. Commonly, a flare-up occurs just prior to the menstrual period, and acne usually improves during pregnancy. Acne is also seen in various endocrine diseases characterized by adrenal or ovarian hyperactivity.

With regard to treatment, lotions containing varying combinations and concentrations of sulfur, resorcinol, and salicylic acid are intended to induce peeling of the skin and ease the evacuation of comedones. There are also some newer drugs that may be prescribed to induce peeling. Ultraviolet is another peeling agent used under medical supervision. Acne sometimes improves during the summer, and this is probably due to the sun's ultraviolet rays. Hormones are occasionally prescribed for acne patients, particularly in females, and under strict medical supervision. Since androgens appear to be largely responsible for increasing the activity of the sebaceous glands and causing acne, the administration of estrogens has been used against acne in women. Estrogen medication is not used in girls who are still growing, and physicians limit its use to severe cases of acne that are resistant to other methods of treatment.

Diet therapy is usually ineffective in acne. Chocolate appears to aggravate acne only infrequently. Elimination of chocolate, shellfish, nuts, cola drinks, milk, and cheese may be of help in some persons. Greasy facial makeups, hairdressings, or pomades can aggravate acne or produce acneform eruptions. In general, if acne requires medication it requires a doctor's attention.

Alopecia (Baldness)

Women coming into the office of a dermatologist often complain, "My hair is falling out!" This complaint is better understood by reviewing the growth of hair.

Hair and its internal sheath arise from actively multiplying cells deep in its root bulb. Each scalp hair independently goes through a cycle of growth and rest. Hair grows actively for a variable period, usually years. This growth period is called the anagenetic phase. Then hair goes through a catagenetic, or transitory period of regression, which lasts only about two or three weeks. A telogenetic phase follows, lasting about two or four months, during which time growth completely ceases until it begins again.

Development of Severe Acne

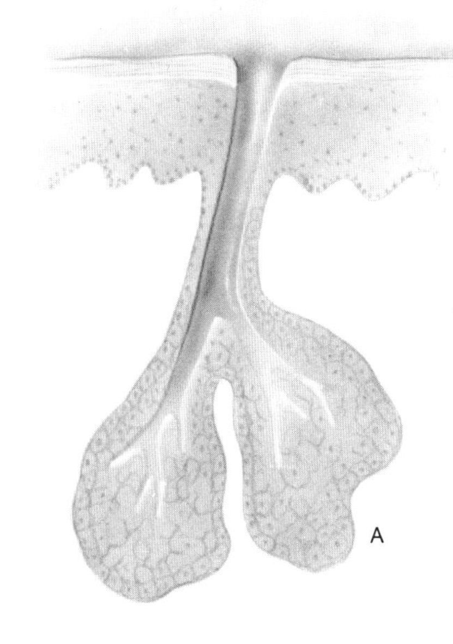

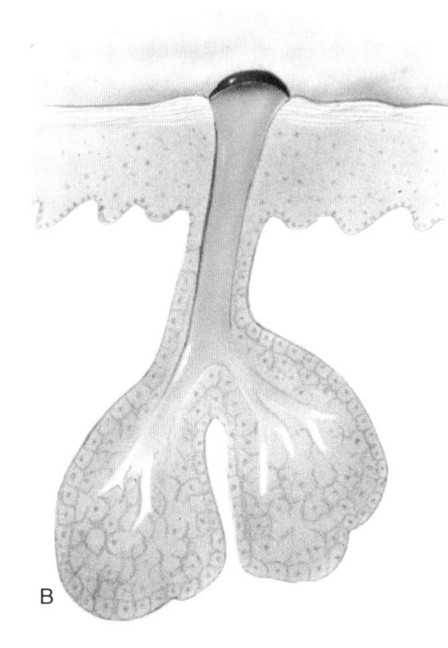

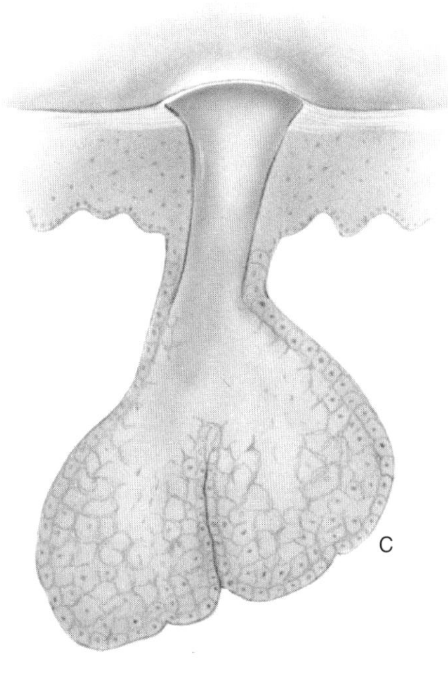

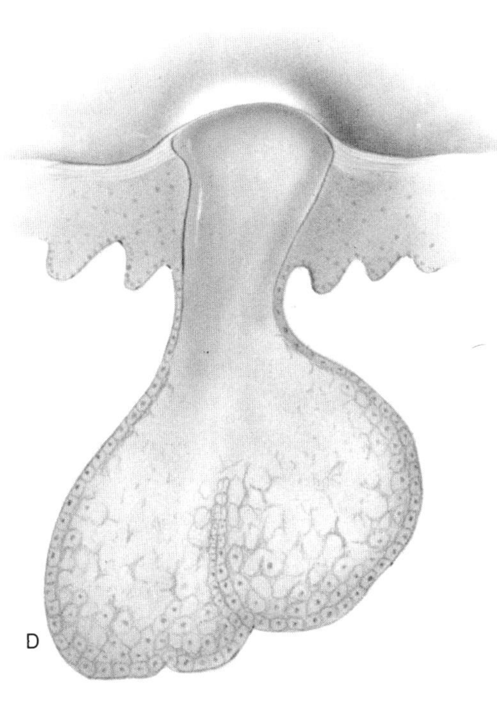

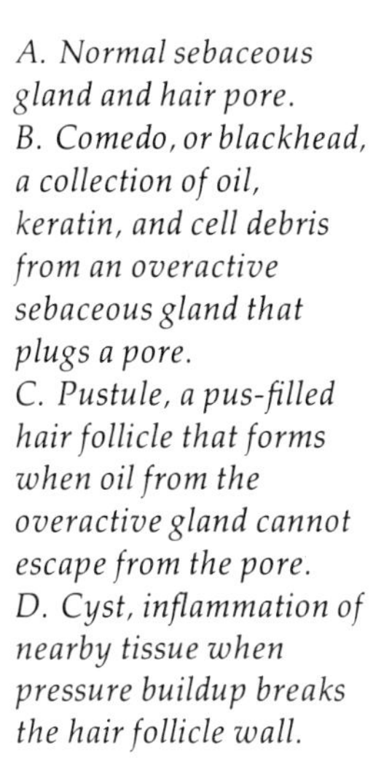

A. Normal sebaceous gland and hair pore.
B. Comedo, or blackhead, a collection of oil, keratin, and cell debris from an overactive sebaceous gland that plugs a pore.
C. Pustule, a pus-filled hair follicle that forms when oil from the overactive gland cannot escape from the pore.
D. Cyst, inflammation of nearby tissue when pressure buildup breaks the hair follicle wall.

The photos above and on the facing page show male pattern baldness, or alopecia.

Hair grows in cycles. Many hairs grow for four or five years, entering a resting phase for about three months, then fall out. After this shedding, a new hair shaft is formed. Ordinarily, about 90 percent of the hairs are in the growing phase and 10 percent in the resting phase.

Normally, a person loses twenty-five to one hundred hairs a day. These are shed from follicles—the skin depressions containing hair roots—in the resting state. Such hairs are less firmly anchored and are lost from combing, brushing, and shampooing. However, these hairs are replaced because the scalp starts growing new ones every day. The rate of loss and replacement varies from person to person and from time to time in the same person. When the rate of hair loss exceeds the rate of new growth, thinness and balding become apparent.

Ordinary baldness, called male pattern alopecia, has many characteristics that make its diagnosis simple. The telltale feature is the pattern of hair loss. It may begin at the crown, but more often the hairline recedes at the temples. Eventually, the frontal hairline assumes the shape of the letter M. By this time, the hair is thinning at the crown, and the thin area progresses forward until it joins the denuded area in front. In extreme cases only a fringe of hair is left around the base of the scalp.

All males tend to lose some scalp hair in the course of time, although some reach old age without any thinning becoming apparent. About two out of five of all white men will have male pattern baldness. It usually begins in the late twenties or early thirties, but in some individuals with a strong family history of baldness, balding becomes obvious shortly after adolescence. In such cases the outlook for hair growth is poor, and most men in this circumstance will usually be quite bald before their thirtieth birthday. This type of baldness is not a skin disease but has been described as a disability or death of individual follicles.

Somewhat similar but less complete balding occurs in women as well as in men. The feminine type proceeds according to a different pattern. Instead of a receding hairline in front, women suffer a diffuse thinning over the crown and front of the scalp. This, however, rarely progresses to complete baldness, though thinning may approach baldness in very old women. The victims have a family history of "thinning" of the hair in female ancestors. Such hair loss has a great variety of causes—infections, systemic diseases, drugs which have toxic effects, mechanical stress, friction, and radiation.

Infections which cause loss of hair can be fungal (e.g., ringworm of the scalp), bacterial (e.g., staphy-

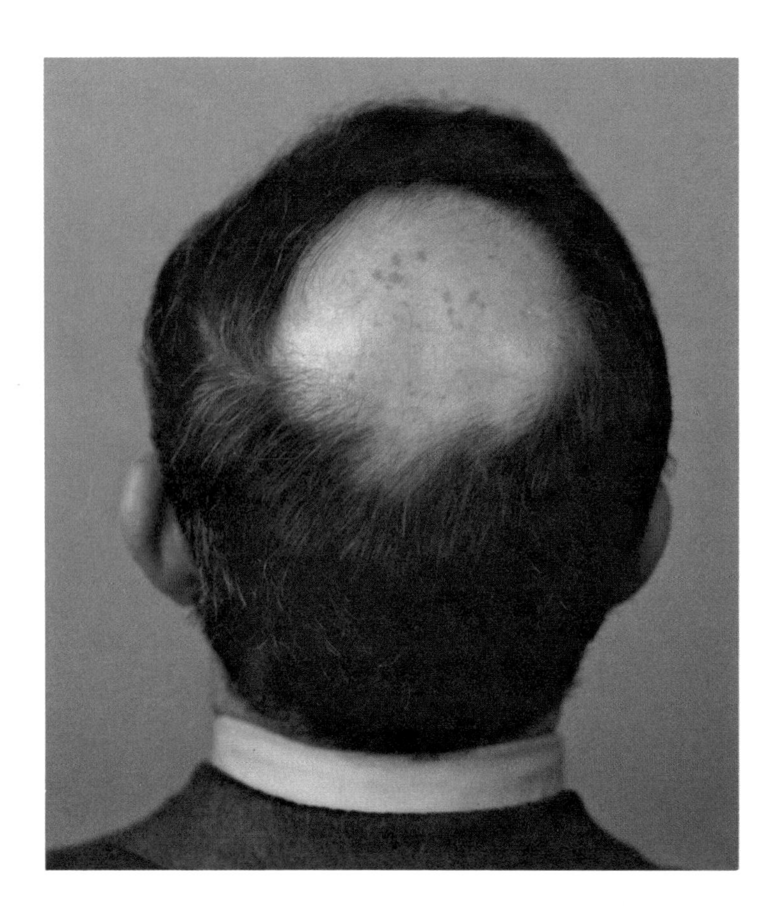

lococcal or streptococcal), or viral (e.g., herpes zoster). Early diagnosis and treatment are important, otherwise the hair loss may be permanent.

Systemic diseases which may lead to baldness include allergic and other reactions to drugs, advanced diabetes, influenza, scarlet fever, syphilis, pituitary disorders, leukemia, and cancer.

Blame for hair loss also extends to toxic effects of drugs and chemicals. Certain substances ingested or applied to the skin or hair can result in loss of hair. This has occurred after accidental ingestion of insecticides and rodenticides containing thallium. The topical use of selenium disulfide, an ingredient of certain shampoos and ointments, may cause temporary hair loss. Permanent waves, when inexpertly used, have destroyed hair follicles, but with products in use today, virtually no risk exists.

Friction against bedding breaks hair off close to the scalp and may cause a bald spot on the back of the head, for example when the head of an infant is turned continuously in one direction. Bald spots of this sort also occur in adults when work gear or rough clothing causes friction.

Radiation from X rays, radium, radioisotopes, nuclear explosions, or nuclear accidents may result in temporary or permanent baldness.

Diet ordinarily has little to do with baldness. However, chronic starvation or severe deficiencies of certain vitamins contribute to dryness, lack of luster, brittleness, and hair loss. On the other hand, grossly excessive intake of vitamin A may cause baldness.

Loss of hair is quite common in women toward the end of pregnancy, after childbirth, and during menopause. In most cases, the majority of these lost hairs will regrow after several months. This situation is due to the fact that all of the hairs rest at the same time and are not replaced immediately. Many people worry that dandruff might lead to baldness, but no cause-and-effect relationship between the two conditions is known.

Photosensitivity

In general, photosensitivity is defined as any deviation from the normal response of the skin to light exposure. Although the sun is no longer considered a deity, "sun worship," or the craze of exposing the skin to the sun day after day to obtain a tan, has become a common custom in many parts of the world. The smooth, heavily tanned skin of the teens and twenties, however, is destined to become the dry, wrinkled, and wizened skin of the forties with premature "liver" spots and "crow's feet."

Sunlight has many effects on the skin. One of the most important, both clinically and cosmetically, is aging. Many people are sun faddists, unaware that exposure to the harmful ultraviolet rays of sunlight causes aging. Gross changes in sun-damaged skin are a dry, coarse, leathery appearance, laxity with wrinkling, and various pigmentary changes. In elderly persons, even in some relatively young white adults, a striking difference frequently appears between light-exposed areas of the body and those protected by clothing. A weather-beaten farmer often appears considerably older than a physician of comparable age. Since skin of blacks has a natural protection in its high melanin content, elderly blacks often appear deceptively young.

Excellent protection from sunburn can be provided by sunscreens. A number of compounds have been used, among them para-aminobenzoates, anthranilates, cinnamates, pyrrones, benzimidazoles, carbazoles, naphthosulfonates, quinine disulfate, and benzophenones.

Exposure to substances that can make a person respond to light in an abnormal way occurs more and more in our daily living routines; these substances lurk in the bedroom (perfume), the bathroom (soap), the kitchen (limes), the garden (insecticides), the pool (suntan lotion), and most important, the medicine cabinet (drugs). Some knowledge of the photosensitizing agents is necessary in order to do the detective work required to diagnose and treat successfully the light-sensitive person.

Skin Cancer

Cancer is more common in the skin than in any other organ, and physicians will diagnose one hundred thousand new skin cancers in the United States annually. Skin cancer constitutes 20 percent of all cancers in men, and 11 percent of all cancers in women. Ninety-five percent will be cured.

Four clinical observations provide evidence that relate skin cancer to sunlight: (1) Skin cancer occurs most frequently on exposed parts of the body. (2) It occurs more frequently among outdoor and rural workers than among indoor and urban workers. (3) It occurs more frequently in light-skinned than in dark-skinned persons, and (4) the greatest incidence of skin cancer is in geographical areas where the sunlight has high ultraviolet-ray content.

Moles and Hemangiomas

Since nevi, or moles, are the most prevalent of all skin tumors, and since many are situated on the face, patients frequently request their removal for cosmetic reasons.

The pigmented nevus, or common mole, is peculiar to man and is the most common tumor found in humans. Everyone has a few nevi somewhere on his body; most persons at the age of twenty-five have about forty of them, and some people have one hundred or more.

The cause of nevi is not known; it is assumed that some are genetically determined, and there is evidence for a hereditary predisposition. Both sexes are equally involved.

Only 3 percent of infants are born with nevi; the remainder of nevi are acquired. By age twenty-five a person will usually have acquired his maximal number of nevi. However, some flesh-colored nevi may appear when a person is about forty years of age.

The pigmented nevus is thought to arise from melanocytes at the epidermal-dermal junction or from nevoblasts (precursor cells), both derived from the neural crest (an embryonic site). Nevi appear in early childhood, probably in the second year of life, when some collections of melanocytes at the epidermal-dermal border become activated and proliferate. The result, a brown or black lesion the size of a pinhead, usually becomes visible after age two. However, nevi may appear earlier or later than age two and in crops instead of singly. After a variable period (usually years), they become elevated above the surface of the skin and assume a variety of surface characteristics—smooth, dome-shaped, with or without a stem, and warty.

In children, some nevi on the trunk, particularly over the shoulder blades, become surrounded by an area of depigmented skin a few millimeters to 2 centimeters in width to form the so-called halo nevus. The appearance of this halo is an almost certain indication that over a period of a few months or a year the nevus will lose its brown pigmentation, turn red, flatten, and disappear, whereupon the area of leukoderma (depigmentation) will resume its normal pigmentation.

Most nevi, especially those on the face, are best removed surgically for cosmetic reasons. After removal, they are examined microscopically for signs of malignancy.

There are three common types of hemangiomas (tumors consisting of blood vessels); namely, the "portwine stain," "strawberry mark," and the cavernous hemangioma. The physician differentiates them by their appearance. Although some of these lesions will lighten or reduce in size spontaneously,

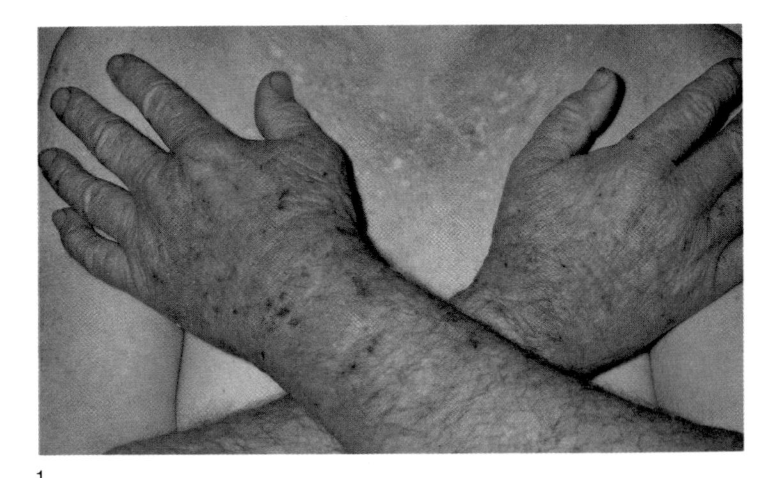

1

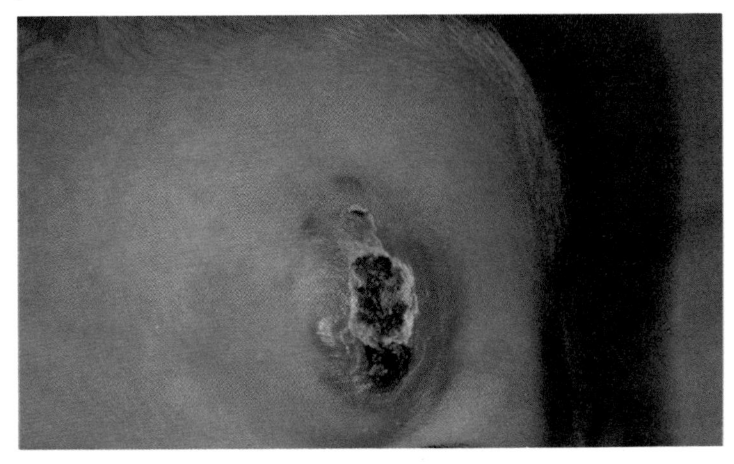

2

they all should be seen by a physician for diagnosis, and at regular intervals thereafter, particularly if any change occurs in shape, color, or consistency of the lesion. Some of them are treated by radiation therapy. Cosmetic preparations which effectively cover the portwine stains are available.

Simple and cavernous hemangiomas present difficult problems in regard to management. Spontaneous involution, or a return to normal, of both types has been documented in numerous cases. These undeniable findings affirm the belief of many physicians that all hemangiomas should be left alone and that within five years involution of these lesions will take place.

On the other hand, the attending physician is faced with several disturbing potentialities should he decide to await spontaneous involution of simple or cavernous hemangiomas. Most lesions are small, as they are seen at birth or shortly thereafter, and it would then be simple to treat these so-called insignificant lesions effectively. Since the physician is armed with the knowledge that most of these lesions will involute spontaneously, he is tempted to regard all lesions as insignificant. However, many of the large ulcerated extensive lesions start as small ones, which would at the beginning lend themselves to effective, uncomplicated, and safe treatment. Many of these will not involute spontaneously but, on the contrary, will tend to grow, sometimes rapidly. Other lesions will persist with growth proportional only to the normal development of the child.

The tendency for hemangiomas to become ulcerated occurs with some frequency. If ulceration occurs, cosmetic defects after the healing of the ulceration become a problem, especially if this takes place on the face, arms, and chest. More disturbing than the consideration of cosmetic defects is the fact that serious infections may result with impairment of the function of the involved area. The physician is also faced with the fact that once the lesions have been allowed to become enlarged therapy will be more difficult. Moreover, the cosmetic result will be inferior to that of therapy performed when the lesion is still small. It is generally conceded by most dermatologists that the earlier hemangiomas are treated the better the result.

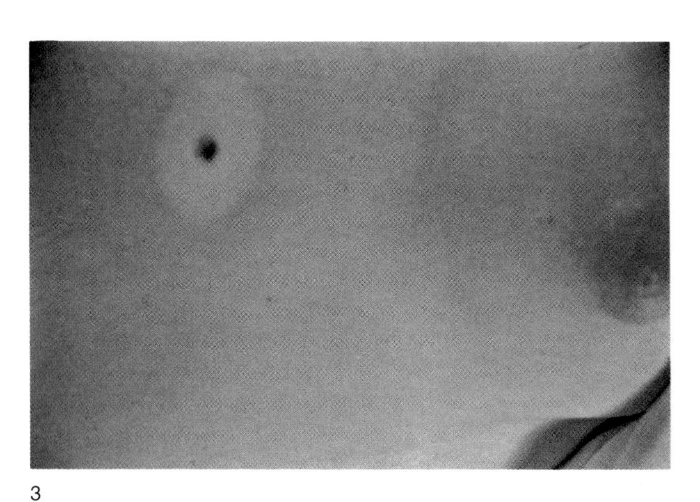

3

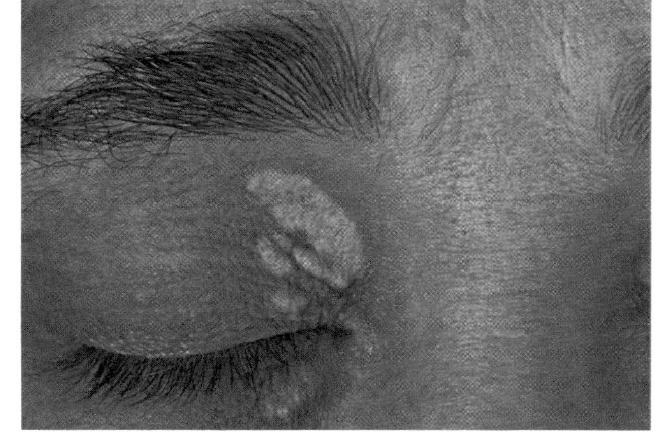

5

The conditions depicted by most of the photos in this series are described in the text—photosensitivity (1), cavernous hemangioma (2), "halo" nevus (3), pigmented nevus (4), and "port-wine stain" (6). Xanthelasma (5), around the eyelids, is caused by fat-filled dermal cells.

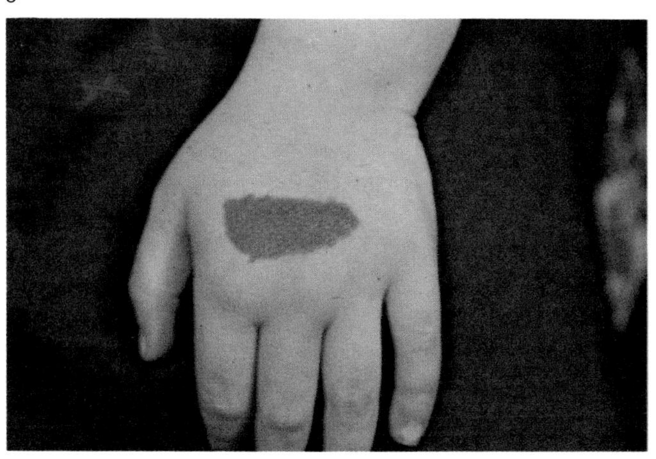

4

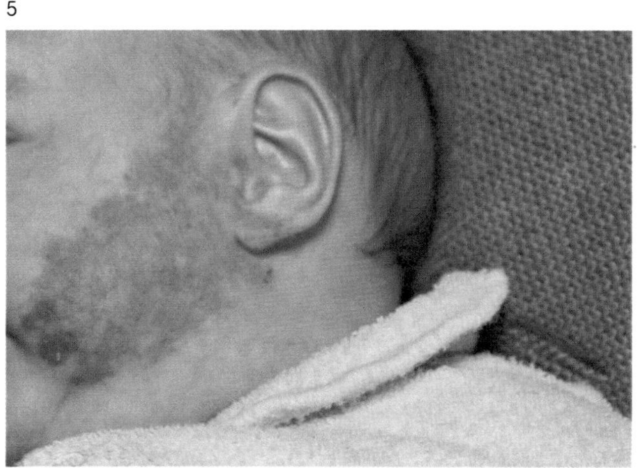

6

Dermatologic Manifestations of Internal Diseases

Changes in the skin may be an early clue to the presence of an internal disease. Pruritus, or itching, and pigmentary abnormalities often accompany internal malignant tumors and blood abnormalities. The accessibility of the skin and mucous membranes to direct observation and biopsy of early lesions is a great convenience. The physician is able to perform skin tests to determine the causes of contact and drug dermatitis and to investigate the allergic status or susceptibility to disease.

Dermatologists have long been interested in the skin manifestations of internal diseases. For example, pruritus with parallel skin abrasions suggests lymphomas, or tumors of lymphoid tissue; carbuncles and yellow skin may reflect diabetes mellitus; pitting and transverse furrowing of the fingernails may indicate infection or an acute heart condition. The accentuation and hyperpigmentation of the normal markings of the palms and of the mucous membranes, accompanied by fatigue, indicate the possibility of Addison's disease, a malfunction of the adrenal glands; the appearance of a raw, raised eruption on the extremities and cutaneous tags on the sides of the neck occasionally reflect the pregnant state of the female.

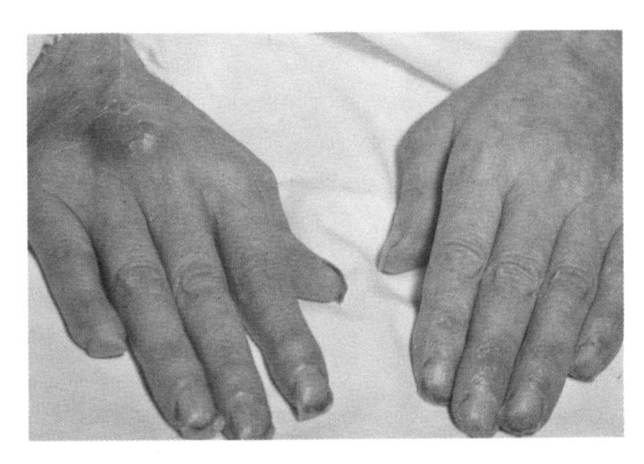

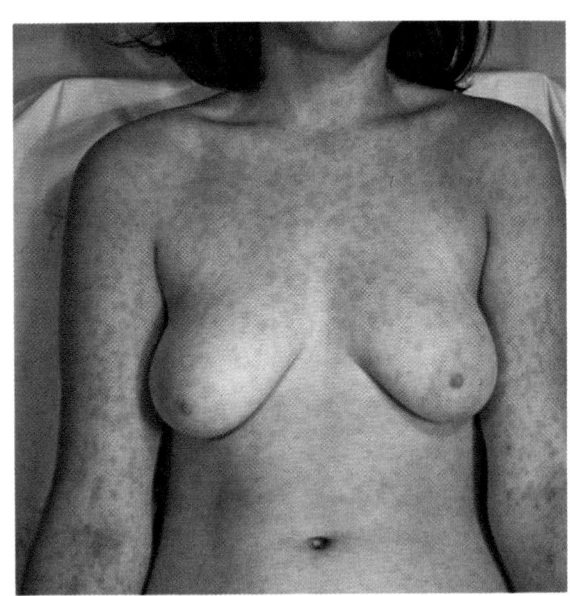

Disease within the body is sometimes indicated by changes in skin appearance. Gout (upper photo) and syphilis (lower photo) are two diseases that leave clues on the skin.

Itching

Toxic substances of many systemic diseases cause itching. This circumstance is true in a diabetic patient, even when visible lesions are not present. In addition, candidiasis, a yeast infection, in diabetes may aggravate itching, and bacterial infections are common. In liver disease such as hepatitis, or in gallbladder disease, itching often precedes visible jaundice and may be difficult to relieve. In kidney disease retained waste products entering the skin generally produce severe itching. Occasionally internal cancer is accompanied by intractable itching even though there may be no lesions in the skin. Generalized itching may also be caused by internal parasites, gout, anemia, other blood disturbances, and hardening of the arteries.

Other Manifestations

Excessive pigmentation of the skin or mucous membranes may represent the local deposit of metals such as silver, bismuth, or gold. Deposition of melanin may likewise be seen in several serious internal conditions.

The physician may note skin changes which by color, location, or consistency indicate that an unsuspected tumor may be present in an internal organ. Sometimes a skin biopsy will lead to the detection and identification of such a hitherto undetected tumor. The scalp is often the site of such a metastasis.

"Spider nevi," networks of tiny veins under the skin surface, may occur during pregnancy. Also, they may develop in great numbers in persons with liver disease, particularly cirrhosis.

A large group of diseases, collagen diseases, have skin lesions which assist the doctor in the diagnosis. The cause appears to be related to sensitization of the body. Skin changes are variable and complex but are evidence of widespread, serious systemic disease.

Blood Abnormalities (Dyscrasias)

Anemia—a decrease in the number of red blood cells and a diminution of their hemoglobin content—is associated with a paleness of the skin and mucous membranes. On the other hand, an excess of these elements of the blood brings exceptional redness.

Deficiencies of certain blood elements may interfere with clotting and give rise to spontaneous bleeding into the skin or appear as a bruise, or purpura. When such a bruise occurs without an injury, it is often serious and requires a prompt search for the cause.

Certain general diseases often show characteristic local signs. For example, the tongue may exhibit an excessive slickness and redness in pernicious anemia. Leg ulcers may be a manifestation of sickle-cell anemia. Skin or gum lesions are common in leukemia and should be reported to a physician.

Signs of Metabolic Disease

Errors of metabolism often bring skin lesions. In one such disease, porphyria, extreme light-sensitivity appears. Patients develop excessive hair, blisters, and heavy pigmentation on the exposed parts. Diabetes, which is also a metabolic disease, likewise produces characteristic signs of deep-seated trouble.

Gout, low thyroid level, and many other generalized diseases cause skin changes. At times, a biopsy is taken to establish the cause of the rash. Xanthomas, which are yellow thickened areas in skin below the eyes and elsewhere, are the most common lesions in this group, and their diagnosis is generally apparent on simple inspection. Every person with a xanthoma requires a thorough investigation. In many such individuals profound changes appear in the chemical content of the blood, with a great increase in cholesterol. Many of these persons are candidates for hypertension, diabetes mellitus, or coronary disease.

The diabetic may have yellowish brown, thickened patches in the skin, particularly on the legs. Ulcers develop at points of pressure. Persons with diabetes mellitus are prone to infections, these often appearing in the skin as boils or fungus infections.

Infections

Impetigo, a staphylococci infection that is readily transmitted from one person to another, usually is superficial and confined to the skin. Occasionally, the infection is not confined and may be followed by nephritis, an inflammation of the kidneys.

Characteristic lesions of syphilis in any of its three stages may appear in the skin. Diagnosis is usually confirmed by laboratory tests.

Leprosy, a common chronic disease in certain parts of the world, affects the skin, usually producing typical appearances and symptoms. These include a heavy thickening of the skin, particularly that of the face. The skin becomes coarsened with heavy cellular growth giving these people a "lion-like" appearance. Fortunately there are now drugs that can control leprosy with relative certainty. Fortunately, too, this disease, though infectious, is not readily communicable.

Tuberculosis is another widespread disease. While the primary site of tuberculosis infection may be an internal organ such as the lung, skin manifestations may be the first clue to a diagnosis. Such infections are serious and require active professional treatment.

These dermatologic lesions are but a few of those commonly occurring. It is important to remember that while a disease may appear in the skin, its changes may be mirroring a generalized infection or one involving a single distant organ.

While it is true that the human body is so well constructed that man can live without some of his internal body organs and can live without major portions of others—e.g., 20 percent kidney function can keep a person in relatively good health—the integrity of man's skin is vital to life. If 70 percent of a man's full thickness of skin is lost through accident he almost surely dies.

For further information on the subjects covered in this chapter, consult ***Encyclopædia Britannica:***

Articles in the Macropædia	*Topics in the Outline of Knowledge*
SKIN, HUMAN	422.I.9
SKIN DISEASES	424.L.7

When a small amount of muscle or other tissue is removed from the body, it can be examined under the microscope for signs of disease. The tissue at the left shows signs of excess urate buildup, and hence, gout.
Photo, Camera MD

Van Doren. *This chapter is about support, protection, and locomotion — right?*

Kessler. *Right. The skeleton provides the framework and the levers for all other organs and tissues. Skull, ribs, and pelvic bones protect the soft tissues. Spine and long bones (arms and legs) are the articulated structures on which the muscles act in locomotion. And the range of motion is, of course, very wide.*

Van Doren. *For example, just standing or sitting.*

Kessler. *Running, or climbing stairs or a mountain.*

Van Doren. *Or playing the piano. What a machine!*

3

Musculoskeletal System

BY WILLIAM J. KANE, M.D.

The ballet dancer pirouetting on the stage, the pitcher hurling a baseball at nearly 100 miles per hour, the housewife pushing a cart in the supermarket, and the student writing his homework: all are using the musculoskeletal system to achieve controlled motion—motion that can be graceful, powerful, rhythmic, and precise. The bones and muscles that constitute this system account for over half the total body weight of the adult human.

The human body is supported by an internal framework, or skeleton, of 206 bones, which account for approximately 12 to 14 percent of the body weight. The skeleton supports the softer tissues surrounding it, and in some cases the bones form protective cages around internal organs. Thus the heavy, bony structure of the skull protects the brain, eyes, sinuses, and the semicircular canals of the inner ear that give man his sense of balance. The rib cage protects the heart and lungs. The intricate network of bones forming the spinal column protects the spinal cord. However, the major function of the skeleton is to serve as a series of levers for the muscles, permitting the body and its parts to move. Two additional functions of the skeleton are to provide storage for such minerals

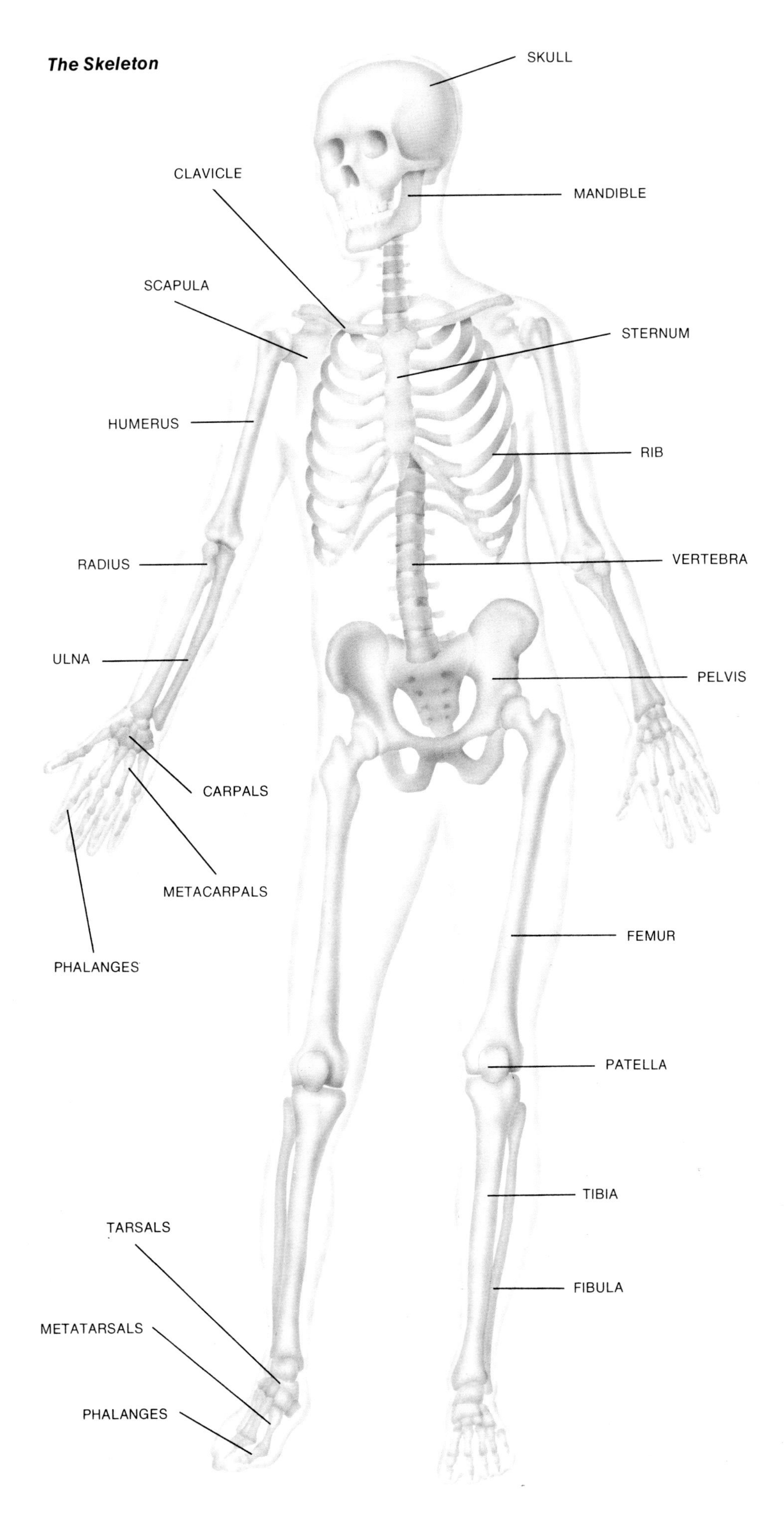

as calcium, phosphorus, magnesium, and sodium that are needed by the body and to serve as a storehouse for the blood-forming tissues of the marrow.

Since a continuous rigid framework would prevent motion, the bony skeleton is divided into separate bones joined by softer tissues that are, ordinarily, sufficiently flexible to allow motion to take place. The connection between two or more adjacent bones is called a joint, and the surface of one bone that comes into direct contact with another at a joint is called a joint surface. The joint surface is covered with a layer of smooth, glistening cartilage which makes motion between two bones much easier. In joints where little motion is required, the adjacent bones are attached by tough bands of connective tissue.

Joint motion is controlled and produced by 434 individual muscles, which make up 45 percent of the adult body weight. The muscles are attached to the bones, but the bones are held in place by all the tissues that surround them. These tissues include the ligaments, fibrous bands that run from one bone to another, and the joint capsules, broad sheets of fibrous tissue that completely surround the joint. A slippery tissue known as synovial membrane forms the inner lining of the joint capsule.

Bony Framework of the Body

A chemical analysis of bone tissue shows that it consists of about 20 percent water and 80 percent solid materials. Thirty-five percent is organic protein and 45 percent consists of minerals such as calcium and phosphorus, which give bone its hard and rigid characteristics. This bony tissue is produced by cells that gradually become incorporated into the tissue they are producing. During the initial stage of bone production, these cells are known as osteoblasts. When bone production slows down, they become less active and are known as osteocytes. Later, as changes in the shape of the bone become necessary because of growth and repair and replacement, these same cells are able to break down bone and are then called osteoclasts. This constant remodeling process allows for the growth of bone, for the repair of bones that have been fractured, and for the processes of aging, which are accelerated in certain disease states. It is an extremely interesting and mysterious process whereby one cell has the capacity to make bone at one point in time and destroy it sometime later.

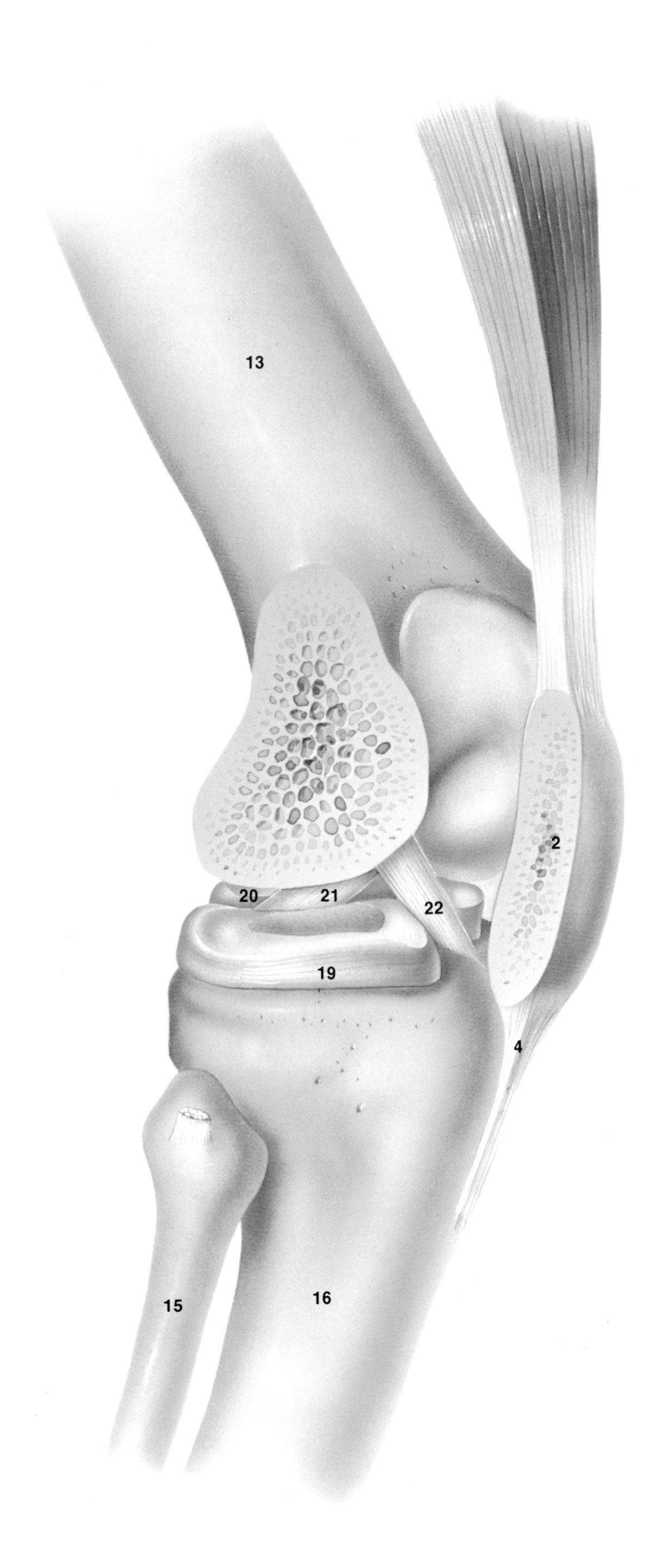
13
2
20
21
22
19
4
15
16

The Knee

1. Skin, 1
2. Patella, or knee cap, 2, 3
3. Quadriceps femoris, 2
4. Ligamentum patellae, 2, 3
5. Vastus lateralis, 1
6. Fascia lata, 1
7. Gastrocnemius, 1
8. Soleus, 1
9. Biceps femoris, 1
10. Peroneus longus, 1
11. Tibialis anterior, 1
12. Extensor digitorum longus, 1
13. Femur, 2, 3
14. Fibular collateral ligament, 2
15. Fibula, 3
16. Tibia, 3
17. Synovial membrane, 2
18. Tendon of popliteus muscle, 2
19. Lateral meniscus, 3
20. Medial meniscus, 3
21. Posterior cruciate ligament, 3
22. Anterior cruciate ligament, 3

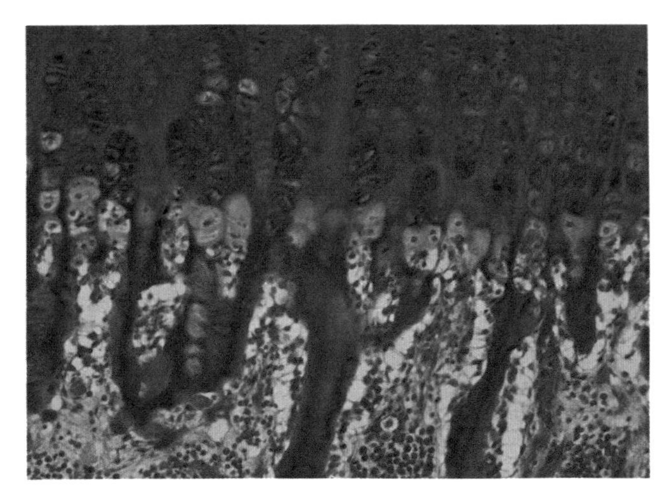

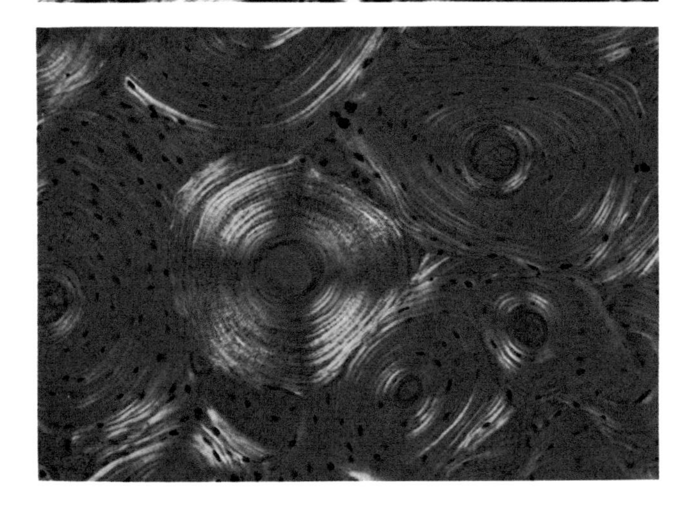

The skeleton of a fetus goes through a cartilaginous stage during which certain cells (upper photo) become converted into cartilage. The cartilage of the skeleton eventually becomes ossified. Under low-power magnification, a cross section of bone shows many holes called Haversian canals and the concentric layers of bone that surround them (middle photo). Under polarized light, the Haversian canals are easily seen (lower photo).

Several factors are responsible for the strength of bone. One is the combination of minerals and salts that bone contains. Another is the fibrous tissue and ligaments that help to support bone and give it added strength. Finally, the basic units of bone, the microscopic tubular osteons, are constructed in such a way as to resist bending equally in any direction. Osteons also provide for storage of materials that the body may need at some future time.

Skeletal Growth

Nowhere are the mystery and excitement of human growth more apparent than in the development of the skeleton. In a sense, an individual has three different skeletons at different ages. The first skeleton is formed of cartilage, which is flexible and adequate for the fetal period when the mother's uterus protects the child and when motion is limited. The second, during the first eighteen months of life, is composed of primitive bone. More rigid than cartilage and more resilient than adult bone, this is adequate for the generally prone infant. As more support and protection are required, the primitive bone is replaced by finer grained bone, which is denser, stronger, and is oriented to the lines of stress of motion.

About twenty-eight days after conception, limb buds appear at the site of the infant's future arms and legs. By four and one-half weeks blood vessels and nerve trunks are growing into the buds. Within the limb buds a spongy network of branching cells develops, and this produces a semifluid jelly that gradually stiffens, forming primitive connective tissue and cartilage.

About the fifth week the semifluid jelly enlarges and separates into areas of differing density. As the fluid becomes thicker and denser, it forms cartilage. As the cells divide and produce new cartilage, the entire mass grows both internally and externally. The internal growth continues as long as the matrix is plastic enough to allow such expansion. Growth on the outside of the cartilage mass occurs by means of the cartilage cells on the surface. These outer cells produce cartilage and are gradually buried within it.

About eight weeks after conception, the general shape of the limb can be seen under the microscope, and even the developing joints can be identified. Subsequently, as the cartilage masses continue to grow, they are gradually surrounded by a ring of fibrous cells that, in turn, is invaded by capillaries, or tiny blood vessels. This invasion causes the destruction of the previously formed cartilage. As the mass of capillaries advances and penetrates deeper into the cartilage, it lays down a type of protein known as osteoid, which becomes more dense and is later impregnated with bone minerals, mainly calcium and phosphorus. Thus a ring of bone is formed around the midportion of the capillary tissue. This becomes the early marrow space, which in turn separates the two narrow, distinct masses of cartilage.

A fibrous capsule, called the periosteum, forms around this entire developing bone. The periosteum separates into two layers, a tough fibrous outer layer and a cellular inner one. The cells of the inner layer, called osteoblasts, gradually produce the calcified protein of bone and are again buried within it. As successive layers are laid down, the compact portion of the bone shaft is formed. In the typical long bone this process is repeated at the bone ends until finally there is a bony mass at each end of the bone separated from the central bony mass by a thin plate of cartilage, known as the growth plate, or the epiphyseal line. Birth occurs at about this stage.

After birth, further growth takes place at the growth plate and continues until about sixteen years of age in girls and eighteen in boys. A person's final height depends on a number of factors. These include genetic inheritance, sound nutrition, the presence of normal hormonal stimulation, normal nerve and muscle function, and, finally, the absence of serious disease states, particularly of the bones and joints.

X rays can provide information about the growth and development of the skeleton and can give evidence about an individual's "bone age." This should approximate his chronological age. A great discrepancy between the two signifies an abnormality that will lead to either delayed or early completion of growth. It is possible to determine unusual bone growth by comparing X rays, usually of the hand and wrist, with X rays that have been obtained through examination of many normal children. Frequently this information is used to determine whether a child will grow any taller. It also assists the physician in deciding on the type of treatment in certain skeletal diseases.

The bones are classified according to their shape when growth is completed. They may be long (the tibia in the lower leg or the humerus in the upper arm), short (the small bones of the wrist), wide (the bones of the skull), or irregular (the vertebrae of the spinal column).

A typical long bone consists of a shaft, or diaphysis, two flaring portions called metaphyses, and the two end portions, or epiphyses. The epiphyses are covered on their blunt surfaces by joint cartilage, and while the bone is growing they are separated from the metaphyses by the growth plates. The tubular outer shaft of the diaphysis is made up of compact bone, while the inner canal and the ends of the bone are composed mainly of spongy bone that provides a framework for the marrow tissue. The shaft is covered with the fibrous periosteum.

The proportions of the skeleton change as the individual matures. In the infant the head represents approximately 25 percent of the total body height, but in the average adult it represents only about 12½ percent. The arms and legs grow proportionately more than the torso, and the torso grows proportionately more than the skull. The spinal column, which at birth has a basically C-shaped curve, gradually changes into a double S curve consisting of a gentle forward curve in the cervical spine about the neck, a gentle backward curve in the thoracic spine around the chest, and a gentle forward curve in the lumbar spine around the low back leading into a gentle backward curve in the pelvic region. Only slight differences exist between the skeletons of the adult male and female, and these stem largely from the childbearing functions of the female; the female pelvis is shallow and broad with a wider pubic arch. The male skeleton also tends to be larger and heavier.

When growth is completed, the cartilaginous growth plates disappear, and no further longitudinal growth of the skeleton occurs. Later, as aging begins, structural changes occur within the

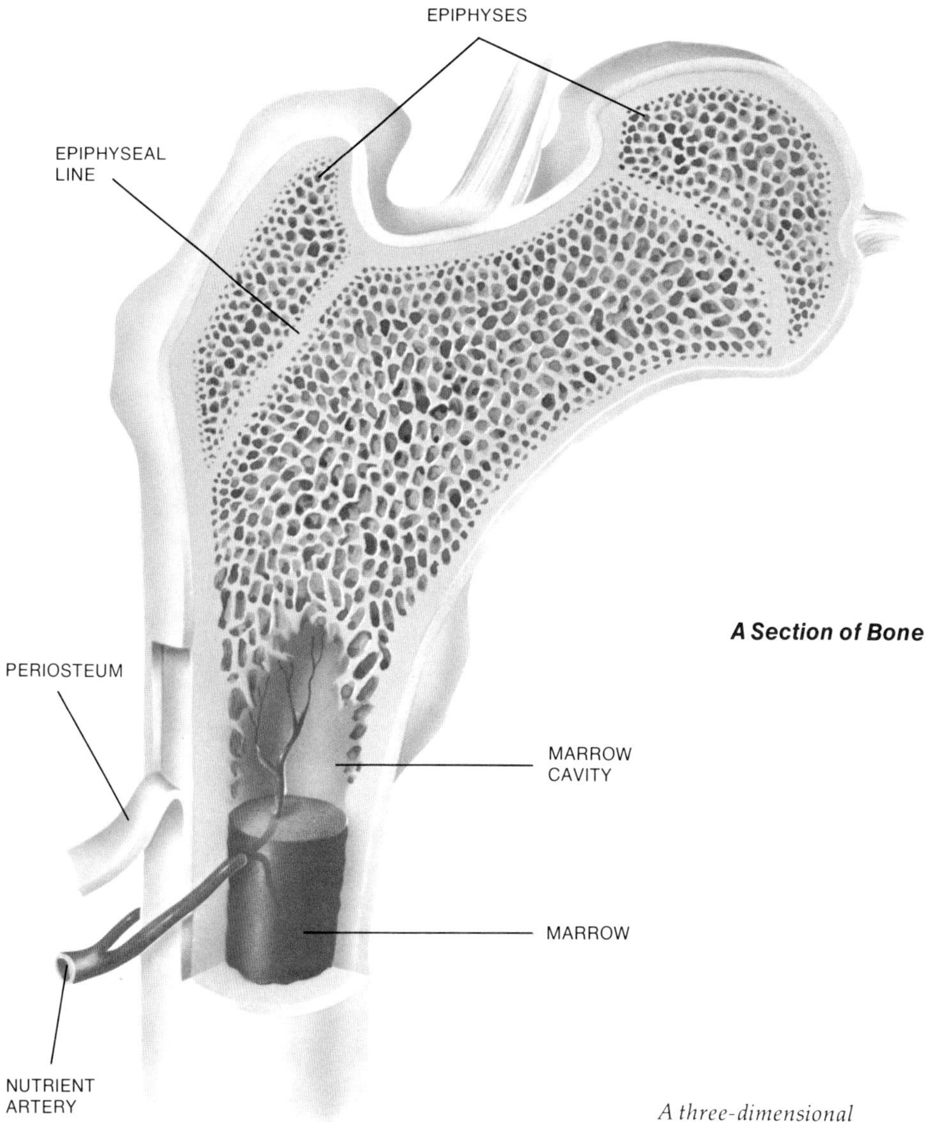

A Section of Bone

A three-dimensional view of the upper portion of the femur, one of the body's long bones, shows its inner structure. The function of the parts is explained in the text. Bone is one of the hardest of all body structures.

Types of Fractures

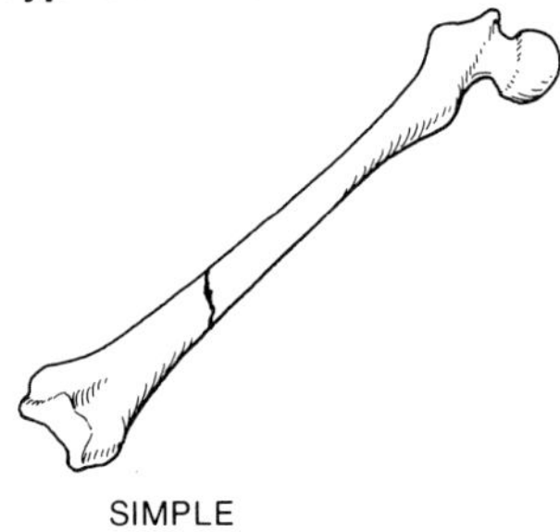
SIMPLE

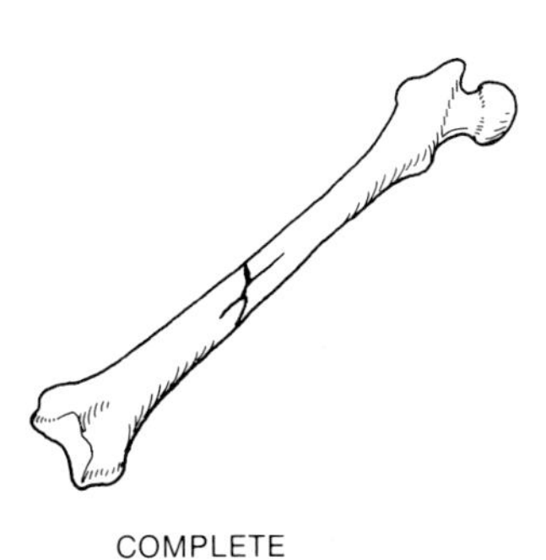
COMPLETE

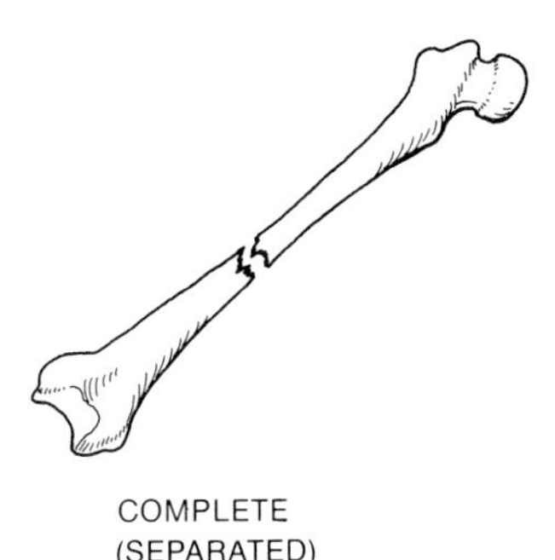
COMPLETE (SEPARATED)

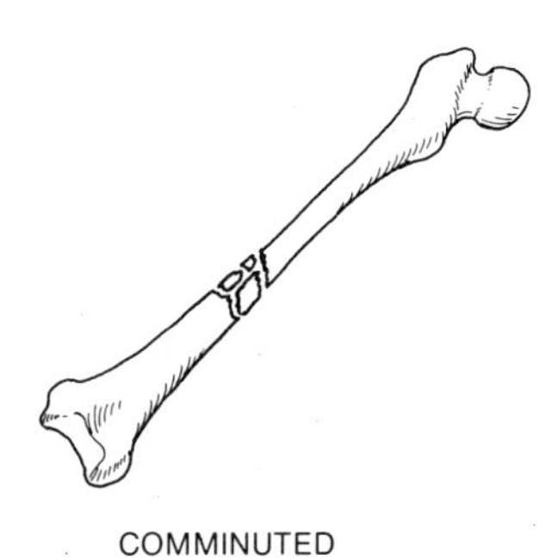
COMMINUTED

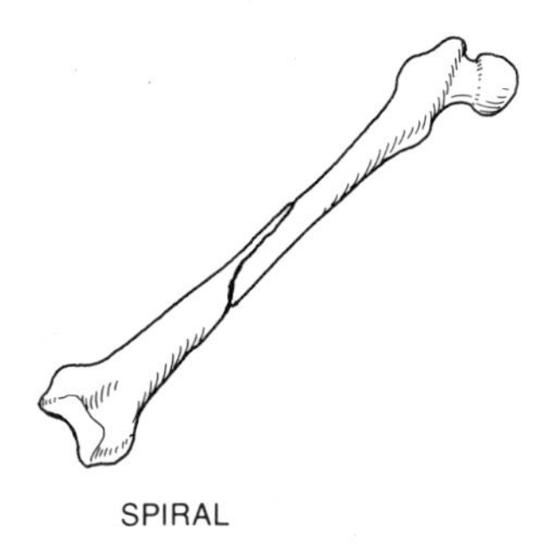
SPIRAL

bones, but no alteration occurs in their general outline. The height of an individual may diminish as the spaces between the vertebrae are reduced through degenerative processes and/or failure to maintain good posture. Occasionally, hormonal and nutritional imbalances may cause softening of the bones, leading to small fractures within the vertebral column and some additional shortening of the individual's stature.

Bone Fractures

A fracture is a break in a bone. Fractures may result from direct violence, such as a blow or a bullet, indirect violence such as an excessive twisting motion, seen for example in skiing injuries, or from a sudden, violent muscle contraction in which the muscle power is greater than the strength of the bone. If the skin is broken and the fracture is exposed to the air, the injury is called an open, or compound, fracture, but if the skin has not been broken, it is called a closed fracture. This distinction is important because an open fracture is much more susceptible to infection and delayed healing. Fractures are also classified by their shape and by the amount of displacement of the ends of the bone fragments. Regardless of the type of fracture, there is always some injury to the soft tissues surrounding the bone. This includes bruising and tearing of muscles and hemorrhage due to ruptured blood vessels. Since nerves are frequently in the vicinity of bones, they too may be injured.

The signs and symptoms that help a doctor to diagnose a fracture include pain, swelling, stiffness, and loss of function, all of which result from the associated soft tissue injury. The fracture itself causes a deformity or change in the shape of the limb and, if the fracture is completely through the bone, allows motion at a place where there should not be any. On occasion there will be a scratching or crunching sound as the bone fragments rub against each other. X rays are commonly taken when a fracture is suspected; they help the doctor ascertain the presence and exact type of fracture. Later, X rays are taken to show the amount of healing that has taken place at the fracture site.

The healing of bony fractures begins immediately after the injury with the clotting of the blood around the bone. Later, this blood clot is replaced by cells that have been produced from the marrow cavity and from the periosteum. These cells grow into the blood clot, and as they grow toward similar tissue being produced by the other bone fragment, they bring along a rich blood supply. Depending upon the type of fracture and the kind of treatment that has taken place, these cells change either into osteoblasts, which will make bone, or into chondroblasts, which will form cartilage. This mixture of cartilage and bone is called callus, and it soon becomes impregnated with calcium and phosphorus to form an immature type of bone. As this bridge between the fragments forms, it provides rigidity to the fracture. X rays taken at this time show whether healing or union of the fracture is occurring in a satisfactory fashion. Later, the immature bone that forms the early callus is gradually transformed by the remodeling action of the cells into more mature bone with a typical bone structure. In the final stage of healing, the callus tissue, which forms an external mass around the outside of the bone, is gradually trimmed back to the usual shape of the bone. After a year's time it is frequently impossible to detect, even by X ray, where the fracture occurred.

Cartilage

Cartilage is similar to bone in that it consists of cells that have been surrounded by protein which the cells have produced themselves. It differs from bone in that the protein is not calcified and, consequently, it has greater flexibility. Another distinguishing feature is that cartilage does not have any blood vessels within it. Cartilage usually occurs in thin sheets or in plates, so it is possible for nourishment to seep through the protein to the cartilage cells (chondrocytes). If the protein became too thick, the cartilage cells would be unable to receive nourishment and would die.

There are three types of cartilage. The most common, hyaline cartilage, is found on the ends of bones where it provides a smooth, slippery cushion for joint motion. Fibrous cartilage is a tough tissue made up of thick, compact bundles of fibers arranged in such a fashion as to provide great tensile strength. It is found, for example, in the discs between the vertebrae. Elastic cartilage is characterized by branching elastic fibers that provide flexibility and resilience and is exemplified by the cartilage of the human ear.

Joints

Joints are connections between two or more bones. A healthy joint will allow painless motion while remaining stable. The shapes of the bones and joint surfaces vary in different joints, depending on whether the joint permits a large degree of movement or serves mainly to give rigid support. Joints that allow little motion are called synarthroses, and the uniting layer of connective tissue may be either fibrous (syndesmosis) or cartilaginous (synchondrosis). For example, the cranial

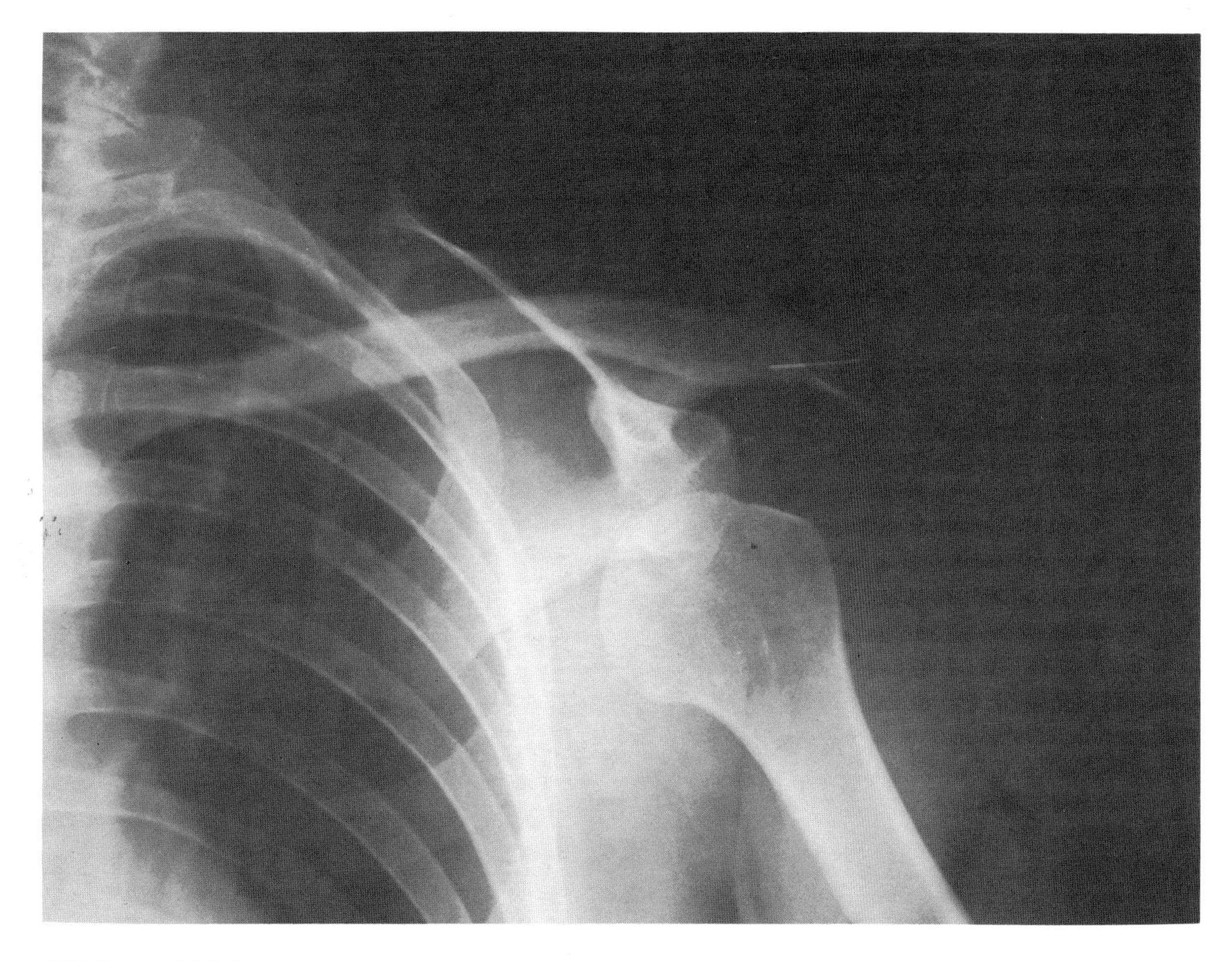

A Dislocated Joint

bones of the skull are connected by fibrous tissue; the vertebrae are connected by cartilage.

Joints that allow considerable movement are called diarthroses. They include ball-and-socket joints, such as the shoulder and hip joint; hinge joints, such as the elbow and knee; pivot joints, such as that between the first and second cervical vertebrae just below the skull; and some other, less common, subgroups. In a diathrosis a space exists between the two bony surfaces, and these surfaces are covered by hyaline cartilage, which provides a smooth, slippery surface cushion. Usually, ligaments run from one bone to the other to prevent excessive motion, and in some joints fibrocartilaginous discs help direct the type of motion. These joints are also lined by synovial tissue; this highly specialized type of connective tissue produces a synovial fluid that serves as a lubricating agent and transmits nutritional material to the surfaces lined by cartilage.

Arthritis

The word arthritis means an inflammation of a joint. This inflammation can be brought about by major or minor injuries, by bacterial infection, or by other causes. One of the most common types of arthritis, rheumatoid arthritis, is believed to be the result of an allergic reaction between the synovial lining of the joint and certain proteins within the person's own body. Arthritis also occurs in gout, a disease characterized by changes in the way a person produces or eliminates uric acid, one of the by-products of digestion.

Modern medical textbooks list more than eighty diseases that may cause arthritis. To some degree, arthritis affects 10 percent of the population of the United States. Rheumatoid arthritis, the most painful and crippling type, affects approximately 1 percent of the population. Because it is often difficult to distinguish between the various types of arthritis on the basis of a general history and physical examination, the physician frequently needs to obtain special X rays and laboratory tests in order to make a proper diagnosis. This is important, since the treatment for the different types of arthritis varies. To a large extent, successful treatment depends on determining the exact type and the degree to which the patient is affected.

Types of Joints

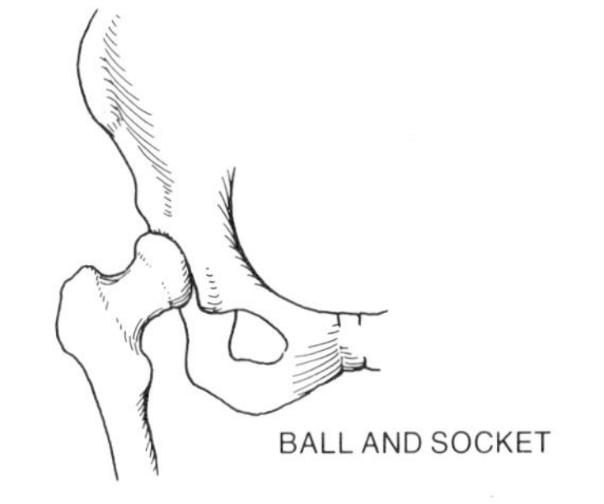

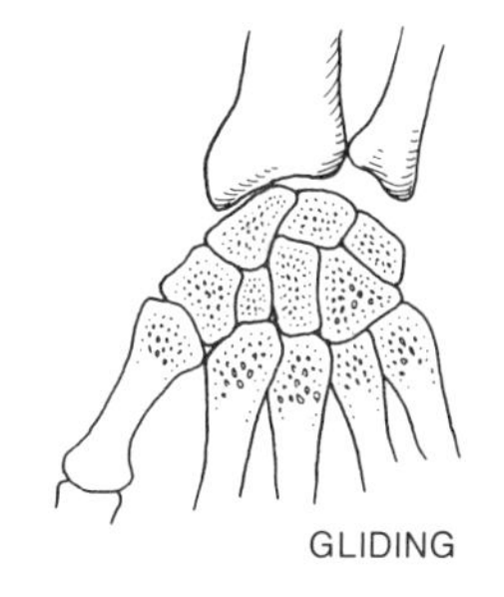

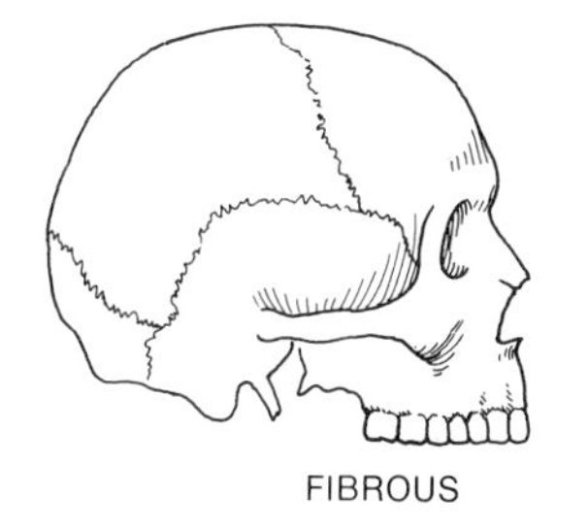

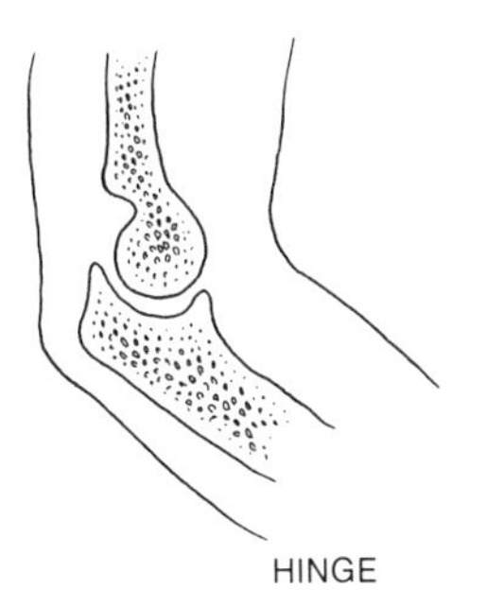

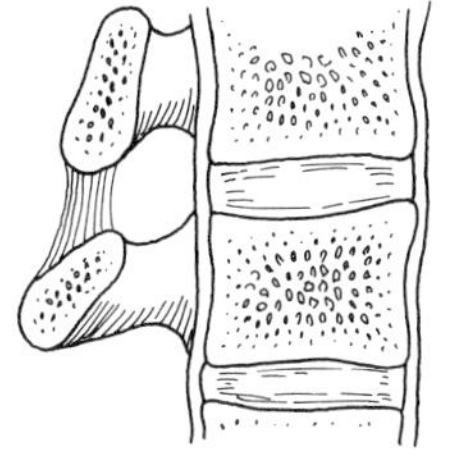

Muscles and Their Action

Some Skeletal Muscles

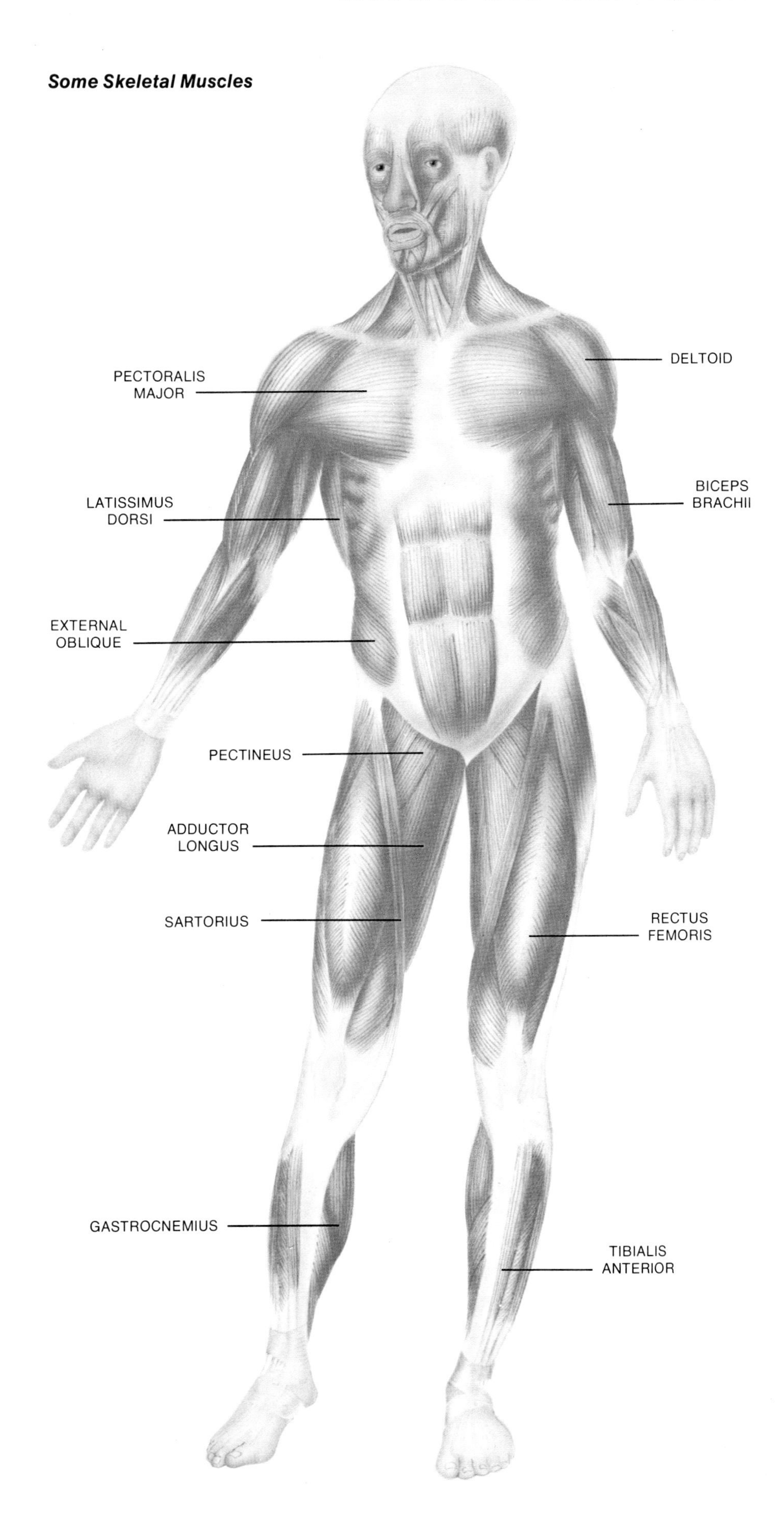

There are three types of muscle in the human body. The chief characteristic of all muscle is its ability to shorten or contract. The contraction of cardiac muscle in the wall of the heart serves to pump blood. The contractions of smooth muscle in the walls of the intestine change the internal diameter of the intestine, thereby forcing material through the intestinal tract. Smooth muscle in the walls of blood vessels serves a similar function.

Skeletal muscle has the basic characteristic of contractility and, in addition, this type is distinguished by the fact that its ability to contract is under the conscious control of the individual. Such muscle, therefore, is called voluntary muscle. The main functions of the voluntary muscles of the human body are movement, the maintenance of posture, and the production of heat.

The movements under voluntary muscle control include not only external motions but also internal motions, such as those a person performs when he takes a deep breath or moves the muscles of his face in order to smile or whistle.

The posture of the body, whether it is in a sitting or standing position, is controlled by the continued partial contraction of many voluntary muscles.

Since muscle cells are extremely numerous (constituting 45 percent of the total body weight) and active, they produce a great deal of the total body heat and are responsible for maintaining normal body temperature, especially when the body is exposed to temperatures lower than its own.

The basic unit of all voluntary muscle tissue is the muscle cell, usually referred to as a muscle fiber because of its long, thin shape. Each muscle fiber contains many myofibrils (thin, threadlike fibers), which in turn are made up of thousands of myofilaments (extremely thin fibers). The myofilaments contain molecules of myosin, actin, and other proteins. Under the microscope, regularly repeating striped areas can be seen in the muscle tissue, variations in lightness or darkness due to the varying components of each of the muscle fibers. These stripes are peculiar to skeletal muscle, which for this reason is often called striated muscle.

Muscle Contraction

Each myosin molecule is surrounded by a number of actin molecules, and there is a sliding action between the actin and myosin that allows the segment of muscle to shorten. When this action is multiplied by all the different subsections of the muscle, a considerable shortening takes place between one end of the muscle and the other. Current theory holds that muscle contraction is triggered by the release of minute amounts of a calcium ion

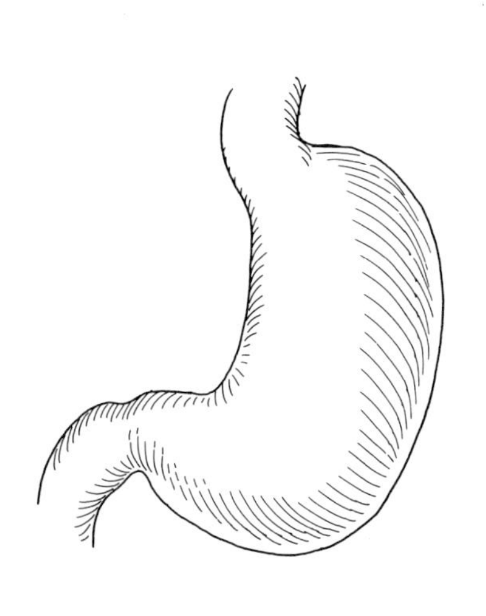

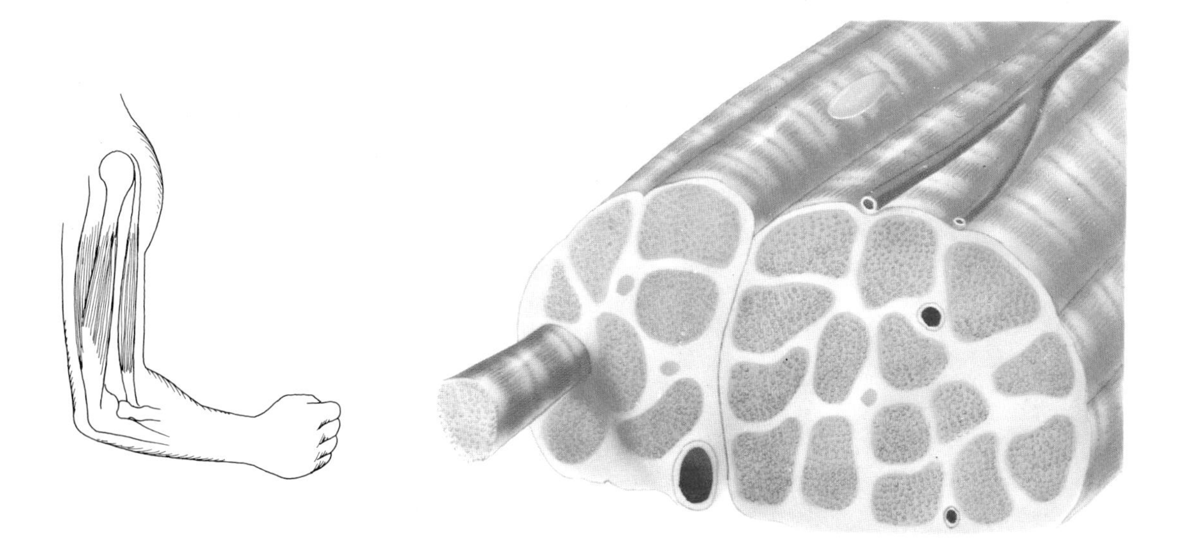

Skeletal, or voluntary, muscle is under a person's conscious control. For example, the muscles of the limbs are skeletal ones. Skeletal muscles are also called striated because of their striped appearance under a microscope.

Smooth, or involuntary, muscle is under the control of portions of the brain with an ancient evolutionary past. Called smooth because they lack stripes, these muscles are involved in bodily functions of which a person is not consciously aware.

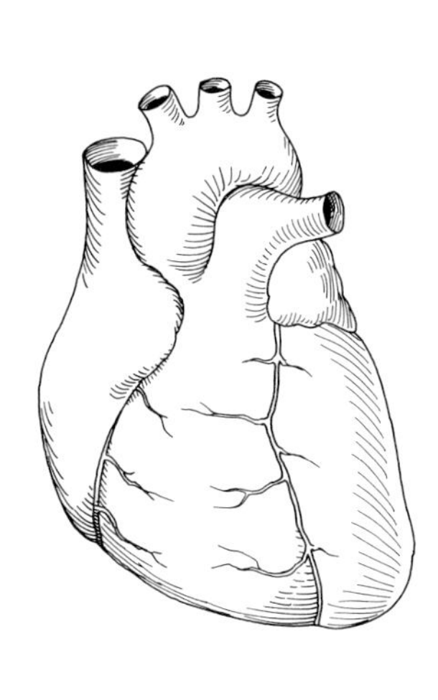

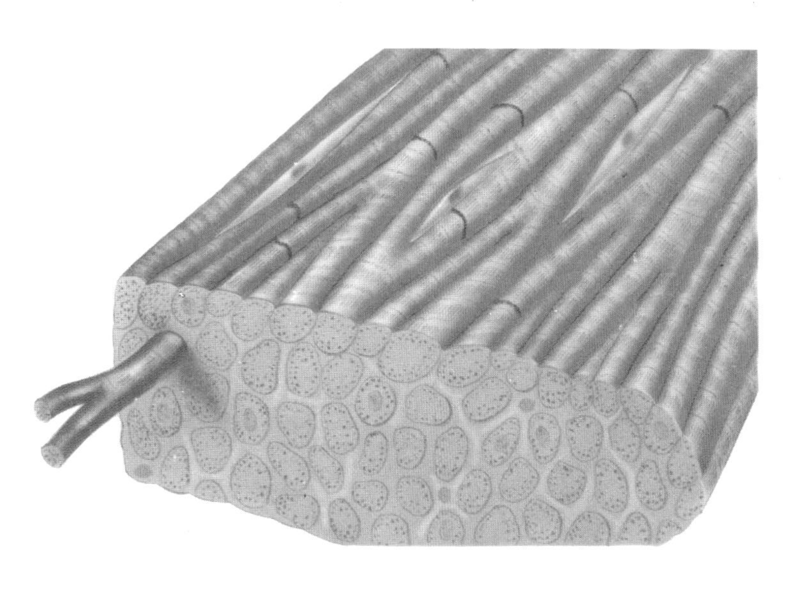

Cardiac muscle, found in the heart, is like skeletal muscle, except that cardiac fibers are interconnected to permit uniform spread of the electrical triggering of a heartbeat.

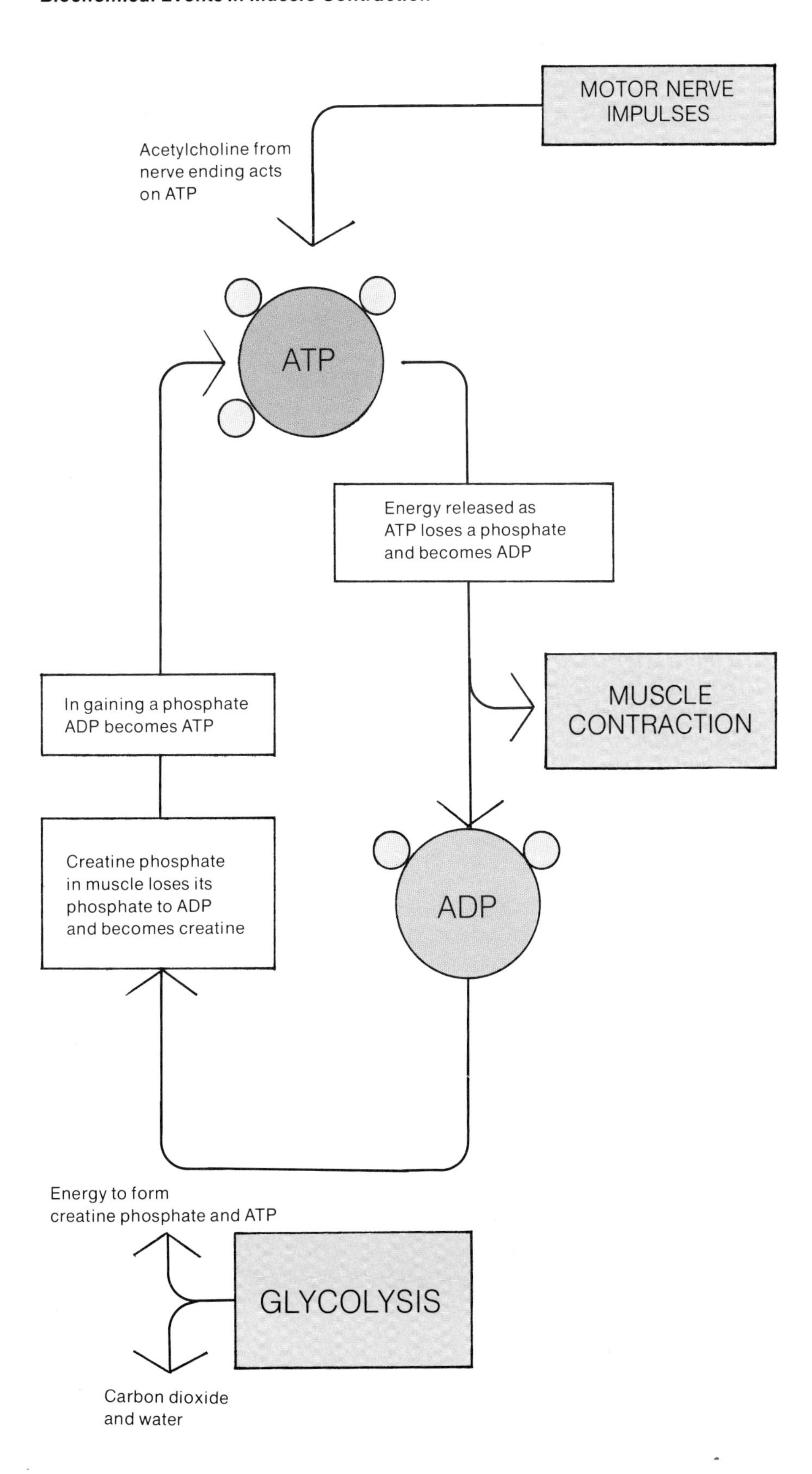

into the fluid surrounding the myofibrils, and that the muscle relaxes when the calcium ion is withdrawn.

The energy needed for muscles to contract is provided by a body chemical known as ATP, or adenosine triphosphate. At present this very complex mechanism is thought to proceed as follows: Nerve impulses that come from the spinal cord down the motor nerves reach the surfaces of muscle fibers through areas of contact known as motor end plates, or neuromuscular junctions. When nerve impulses reach the motor end plate, minute amounts of a chemical called acetylcholine trigger the breakdown of ATP. This releases energy the body can use.

Resting muscles produce more ATP than they need, and some of this ATP is combined with creatine to form creatine phosphate. This acts as a reserve supply for future energy needs. Thus, during vigorous exercise, the muscle has stores of both ATP and creatine phosphate that can be turned into energy. These complex biochemical reactions go on constantly, and since they produce heat, they also serve to maintain body temperature.

In addition to muscle tissue, skeletal muscles contain important connective and nerve tissue components. These muscles vary a great deal in size and shape, depending on their particular function and work load. The entire muscle is enveloped by a sheath of connective tissue called the epimysium, and, within the muscle, individual bundles of muscle fibers are surrounded by portions of sheath called the perimysium. Additional connective tissue compartments surround each individual muscle fiber. This connective tissue is continuous with the fibrous tissue of tendons that transmit the contraction of a muscle to distant sites. It is in this fashion that the motion of the fingers, for example, is controlled by the contraction of muscles in the forearm.

A nerve that sends impulses to a skeletal muscle is called a motor nerve. Motor nerve fibers are axons (that part of a nerve cell that carries an impulse outward) of the nerve cells in the spinal cord. These cells in the spinal cord receive their information and messages from the brain and from other parts of the spinal cord and are responsible both for the voluntary contraction of muscles and for the reflex twitch of muscles seen, for example, when the doctor strikes a patient's knee with a rubber hammer. If a motor nerve is cut as a result of injury or is no longer able to conduct normal impulses, as occurs in certain types of neuritis, the muscle it controls will no longer be able to contract, either reflexly or voluntarily. Such loss also occurs if the spinal cord cells responsible for sending impulses to the muscles are damaged by a disease such as polio. In such cases, the muscles become inactive and gradually waste away, or atrophy. On the other hand, if a person exercises his muscles regularly over a long period of time, the individual muscle fibers—and, hence, the entire muscle—will increase in size. This development is seen in athletes, especially those who do weight lifting and running.

Weakened Muscle

If the muscles are not used for a time, they tend to weaken and must be put to work to regain their normal level of strength and endurance. This is known as rehabilitation or reconditioning of the muscles. It is a necessary part of the treatment of injuries, since weak muscles are very likely to result in re-injury. A familiar example of reconditioning is the spring training camp for baseball players who have been away from their sport during the off season. Similarly, a person who has had his arm or leg in a cast finds, when the cast is removed, that his muscles may have atrophied as a result of enforced immobilization and must be exercised before he can use them.

Frequently, cramps occur in muscles that have not been sufficiently reconditioned or rehabilitated if excessive demands are made on them. Muscle cramps may also occur after injuries, either to the muscles themselves or to nearby tissues and organs. This tightening, or spasm, is usually a protective device that more or less immobilizes the injured portion of the body in an effort to prevent further injury or strain.

Motor End Plate and Associated Structures

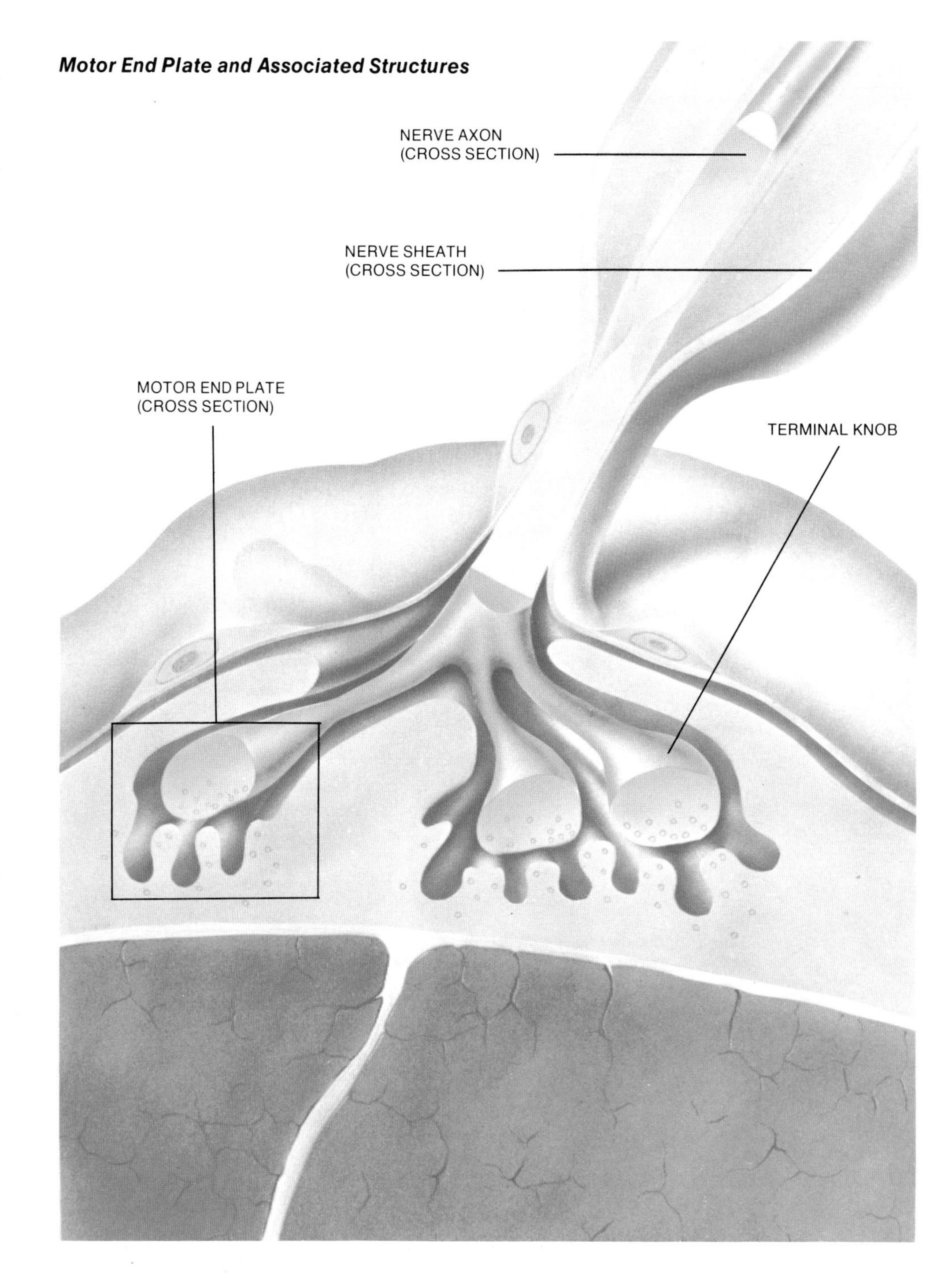

When a nerve impulse down an axon reaches the motor end plate, it causes the release of synaptic vesicles in the terminal knob of the nerve. These vesicles pass into adjacent muscle tissue and release a chemical that, in turn, triggers ATP breakdown and subsequent muscle contraction.

For further information on the subjects covered in this chapter, consult **Encyclopædia Britannica:**

Articles in the Macropædia	*Topics in the Outline of Knowledge*
BONE	422.I.2
BONE DISEASES AND INJURIES	424.L.1
BURSA	422.I.5
CONNECTIVE TISSUE, HUMAN	422.I.3
CONNECTIVE TISSUE DISEASES	424.L.2
JOINT	422.I.4
JOINT DISEASES AND INJURIES	424.L.3
MUSCLE DISEASES	424.L.6
MUSCLE SYSTEM, HUMAN	422.I.7
SKELETAL SYSTEM, HUMAN	422.I.1

The rodlike structure of enamel, the hardest part of a tooth, is shown in this striking micrograph. Enamel covers the exposed part of a tooth and protects its softer inside portions. (By permission of Microscope Publications, Ltd., from Wiemann, M.R. and Besic, F.C.: *The Microscope* 21:2; 94, 1973.)

Van Doren. *There's a lot to say about teeth.*

Kessler. *And all of it fascinating.*

Van Doren. *I suppose the most important lesson I learned from this chapter is—go to the dentist!*

Kessler. *Absolutely. The dentist has a lot to offer. In hardly any other part of the body is it so true that an ounce of prevention is worth a pound of cure.*

Van Doren. *That fine old saying is a tribute to dental progress as well as a lesson to me.*

Kessler. *You're right. Almost everyone who seeks dental care is helped—I wish that were as true in medicine.*

4

Dental System

BY NORMAN H. OLSEN, D.D.S.
AND CHARLES R. MARTINEZ, D.D.S.

The health of the mouth is important to one's general health, personality, and appearance. Disease and infection of the mouth not only detract from the appearance and disposition of a person but can also be related to general illnesses. The following examples illustrate how the mouth can be an index or warning of impending or existing infections.

Both measles and scarlet fever cause early, visible changes in the mouths of children. One of the first indications of measles can be the presence of bluish white spots (Koplik spots) on the inner cheeks two to three days before the characteristic rash appears. Scarlet fever is almost always associated with a red swollen tongue that takes on a strawberrylike appearance.

One of the earlier signs of infectious mononucleosis in an adolescent can be the presence of red spots on the soft palate or throat.

Bleeding and aching gums can be associated with viral infections and blood diseases.

A burning or sore tongue can be a symptom of diabetes or anemia.

Not only does the mouth foretell some of these general diseases, but an unhealthy mouth can also intensify or prolong an illness. For example, a mouth disease that is present prior to the onset of another illness may not only worsen but may also contribute to the severity of the general ailment. The mouth can act as a focal point for bacteria and viruses that become resistant to treatment.

Teeth, gums, and their related structures are important because they are the chief components of the mouth. It is virtually impossible to have an attractive mouth and face without teeth, and, more importantly, it is difficult to maintain a healthy, functioning body when teeth are diseased or infected.

One of the major functions of the mouth is chewing, which helps prepare food for digestion. This is done by the teeth and their supporting structures. The most meticulously prepared food can be contaminated by diseased teeth before it enters the stomach. The constant swallowing of pus and decay from diseased and infected teeth and gums can be an important factor in digestive problems, often resulting in nausea and a decreased enjoyment of food. For these reasons it is not surprising that a physical examination is considered incomplete without a thorough dental examination.

The objectives of this chapter are to acquaint the reader with the various structures of the mouth and the importance of healthy teeth and gums in preserving a healthy, functioning mouth. In addition, consequences of dental neglect during childhood and adulthood will be stressed.

Before proceeding with a discussion of the teeth, the mouth should be mentioned briefly. The mouth can be conveniently divided into upper and lower parts. The roof of the mouth is immovable and is formed by the upper jaw (maxilla), upper teeth, and the hard and soft palate. The floor of the mouth is movable and is formed by the lower jaw (mandible), lower teeth, and tongue.

Structure of Teeth

Teeth, the hardest of the body's tissues, are among the most important components in the mouth and are dependent on personal care and dentistry for their health. A tooth consists of an upper crown portion, which is that part visible in the mouth, and a lower root portion, which is not visible because it is embedded in the bony socket of the jaw. The gums (gingivae) cover both the bony socket and root portion of the tooth.

Enamel

The outer covering of the crown is called the enamel, and it varies in color from white to yellowish depending on its thickness. Enamel is normally white and translucent, whereas the underlying dentin is brownish yellow. Therefore, if the enamel is thick, less of the brownish dentin is visible, and the tooth appears whiter. Conversely, if the enamel is thin, more of the brownish dentin is visible, and the tooth appears more yellowish.

Enamel is considered dead tissue. Once the crown of the tooth is formed, no more enamel can be produced. If the enamel is broken or damaged by trauma or decay, it is unable to heal itself by forming new enamel and must be replaced by a dental filling.

Enamel provides the tooth with a hard outer surface so that the chewing of food is possible. In addition, it serves as a protective coating for the underlying softer structures. Enamel is analogous to the skin of fruits. The portion beneath the skin remains protected as long as the skin is intact, but once the skin is damaged the underlying parts then become vulnerable to injury.

Dentin

Dentin is the second layer of the tooth, beginning immediately below the enamel. Extending from the crown to the root, it provides bulk to the tooth. Dentin is softer than enamel and is considered alive because it has the ability to repair itself: if trauma or decay destroys the dentin, new dentin can be produced. However, this reparative ability is limited; it will not take place if trauma or decay is extensive.

Since dentin is alive, it is capable of producing sensations. Once its protective enamel is lost, a tooth will respond to hot and cold liquids and sweet foods. This phenomenon is experienced during early decay of a tooth when a small portion of the enamel is destroyed, thereby exposing dentin. The sensitivity to these stimuli results from the proximity and communication of the dentin with the nerves of the underlying pulp.

Cross Section of a Tooth

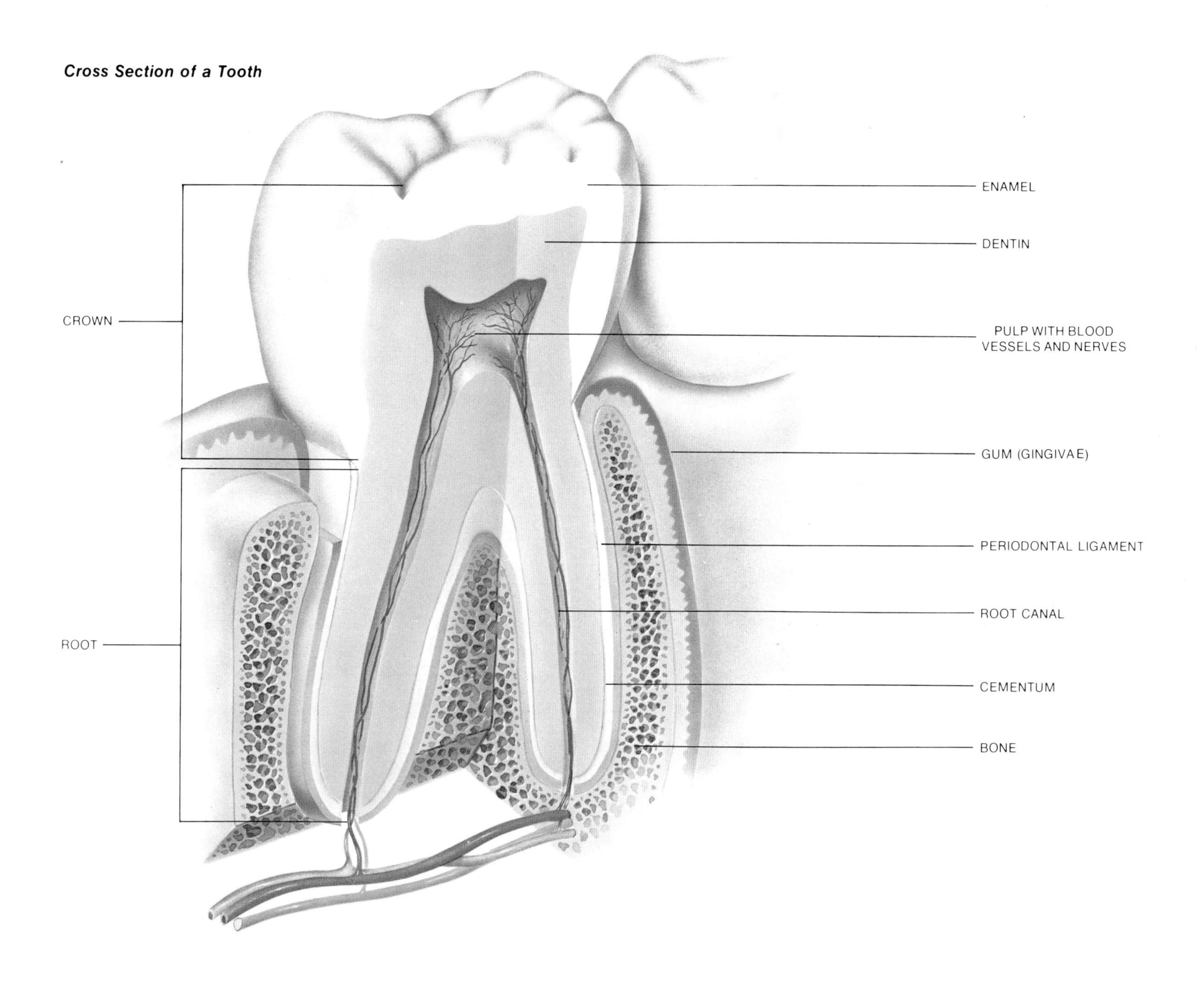

Pulp and Cementum

The pulp keeps the tooth alive. Consisting of blood vessels, nerves, and lymphatics, it provides nutrients to the dentin and initiates the repair of destroyed dentin. The pulp in the crown portion of the tooth is called the pulp chamber, and that in the root is called the root canal.

Pain fibers are located in the pulp. As long as the enamel and dentin remain intact, the nerve-rich pulp is protected. However, when trauma or decay and bacteria remove sufficient enamel and dentin to expose the pulp, the nerves there become irritated and tremendous pain results. This is why toothaches can be so excruciatingly painful.

Cementum is the substance that covers the dentin in the root portion of the tooth. The tooth is attached to the jawbone by ligament fibers.

Periodontal Ligaments

Periodontal ligaments are small fibers that extend from the cementum into the bony walls of the tooth socket. They hold the tooth in the socket and act as a shock absorber during chewing.

Function of Teeth

Man is endowed with two sets of teeth. The primary teeth meet his needs during infancy and childhood, and the permanent teeth are intended for use during the remainder of his life.

There are twenty primary teeth, ten in each jaw. They include (in each jaw) two of each of the following: central incisor, lateral incisor, cuspid, first molar, and second molar.

At maturity an adult has thirty-two permanent teeth, sixteen in each jaw, including two each of the following (in each jaw): central incisor, lateral incisor, cuspid, first bicuspid, second bicuspid, first molar, second molar, and third molar (wisdom tooth).

The incisors and cuspids cut and tear food; the bicuspids crush it; and the molars grind it.

Primary Teeth

Formation of the primary (deciduous) teeth begins about the fourth month of fetal life, and at birth the crowns are nearly complete. Thus it is important that pregnant mothers give strict attention to their diets. Sufficient amounts of calcium and vitamin D are essential for well-developed teeth.

Normally, the lower teeth erupt, or appear, earlier than the upper ones. The average eruption dates of primary teeth are as follows:

CENTRAL INCISORS	*5–6 months*
LATERAL INCISORS	*7–10 months*
CUSPIDS	*12–16 months*
FIRST MOLARS	*14–20 months*
SECOND MOLARS	*24–36 months*

The primary teeth have been called a variety of names: temporary, milk, or baby teeth. These are poor terms because they erroneously suggest that these teeth are only useful for a short period of time. This implication fosters the belief that these teeth do not need attention because they will soon be replaced by the permanent ones. No consideration is given to the fact that the primary molars are not replaced by their permanent successors (bicuspids) until ten to eleven years of age. Thus, the primary teeth function for a relatively long period of time.

All of the primary teeth are erupted by about two-and-a-half to three years of age. From three

Arrangement of Primary and Permanent Teeth

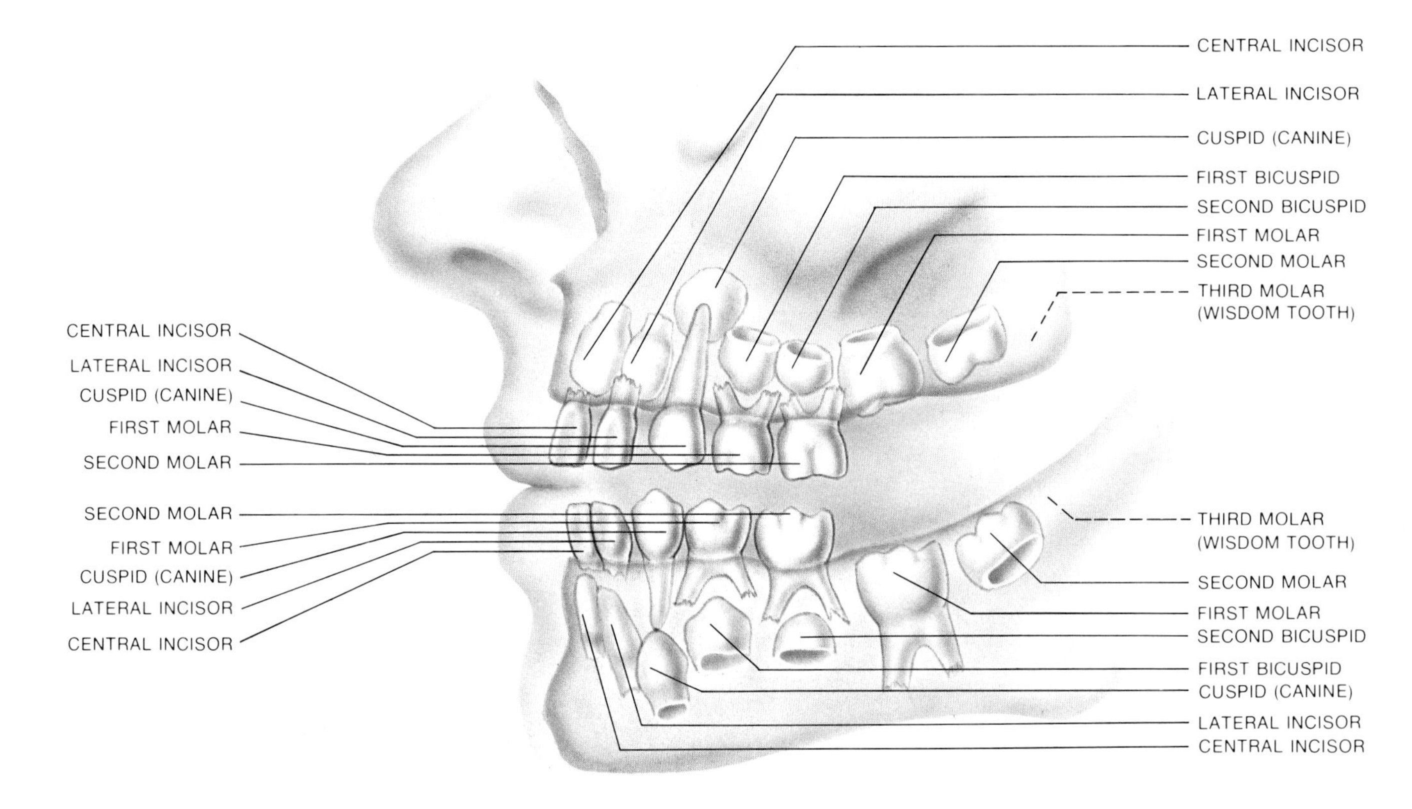

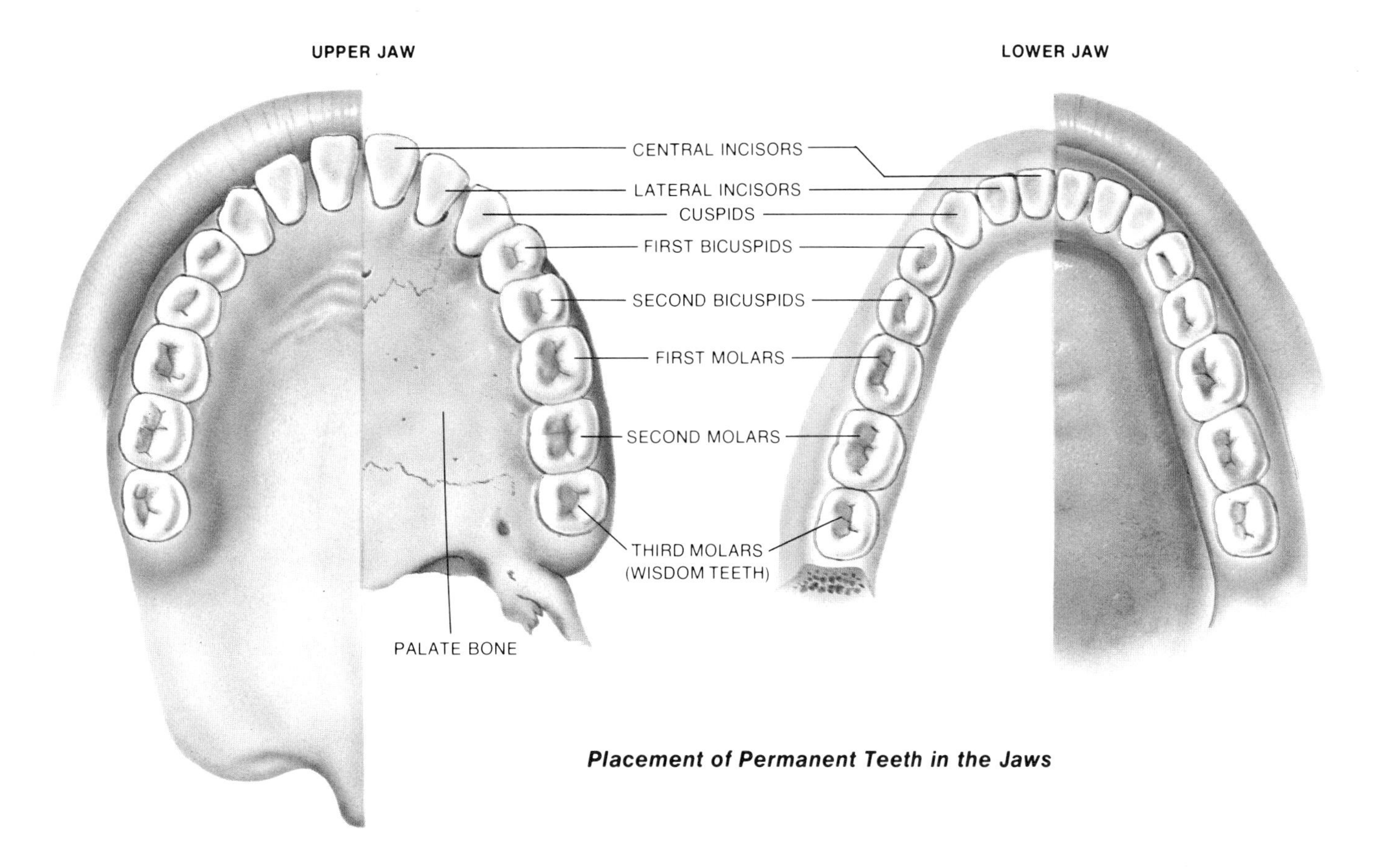

Placement of Permanent Teeth in the Jaws

years on, the roots of these teeth normally begin to resorb, or dissolve, while all the permanent teeth (except the molars) are forming beneath them. Once the permanent tooth is ready to erupt, the primary roots have all been resorbed, and the crown of the tooth falls out. Because only the crown falls out, the belief arose that primary teeth have no roots. This is definitely not true. The roots were present but have been resorbed by the erupting permanent tooth.

Importance of Early Oral Hygiene

Three years of age is the latest at which a child should be first brought to a dentist. At this time the dentist can examine the teeth to see if they have erupted properly and are in normal alignment. The parents can be instructed in proper toothbrushing techniques and counseled as to the types of foods that are least likely to cause decay. More importantly, any teeth with decay can be detected and a dental restoration undertaken at this time.

This early detection of decay is very important. The crowns of the permanent teeth are not fully formed until four to six years of age. Any infection or abscess of the primary teeth can damage the developing permanent tooth. This may result in a permanent tooth with poorly formed enamel that is weak in structure and very susceptible to decay.

In addition to endangering the crown of the developing permanent tooth, undetected decay in primary teeth can cause a toothache. The result is an uncomfortable child who is not able to eat the nourishing foods vital to proper growth and development. It is unfair to subject a child to such needless pain and discomfort when this situation can be avoided.

Permanent Teeth

The primary teeth act as a guide or a template for the erupting permanent teeth. When all the primary teeth are present, the jaws will grow evenly, and the underlying permanent teeth will be guided into a normal position in the growing jaws. If a primary tooth is lost early because of decay or infection, serious consequences may result.

Once a tooth is lost, the teeth on either side of the space will drift or move into the gap. If this happens, there will be no room for the permanent tooth, and it will erupt out of line with its intended position in the jaw. This is the beginning of irregularity in the rest of the mouth and is the most likely cause for the necessity of orthodontic treatment later.

If a primary tooth is lost, early intervention by a dentist can prevent the tipping and moving of the remaining teeth. A space maintainer appliance can be placed where the tooth is missing. This appliance keeps the teeth spaced normally and allows the normal eruption of the permanent tooth. But this appliance must be placed soon after a tooth is

Gross Structure of Individual Teeth

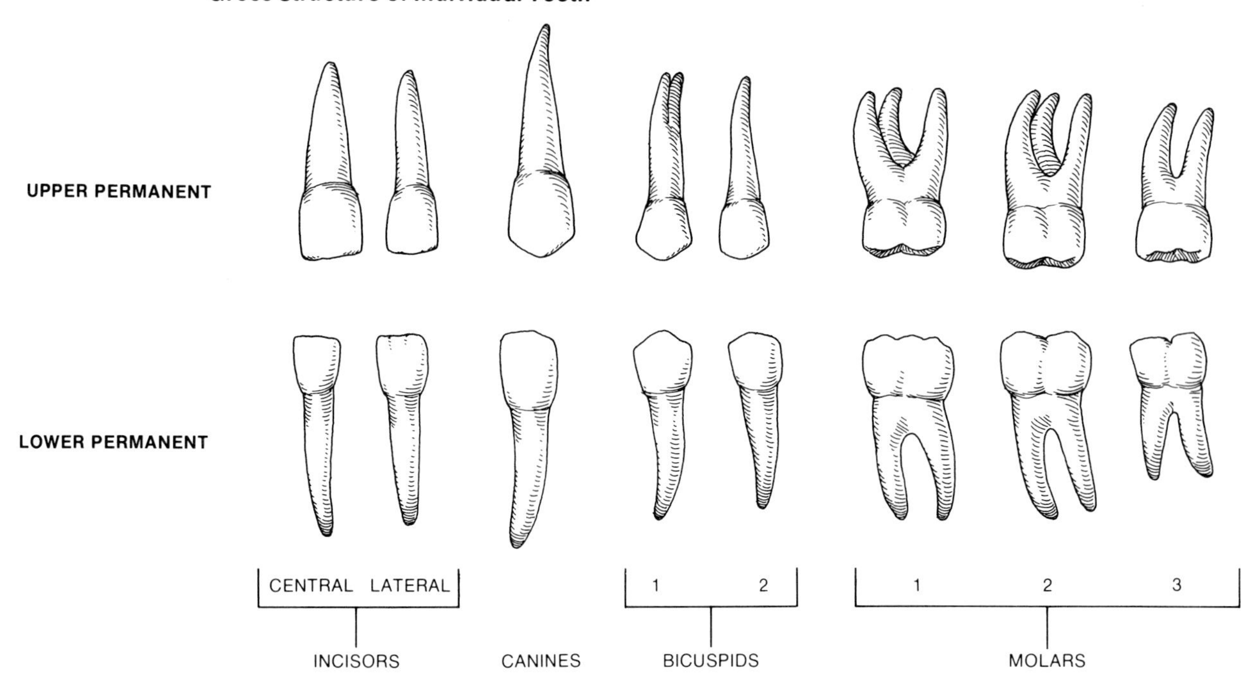

lost because the adjacent teeth move very quickly. Early and frequent dental appointments are, therefore, strongly urged.

The average eruption dates of permanent teeth are as follows:

CENTRAL INCISORS	*6–7 years*
LATERAL INCISORS	*7–9 years*
CUSPIDS	*10–12 years*
FIRST BICUSPIDS	*10–11 years*
SECOND BICUSPIDS	*10–11 years*
FIRST MOLARS	*6 years*
SECOND MOLARS	*12 years*
THIRD MOLARS	*18–22 years*

The usual pattern is for the lower teeth to erupt before the upper teeth. The above dates are averages and should not be considered strict guidelines. It is normal for some children to erupt both the primary and the permanent teeth slightly earlier than the average dates, and for some children to erupt teeth slightly later. In general, if the primary teeth erupt earlier than the average, the permanent teeth will also erupt slightly earlier, and vice versa.

The permanent teeth develop slightly behind the primary teeth. One common situation that can arise because of this is two rows of teeth in the lower incisor area of a seven- or eight-year-old child. This indicates that the roots of the primary teeth have not resorbed, making it very unlikely that they will fall out by themselves. A dentist should be consulted as soon as possible so that he can remove these primary teeth.

By twelve to thirteen years of age, all the primary teeth have been lost and the jaws are filled by twenty-eight permanent teeth. The third molars (wisdom teeth) will not erupt until much later (eighteen to twenty-two years of age) and occasionally not at all.

The permanent molars are different from the other permanent teeth in that they do not develop beneath primary teeth. Room for them is made by growth of the upper and lower jaws.

The first permanent teeth to erupt are the first, or six-year, molars. Often mistaken as primary teeth, they are frequently allowed to decay until repair is too late. This is particularly unfortunate because the first molars are the most important teeth in the mouth. The eruption of these teeth in a normal position is essential in determining the future shape of the jaws and whether the other permanent molars will be crowded or irregular.

Once all the permanent teeth have erupted, the challenge is to keep the teeth and their supporting structures healthy. In order to maintain this health, irreversible dental disease must be controlled. The two diseases that must be combatted are tooth decay and gingival disease. They can be controlled, but only through the willing cooperation of the individual patient.

Tooth Decay (Caries)

Tooth decay (caries) is a unique disease. Once it begins, it does not stop unless the process is intercepted by a dentist. Because enamel is dead tissue, it can not heal itself after it has been destroyed by caries. The lesion will only become larger and larger until the entire tooth is destroyed.

Dental caries involves three variables: the tooth, bacteria, and substrate (food). The interaction of all three is necessary before caries will begin. The following equation illustrates this interaction:

bacteria + food = acid → dissolution of enamel (caries)

However, dental plaque must be present on the tooth before the bacteria can digest the food and form the acid. Dental plaque is considered the first step in the initiation of both dental caries and gingival disease. Plaque's effect on gingival disease will be discussed later.

Plaque Formation

A film (pellicle) forms on the teeth all of the time. This pellicle is produced by saliva and in itself presents no problem. But if this film is not removed from the teeth at least once a day, the bacteria in the mouth begin to stick to it. When large numbers of bacteria become attached to the pellicle, it becomes thicker and more difficult to remove. A combination of the bacteria and the pellicle is called dental plaque.

Carbohydrates (starches and sugars, particularly sugars) allow plaque to form very fast. In addition, the bacteria will digest sugar-containing foods eaten by an individual. A product of this digestive process is an acid. If the acid forms for long periods without being removed by toothbrushing, it will begin to dissolve the tooth beneath the layer of plaque. This is the beginning of dental caries.

The caries will continue to dissolve the enamel until pieces of it break away from the tooth. The process then continues to the dentin. If caries remains untreated by a dentist, it will finally reach the pulp. When that happens, it causes a toothache.

Methods of Preventing Caries

An individual has little control over the bacteria present in his mouth, and so prevention must be aimed at plaque removal by means of toothbrushing and diet control. In addition, the tooth can be made less susceptible to acid dissolution by being treated with fluoride.

Toothbrushing and Dental Flossing. The most effective method of removing plaque from the teeth is by using a toothbrush and dental floss. Toothbrushing is best accomplished with a medium-soft nylon-bristled toothbrush. The brush is placed on the outer surfaces of the teeth with the bristles on the gingivae (gums). The brush is then moved downward as if to pull the gingivae over the teeth. This is repeated on the outside and inside of all the teeth. The brush is held slightly differently when brushing the tongue side of the lower teeth.

How a Tooth Cavity Forms

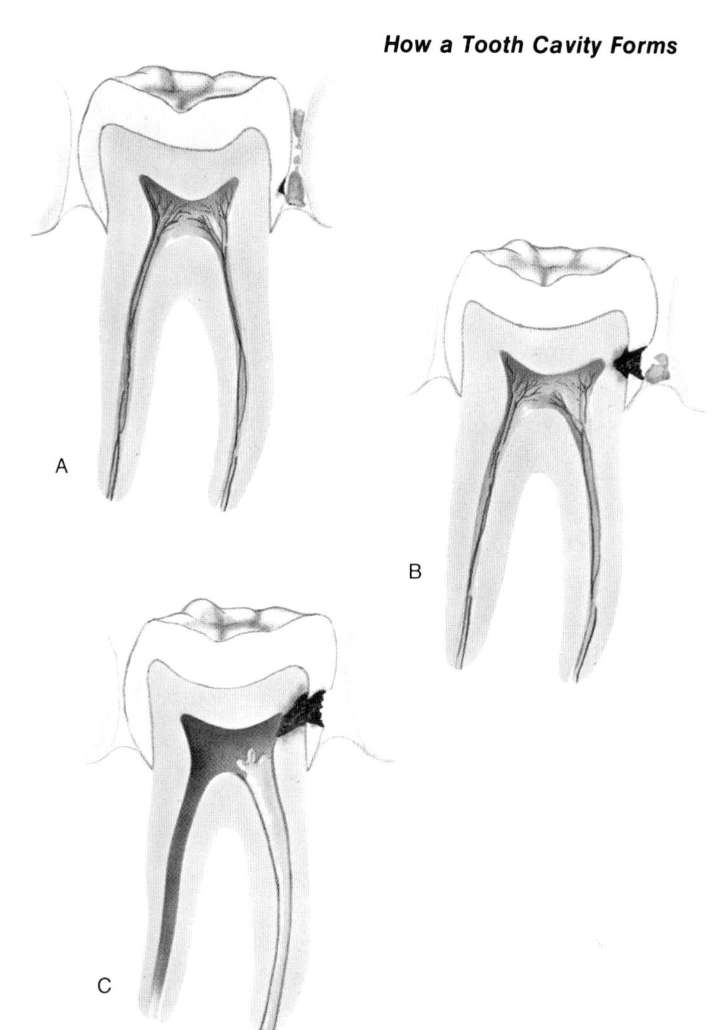

A. Food acted upon by bacteria held together by plaque produces enamel-dissolving acid. B. The dissolution, or caries, proceeds through the dentin and into the pulp. C. Nerves in the pulp are irritated and cause pain; pus forming from the infected pulp drains through the canal and forms an abscess.

Toothbrushing will remove the plaque from all surfaces of the teeth except those surfaces between the teeth. For these areas dental floss must be used. The floss is placed between the teeth and moved up and down polishing the side of one tooth and then the other. This must be done carefully to avoid damaging the gingivae. Flossing is best taught by the dentist and should not be attempted until properly demonstrated.

Diet Control. Of all foods, sugars have the most harmful effect when they interact with plaque. The bacteria on plaque digest the sugars, and the end

How to Brush the Teeth

THE UPPER TEETH

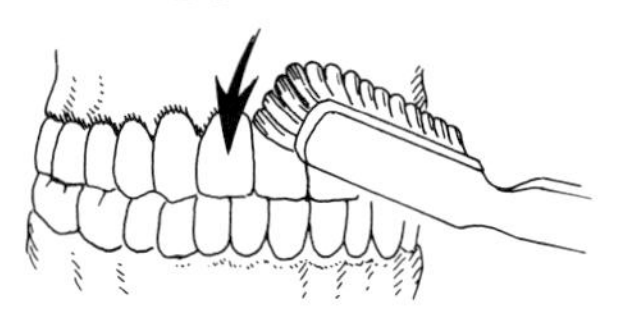

THE LOWER TEETH

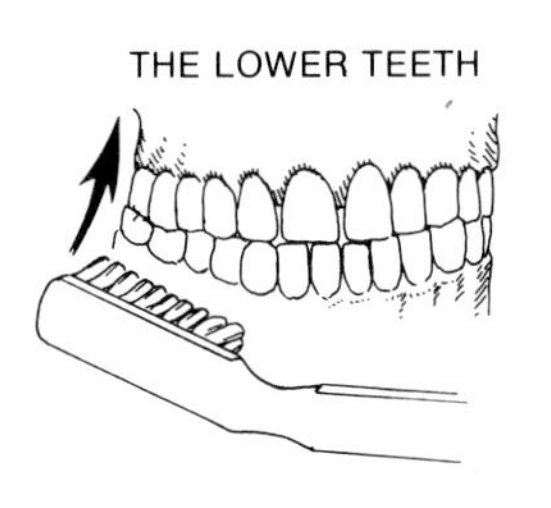

THE INSIDE OF THE BACK TEETH

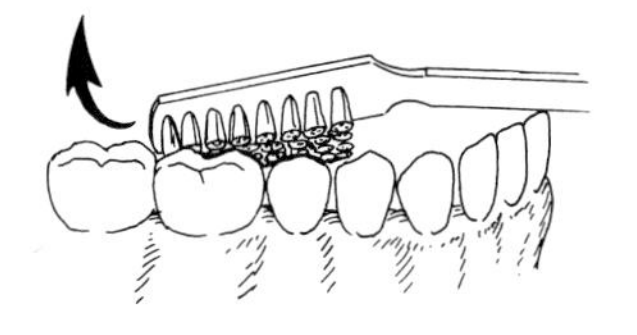

THE INSIDE OF THE FRONT TEETH

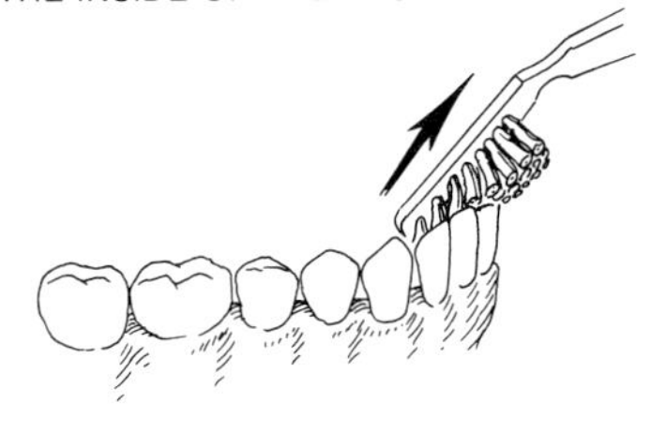

How to Use Dental Floss

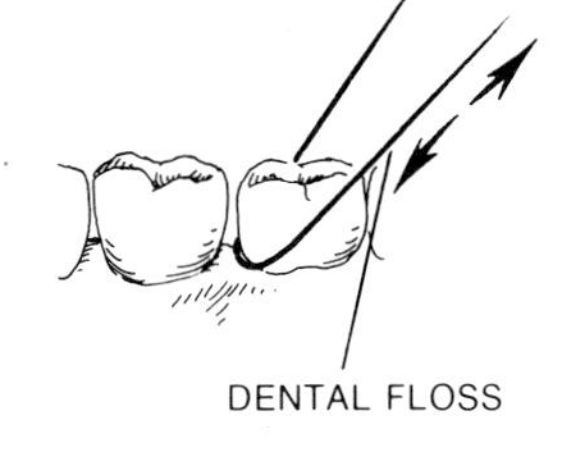

product is not only acid but also very sticky. This substance, dextran, allows more food and bacteria to stick to the plaque. This causes the plaque to become very thick, which, in turn, allows more bacteria to be present. When more bacteria are present, more acid is produced; the more acid produced, the more likely it is that the acid will dissolve the tooth.

A dentist can perform a diet analysis of each patient and determine which of the foods eaten are high in sugar content. Substitutions can be recommended that contain little sugar and thus are less likely to form excessive amounts of plaque.

Fluorides. The use of fluorides is the most effective method of making the teeth more resistant to dissolution by bacterial acid. Fluorides are most beneficial if they are ingested in drinking water during tooth formation—from birth to ten years of age. During that time the fluoride becomes part of the tooth, producing a tooth more resistant to acid dissolution. Studies have shown that such ingestion of fluorides can reduce caries by 40 to 60 percent.

At this time, research has not demonstrated the effectiveness of giving fluorides to women during pregnancy to help in developing healthier teeth in the child. Therefore, fluorides appear to have their main effectiveness on the permanent teeth.

Other methods of using fluorides to prevent tooth decay include putting fluorides on the tooth by brushing with a fluoridated toothpaste, and by having the dentist put topical fluoride on the tooth every four to six months. There is a trend toward self-administration of topical fluoride by the patient at home. This should be done only under a dentist's supervision and direction. The effectiveness of this approach is still being studied.

Myth of "Hard" Teeth

One point should be made about a myth that has persisted for many generations. There is no such thing as hard teeth or soft teeth. All have the same basic chemical composition, and, barring a systemic infection during the development of a tooth, all will be of the same hardness. A tooth that has been injured during development will be damaged, but this is usually seen only in isolated situations.

The brownish spots seen on the front teeth of some children indicate damage during tooth formation. There are two main causes of this type of defect: systemic infections and trauma. If a child has a high fever or serious illness during tooth formation, the enamel will be malformed prior to eruption of the tooth. When the tooth erupts, the saliva and bacteria will stain the defective area, resulting in brown or yellowish black spots. Trauma can cause a tooth to have a similar appearance. If the primary tooth is hit or pushed into the gingivae, the root may be pushed against the developing permanent tooth. This may damage the enamel, and when the tooth erupts, the defective enamel will be discolored.

Result of Untreated Caries

If caries is untreated, it will eventually progress to the pulp. When that has happened, there are only two ways of treating the tooth: extraction or removal of the infected pulp tissue. The latter procedure is called root canal therapy. In this treatment the pulp is removed and a filling is placed in the root canal. It is recommended because the tooth can be kept and the integrity of the mouth retained. The natural tooth will still be present, though it will no longer be alive.

If the tooth is extracted, it must be replaced with a false tooth in order to maintain the integrity of the mouth. If the space is left open, the adjacent teeth will begin moving into the space. The opposing teeth will also elongate and interfere with the chewing efficiency of the teeth. Both of these situations can lead to gingival disease or joint problems, which will be discussed later.

A missing tooth can be replaced by either a fixed bridge or a removable partial denture. With a fixed bridge, the artificial tooth is supported by two or three of the adjacent remaining teeth and no stress is placed on the gingivae or palate. The advantages of the fixed bridge are: first, it is cemented onto the teeth and will not come out unless it is broken, and second, there is less ultimate damage to the soft tissue. The disadvantage is that it requires the cutting of some of the adjacent sound teeth so that the crowns (caps) that hold the false tooth may be cemented onto the teeth.

With a partial denture, the false tooth is attached to a plastic or metal framework. This framework is then supported by the gingivae or palate and is held in the mouth by clasps on the teeth adjacent to the space. The advantage of this appliance is that the adjacent teeth do not have to be cut. The disadvantage is that it puts stress on the soft tissue and the remaining teeth. In addition, the unsightly clasps are usually visible.

The gingivae (gums) function as a protective coating for the jawbones. This is important because the jawbones hold the teeth by means of periodontal ligaments. Stated simply, once the gingivae are destroyed, the bone will be destroyed and the teeth will eventually fall out.

An infection of the gingivae is called gingivitis. If the infection also includes the bone, the condition is called periodontal disease (pyorrhea).

Periodontal disease primarily affects adults, causing 90 percent of the tooth loss in persons over thirty years of age. Gingivitis can affect anyone, from children to adults. However, in children it rarely progresses to periodontal disease, most probably because children can combat infection better than adults.

Periodontal disease follows a basic pattern. The first phase is gingivitis. When gingivitis is present, the gingivae become red and swollen and bleed easily. The first sign of this condition is bleeding gingivae after eating, toothbrushing, or flossing. Healthy gingivae do not normally bleed during these activities.

Gingivitis

Gingivitis is a direct result of plaque accumulation between the gingivae and the teeth. The waste products (toxins and acid) of the bacteria present in plaque are irritating to the gingivae. In addition, if the plaque is not removed in two to five days, it may become calcified. Calcified plaque, more commonly known as calculus or tartar, is also irritating to the gingival tissues.

Calculus and plaque lodge between the teeth and gingivae. When this has happened, toothbrushing or coarse foods will cause the gingivae to rub over the plaque and calculus. This continual irritation causes the gingivae to become swollen and inflamed, and the process will continue until an infection—gingivitis—begins.

Gingivitis requires treatment by a dentist because calculus beneath the gingivae is not easily removed by toothbrushing. In addition, since the gingivae are sore, the patient is less likely to continue brushing the teeth. As a result, more plaque will accumulate and a vicious cycle begins. At this time, the dentist can remove the calculus and instruct the patient in the procedures that will alleviate the soreness and restore the gingivae to health.

Untreated gingivitis will progress to the next stage. The infection will spread even farther below the gingivae, eventually reaching and dissolving the underlying bone. Once the bone is involved, the condition is called periodontal disease.

Gingival and Periodontal Disease

Progression of Gingivitis to Periodontal Disease

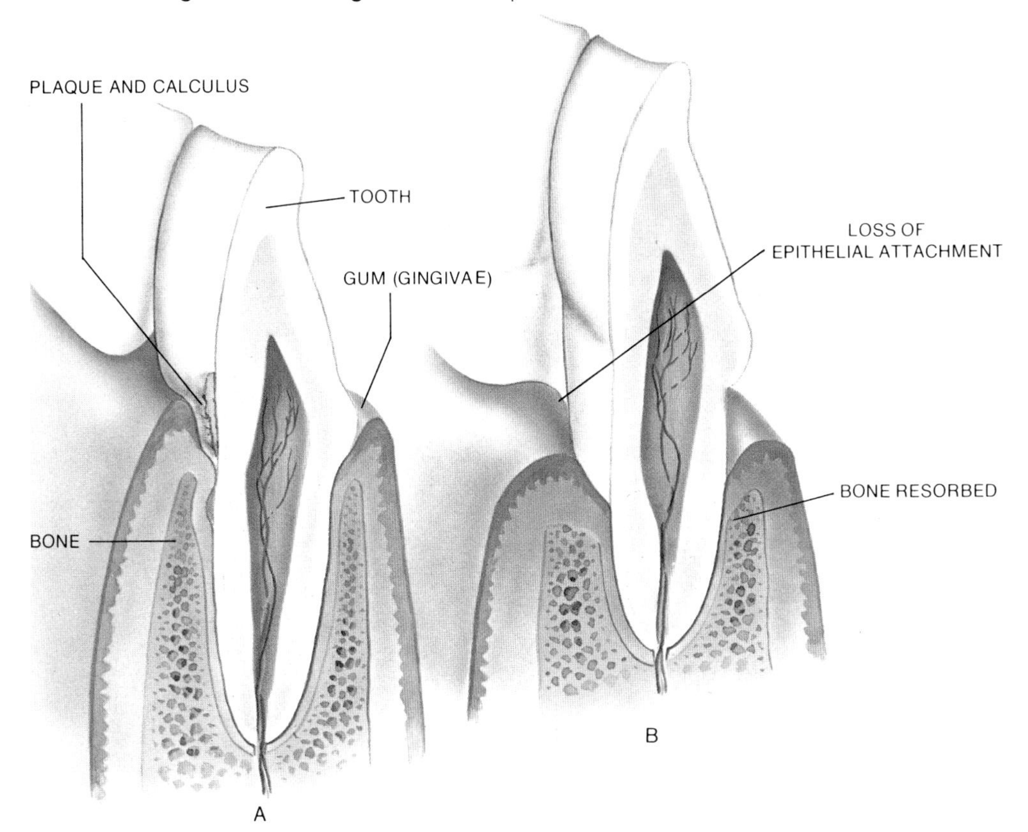

A. Gingivitis—plaque and calculus between tooth and gum causes gum to become swollen, sore, and infected.
B. Periodontal Disease—infection spreads to the bone, bone resorbs, and tooth loosens.

Periodontal Disease

As the bone is gradually dissolved, fewer periodontal fibers will be left to hold the tooth in the socket, and the tooth will become loose. This infection will continue until so much of the bone is destroyed that the teeth literally fall out. If treated early enough, the infection can be stopped, but if too much bone has been lost, the teeth will have to be extracted.

Some situations present in the mouth can worsen or even initiate periodontal disease. As mentioned earlier, if a tooth is lost, the adjacent teeth will begin to tip and move in an attempt to close the space. When teeth are tipped, more protection is afforded to plaque and bacteria, and removal of plaque consequently becomes more difficult. In addition, because the teeth are tipped they do not meet the opposite teeth normally. Instead of the upper and lower teeth meshing evenly together, only a few teeth contact one another during chewing. Those teeth that make contact have to absorb all the forces generated by the chewing. This pressure eventually loosens the teeth, making them much more susceptible to any inflammation.

Effects of Malocclusion

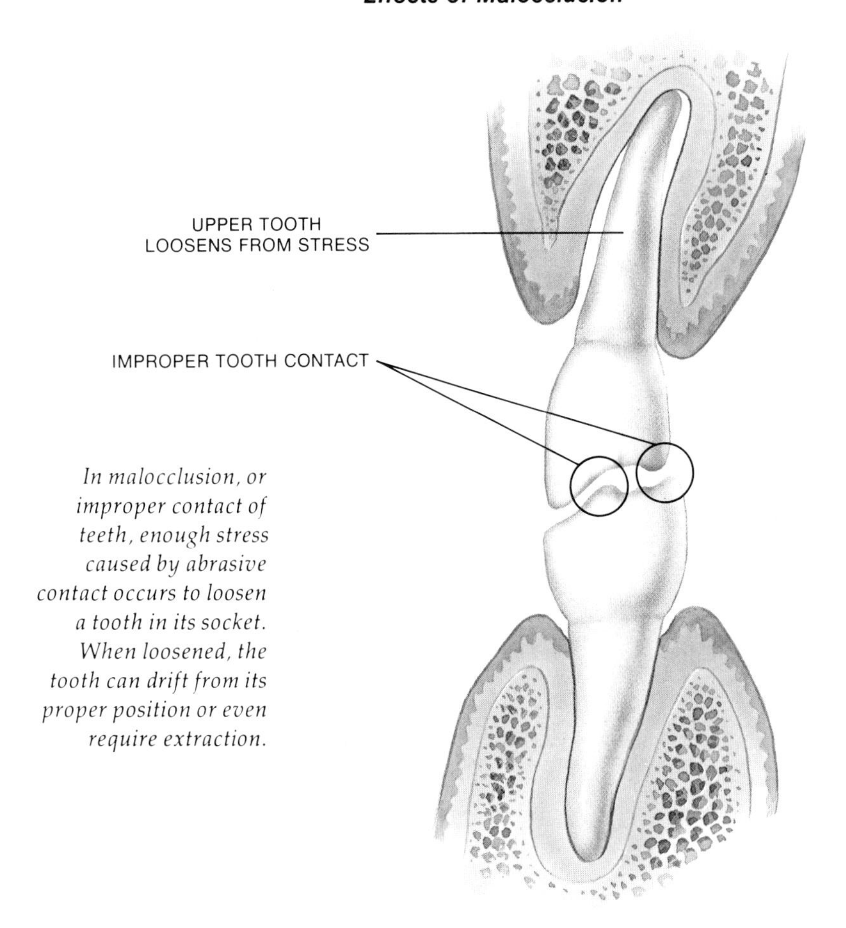

In malocclusion, or improper contact of teeth, enough stress caused by abrasive contact occurs to loosen a tooth in its socket. When loosened, the tooth can drift from its proper position or even require extraction.

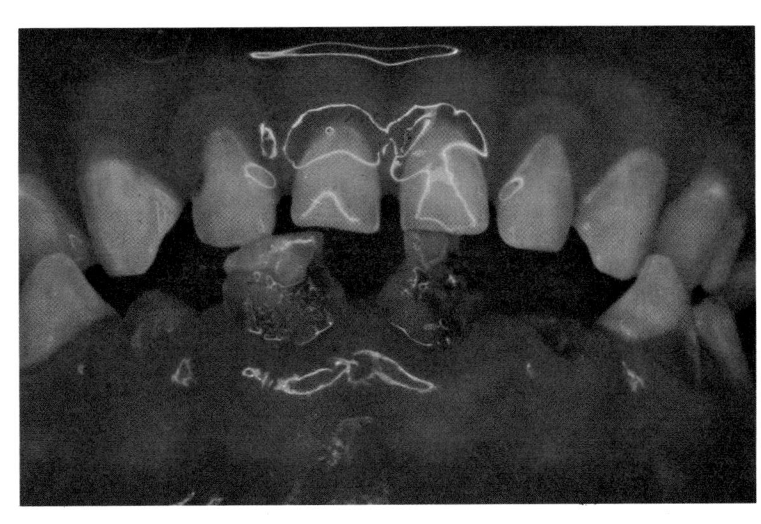

The photo illustrates a case of advanced periodontitis. The teeth have begun to drift, and the gums, now discolored, have receded because of underlying bone resorption.

Another source of irritation is rough margins around dental restorations. These act much the same way as calculus. Frequent visits to the dentist will allow him to detect and repair these defective restorations.

Prevention of Periodontal Disease

The prevention of periodontal disease depends on an individual's diligent removal of plaque between the gingivae and the teeth. This requires a different type of toothbrushing technique than that employed for caries prevention. Actually, it is best to brush the teeth and then the gingivae, treating each as a separate area.

A soft nylon toothbrush should be used for brushing the gingivae. It should be placed so that the bristles touch both the gingivae and the tooth at approximately a forty-five degree angle. The brush can then be either gently vibrated back and forth or moved in a circular motion. These motions will both massage and stimulate the gingivae, thereby increasing the blood circulation. In addition, the plaque will be effectively removed from between the tooth and gingivae.

This method of brushing is best taught by a dentist and should not be attempted until the dentist has cleaned the teeth of all plaque and calculus. If the patient attempts this brushing technique with calculus beneath the gingivae, he will increase the inflammation.

The water pick has received much attention as a means of removing plaque. However, its effectiveness for this purpose has not been unequivocally demonstrated in research studies. It is more useful as a means of removing lodged food particles from between the teeth. A water pick should not be used without first consulting a dentist. The bombardment of the water stream on diseased gingivae will do more harm than good, because it acts as an additional source of irritation. The gingivae must be healthy before this device is employed.

Possible Jaw Problems

The term occlusion refers to the relationship of the upper and lower teeth to each other when the jaws are closed. Normally the cusps of each tooth interlock with those of the opposing tooth evenly, much like gears on a wheel. In this way, the teeth are supported by each other.

The muscles of the face open and close the jaws, thereby allowing the teeth to lock and unlock during chewing. The lower jaw is the only bone of the face that moves. The upper jaw is immovable, and so chewing occurs only as a result of movement of the lower jaw.

This movement is possible because the jaw is attached to the skull in the ball-and-socket type of arrangement. The socket is formed by the skull, and the ball is formed by an extension of the lower jaw. When the lower jaw moves, the ball rotates in the socket. This arrangement is called a joint and is similar to the knee and elbow joints. Named the temporomandibular joint, it is designed so that the ball fits in the socket in a precise position. This position changes as the jaws grow and is primarily determined by the position of the teeth and muscles of the face.

When all of the teeth are in normal alignment, the ball rotates smoothly in the socket. If the teeth are lost or in poor position, the relationship of the ball and socket changes. The improperly positioned teeth push the ball back farther in the socket. The muscles that move the lower jaw then become stretched and attempt to move the ball back to the position it normally occupied. But since the maligned teeth prevent this, a tug-of-war between the muscles and the ball and socket results. The patient then begins to experience distress in this area, which is located just in front of the ear. The muscles become sore, and it becomes painful to open and close the mouth. This marks the beginning of temporomandibular joint problems. If left untreated, a degeneration of the socket may result (osteoarthritis). This is very painful and can eventually lead to permanent destruction of the joint.

III-Effects of a Missing Tooth

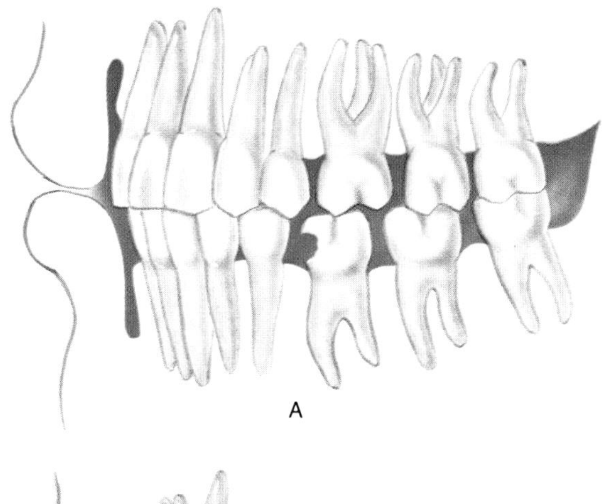

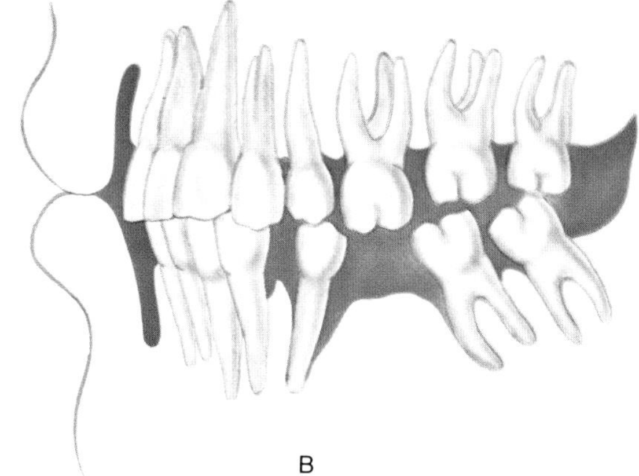

A. Caries can so erode a tooth that it must be pulled.
B. If a missing permanent tooth is not replaced by a false one, adjacent teeth will tip into the gap. As a result, chewing is affected, and gingival disease can form.

Preventive Dentistry

Probably the most important fact concerning disease of the teeth and their supporting structures is that the disease can be prevented. This prevention requires the equal cooperation of the patient and the dentist. The dentist can restore or repair damage that has occurred and, more importantly, guide the patient in procedures that will prevent the problem from reoccurring.

But after he has done this, the responsibility for maintaining a healthy mouth shifts. It is now up to the patient to follow the advice and instructions of the dentist. In our present society, with its sugar-rich diet and generally casual care of the teeth and gums, the dentist and patient must work together to control tooth decay and maintain the patient's mouth in a sound, healthy, and attractive condition.

For further information on the subjects covered in this chapter, consult ***Encyclopædia Britannica:***

Articles in the Macropædia	*Topics in the Outline of Knowledge*
TEETH AND GUMS, HUMAN	422.E.1.ii

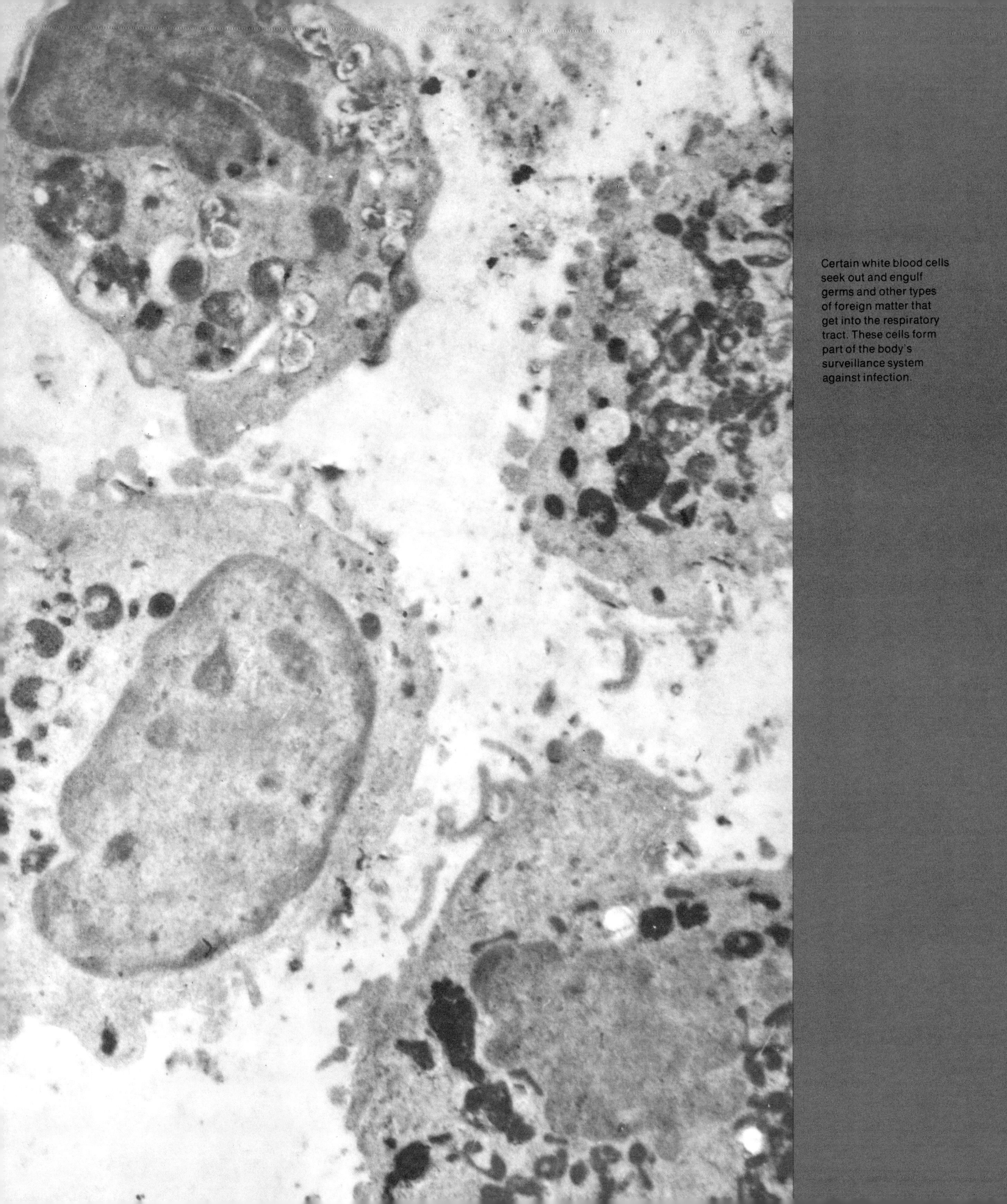

Certain white blood cells seek out and engulf germs and other types of foreign matter that get into the respiratory tract. These cells form part of the body's surveillance system against infection.

Van Doren. *Dr. Cugell, who wrote this chapter, has evidently seen a lot of patients who have trouble breathing.*

Kessler. *Naturally, he has spent his professional life doing just that. You're referring to his warning — a rather frightening one — about smoking?*

Van Doren. *I must say I've decided to try to stop — again. But the chapter is also about other things besides smoking.*

Kessler. *Yes. To me, its most important message is that, although a patient's focus is often on acute problems, there are many occasions on which a patient's long-term problems are equally crucial.*

5

Naso-Respiratory System

BY DAVID W. CUGELL, M.D.

Energy to support the function of individual tissue cells, entire organs, and the body originates not only in the food we eat but also from the air we breathe. An adequate oxygen supply is just as essential as proteins, calories, etc. Foodstuffs can be consumed intermittently and their energy-giving substances stored for subsequent use, but we cannot build up internal stores of oxygen for later use—a continuous supply is necessary. If there is an oxygen deficit because: (1) it is missing from the air we breathe; (2) there is a mechanical interference with breathing; or (3) diseases prevent absorption of oxygen, then death soon follows. Of course, starvation is just as lethal, but death takes a lot longer. The consumption of foodstuffs is associated with the production of waste products, and, in similar fashion, the utilization of oxygen as an energy source also results in the production of a waste product—carbon dioxide. The disposal of waste products, whether they originate from the utilization of foodstuffs or oxygen, is every bit as important as the intake and use of energy sources. For example, when kidney disease prevents excretion of urine, death soon follows from retention of toxic materials—unless a suitable kidney substitute, such as a renal dialysis machine, is available. Likewise, advanced lung disease, or cessation of breathing irrespective of cause, results in an accumulation of carbon dioxide that can rapidly prove fatal.

The foods we eat travel a circuitous route through the gastrointestinal system before the energy they contain can be utilized. Preliminary preparation occurs in the mouth as a result of chewing and the addition of saliva with its digestive enzymes. After a lengthy trip to the stomach via the esophagus, which is primarily a simple muscular tube that serves to transport foods from the mouth to the stomach, considerable additional modification occurs. Partially digested food moves out of the stomach into the small intestine where the transfer of energy-giving materials into the bloodstream occurs.

The air we breathe also traverses as complex a pathway as food. The air first enters the nose and throat where it is prepared in a preliminary manner; then it passes through the trachea and the major bronchi—primarily inactive pathways—to the smallest subdivisions of the lung, where oxygen is absorbed and carbon dioxide is excreted. Unlike the gastrointestinal system, which is a "one-way street" under normal circumstances, there is a back and forth flow of gas through the naso-respiratory apparatus with inspiration serving to refresh the oxygen supply in the lungs and expiration disposing of the accumulated carbon dioxide.

What specific structures comprise the naso-respiratory system, and what functions do they serve? Each will be considered in turn together with its unique anatomical features and specific functions.

The Nose

Despite its prominent location and cosmetic significance, the nose is not an essential organ. Patients in whom the upper air passages have been completely bypassed because of the need to create a hole in the neck for breathing—a procedure sometimes required when there is a malignancy of the upper airway—have no breathing difficulty. This procedure is called a tracheotomy.

The Nasal Cavity

The external nares—the openings into the nasal cavity—are deceptively small. The nasal cavity extends from the external nares into the head as far back as the rear upper teeth, up to the bridge of the nose, and back under the eye sockets. It is divided into two sections by a centrally placed cartilage plate called the nasal septum. This forms the bridge of the nose. Hair in the nose traps some of the particulate material. All of the surfaces in the nasal cavity are covered with an outer cell layer, or epithelium, that contains cells with distinct features: (1) ability to produce mucus, the output of goblet cells, and (2) mobile, filamentous protrusions called cilia. Surface cells with these characteristics exist throughout the naso-respiratory system. Microorganisms and particulate material suspended in the inspired airstream are trapped on the mucous lining, and then they are gradually propelled outward by the coordinated, wavelike motion of the cilia protruding from the cell surface. The motion of the cilia in clearing inhaled materials is particularly important in the lower respiratory tract and will be described further in connection with a review of the various functions of the lungs.

The lateral walls of the nasal cavity have three major convolutions—the superior, middle, and inferior turbinates—thereby greatly increasing the surface area of the nasal cavity. Openings into various sinus cavities are present in these lateral walls. When a person catches cold, or has an allergy such as hay fever, or some type of sinus infection, the passages to the sinuses may become obstructed, preventing drainage of normal secretions. Exaggerated examples of this have been the subject of many television commercials. There is a rich network of small blood vessels immediately beneath the mucous membrane surface. The large area of the nasal cavity and its blood vessel network facilitates the rapid "conditioning" of inspired air. Moisture and heat are added to the inhaled airstream so that by the time it leaves the nasal cavity and reaches the throat it has a water vapor saturation of at least 75 percent and has been

The Respiratory System

1. Superior lobe of right lung, 1
2. Middle lobe of right lung, 1
3. Inferior lobe of right lung, 1
4. Superior lobe of left lung, 1
5. Inferior lobe of left lung, 1
6. Trachea, 1, 2
7. Pleural covering, 2
8. Cardiac notch, 2
9. Right bronchus, 2
10. Left bronchus, 2
11. Bronchial tube, 2
12. Lung lobules, 2
13. Left pulmonary artery, 2
14. Left pulmonary vein, 2
15. Diaphragm, 2, 3
16. Lung lobule (greatly magnified), 2
17. Thyroid cartilage, 3
18. Larynx (surrounded by thyroid cartilage), 3
19. Cricoid cartilage, 3
20. Rib, 3
21. Vertebra, 3
22. Intercostal muscle, 3
23. Pulmonary vein branch, 3
24. Pulmonary artery branch, 3
25. Alveolus (air sac), 3

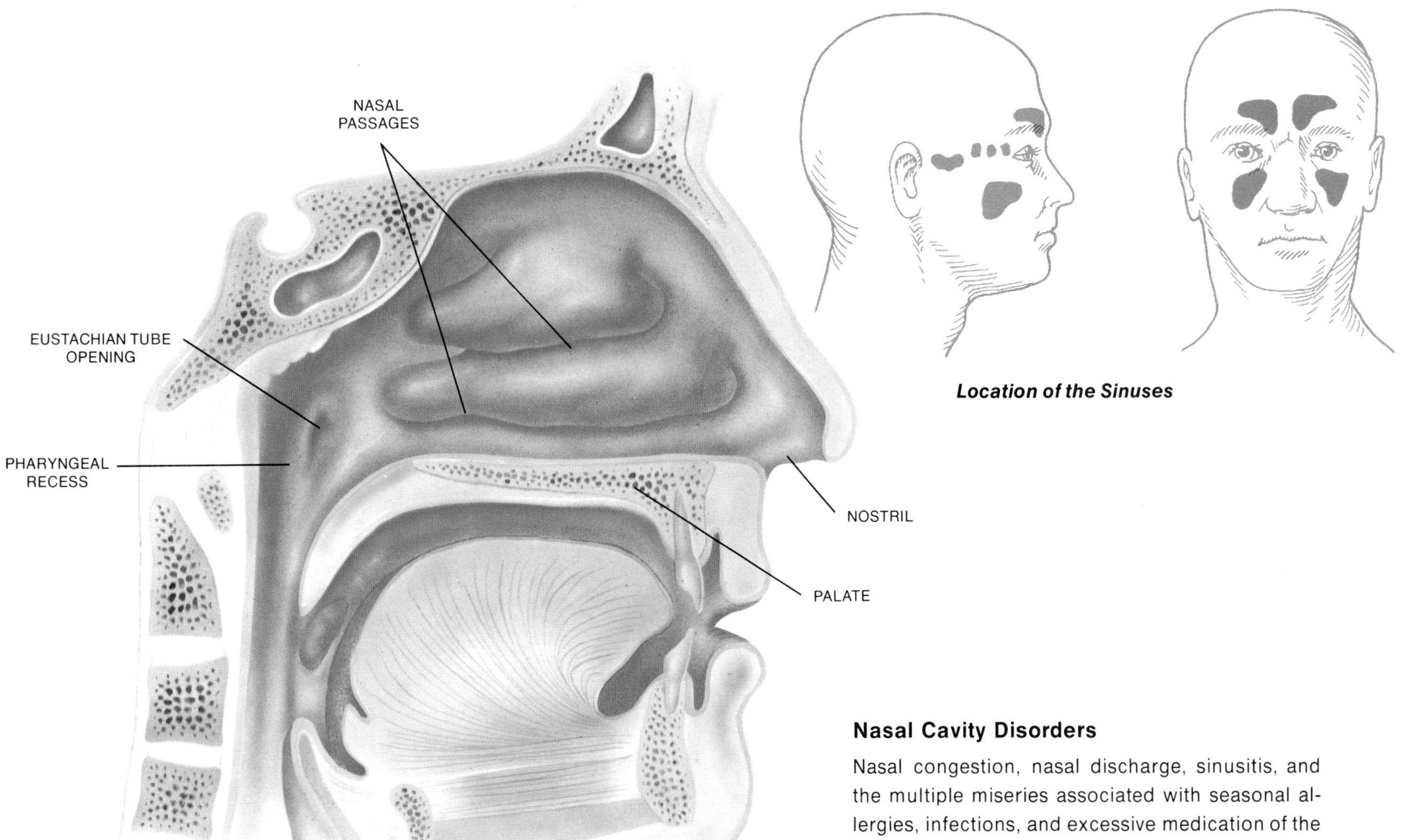

Location of the Sinuses

heated to approximately body temperature. The two major contributions of the nose to breathing—conditioning of the inspired airstream, and removal of particulate material—are both facilitated by virtue of the anatomy of the nasal cavity and the peculiarities of the cellular layer with which it is lined.

Odor detection occurs at the apex, or top, of the nasal cavity. Small branches of the olfactory nerve enter through multiple openings in the bony confines and are distributed in a network throughout the mucous membrane in the upper third of the nasal cavity. Because the inspired airstream is normally swept through the lower portion of the nasal cavity, where there are few if any branches of the olfactory nerve, odor detection is greatly enhanced by sniffing as this maneuver redirects air flow upward toward that region where odor detection nerve fibers are more plentiful.

Nasal Cavity Disorders

Nasal congestion, nasal discharge, sinusitis, and the multiple miseries associated with seasonal allergies, infections, and excessive medication of the nasal cavity are among mankind's most common afflictions. Whether they warrant designation as diseases or should be classified only as nuisances might be argued. Everyone has experienced some or all of the symptoms of upper respiratory tract congestion at one time or another, and some persons are plagued continuously. Chronic sinusitis or sinus infection, with secretions periodically running down the back of the throat, is so common as to be an almost universal experience. The incidence of this "postnasal drip" is very high in urban dwellers who live in cold climates. Those who live in warmer regions of the world and in less congested areas suffer somewhat less. How annoying or disabling chronic sinusitis may be depends in large measure on an individual's threshold of annoyance. Some persons tolerate considerable deviations from ideal health with little concern or interference in daily life. Others cannot tolerate minimal discomfort and seek help. Of course, serious infections that affect the whole body and produce an elevation of body temperature or other generalized symptoms require prompt, expert medical attention. But a postnasal drip that is primarily a mild annoyance is probably best left alone. The sufferer can take comfort in knowing that much of mankind is similarly afflicted.

The Throat, Windpipe and Associated Structures

The throat, or pharynx, is approximately five inches long in the adult and about one-and-a-half inches in width at its widest point. It extends from the base of the skull, where it communicates with the nasal cavities, to the level of the sixth vertebra where it becomes continuous with the esophagus.

Throat Structures

The pharynx contains the larynx, which enables us to talk, some specialized tissue—the tonsils—we can just as well do without, and the Eustachian tubes—channels connecting the ears with the pharynx. Because of the close proximity of these tubes to the back of the nasal cavity, inflammation of the entire nasal cavity during the course of a bad cold can result in stuffed-up ears and earaches. Perhaps the best known structure within the pharynx is the uvula, or soft palate, because it can be clearly seen in a mirror. The uvula hangs from the back of the upper jaw where the mouth cavity and pharynx meet. Some individuals have an inordinate preoccupation with the uvula and inspect it frequently. During a bad cold it may become swollen and inflamed, but otherwise it just sits there, bothering no one, doing nothing, and, for all practical purposes, can be ignored.

In this same location (base of the tongue, uvula, upper pharynx) there is a ring of lymphatic tissue of which the tonsils and the adenoids are a part. In the course of a cold, this tissue may become inflamed, reddened, and enlarged. Although chronic infection of the tonsils was once thought to contribute to sore throats and colds leading to widespread removal of tonsillar tissue, particularly in children, tonsil surgery is now rather infrequent and is performed only when there are recurrent, severe tonsillar infections. Further down the larynx, shortly before the esophagus begins, is a most important structure—the epiglottis. Because both air and food traverse the midportion of the pharynx a mechanism is necessary to direct them into their respective conduits. During the act of swallowing the epiglottis shuts off the entrance to the larynx and trachea. During inhaling, swallowing is quite impossible.

Apart from infections that are confined to the tonsils, the pharynx becomes reddened and inflamed in most diseases, both serious and unimportant, that affect the nose. A sore throat may also be part of very serious diseases that affect the entire body, such as "strep throat," diphtheria, mononucleosis, and others. Interference with the function of the tongue, epiglottis, or larynx can occur in diseases of the nerves or muscles of these structures. Tumors in the chest may result in vocal

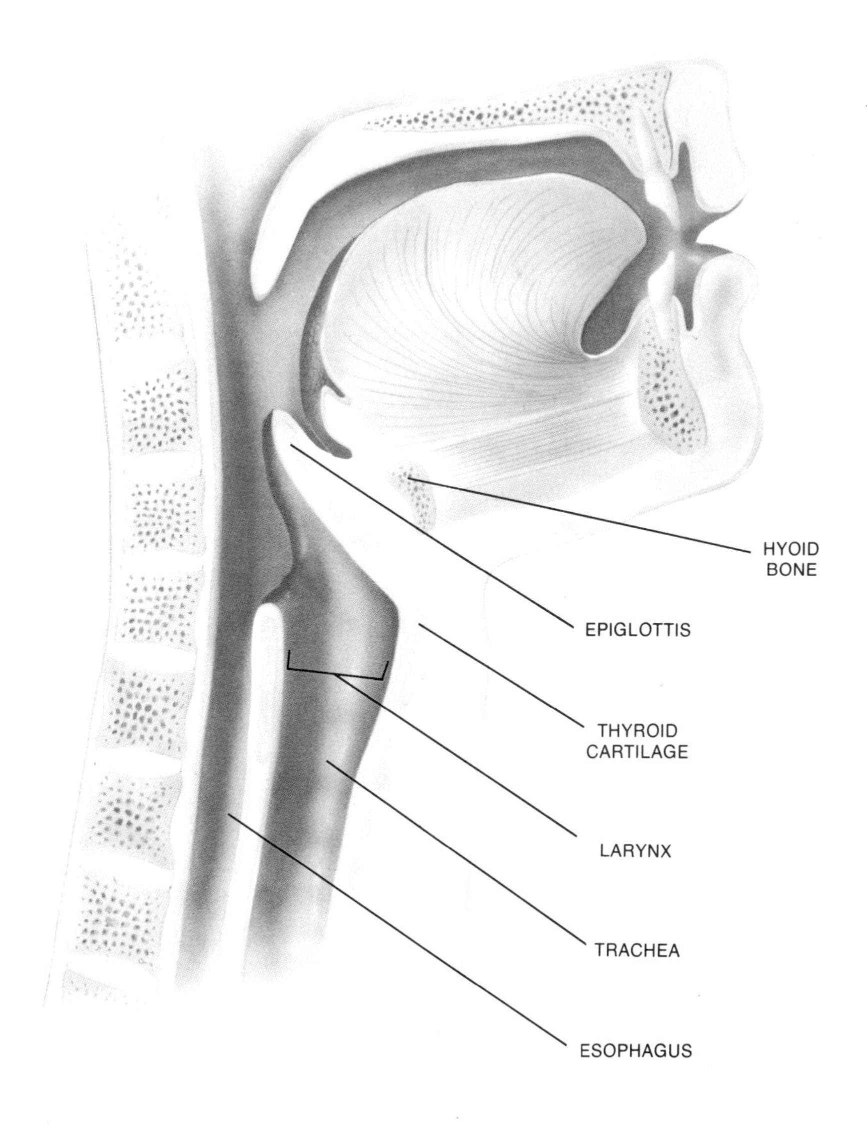

The Larynx

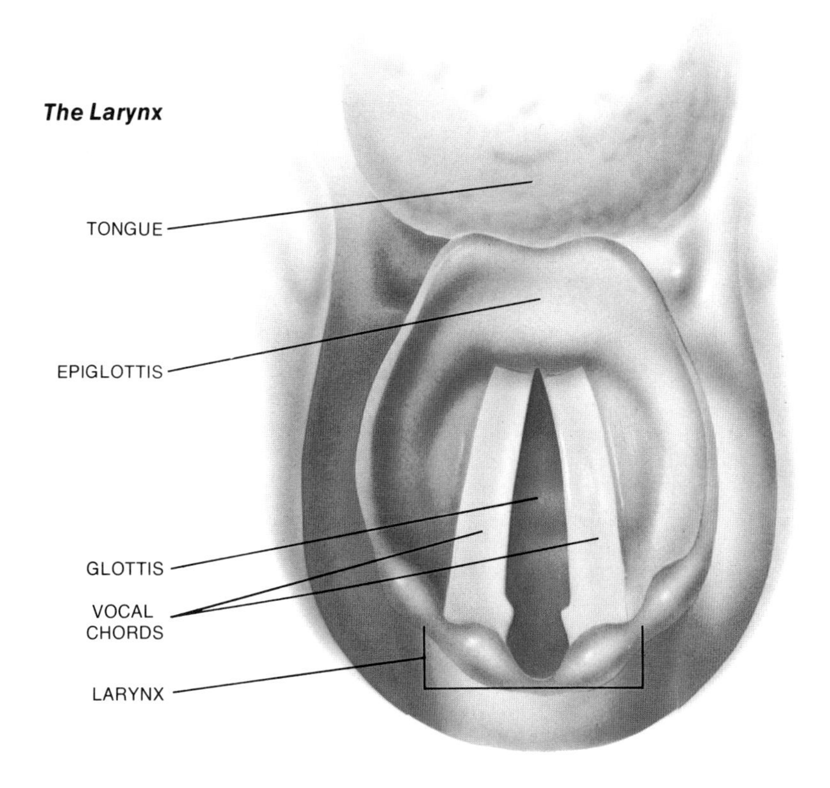

cord paralysis. In these circumstances there may be difficulty in swallowing. Food or liquid may be aspirated into the lungs, leading to severe infections and considerable breathing difficulty.

The Larynx

The larynx, commonly called the voice box or Adam's apple, is located in the front and middle of the neck. It extends from the base of the tongue at its upper end to the trachea, or windpipe, at the lower end, opposite the top of the breastbone. The larynx is made up of various cartilaginous plates, all interconnected by thin layers or bundles of muscle tissue. Tissue folds extend into the main air passage from the walls of the larynx and form the vocal cords. In men the larynx is larger than in women so that sounds produced by vibrations of the vocal cords as air passes across them are at a lower pitch. Respiratory infections, which result in swelling of the mucous membranes of the larynx, modify the vibrations of the vocal cords, creating a change in the character of the voice, or "hoarseness."

The walls of the larynx are liberally supplied with sensory nerves so that any foreign material lodging in this region (food "going down the wrong way") provokes a coughing spell. Infections or tumors in this area may produce a persistent cough. Prolonged, excessive use of the voice often occurs with singers, announcers, and public speakers, and is a common cause of hoarseness and diminished vocal intensity. Usually, such changes are temporary. Paralysis of the vocal cords can occur as a result of various neurological diseases. Cancer can develop on the vocal cords. The treatment for cancer generally consists of surgical removal of laryngeal structures, including the vocal cords. In some cases where cancer of the vocal cords is detected early, minor surgery or radiation therapy has proved effective. With paralysis of the cords, or following surgical removal, normal speech is no longer possible. Some patients may become very adept at "esophageal speech." Air is swallowed or accumulated within the esophagus and then released against taut tissue folds at the upper end of the esophagus, creating sounds that, with practice, serve as a passable voice. Various electromechanical devices have been developed for use as laryngeal substitutes. A vibrator applied to the

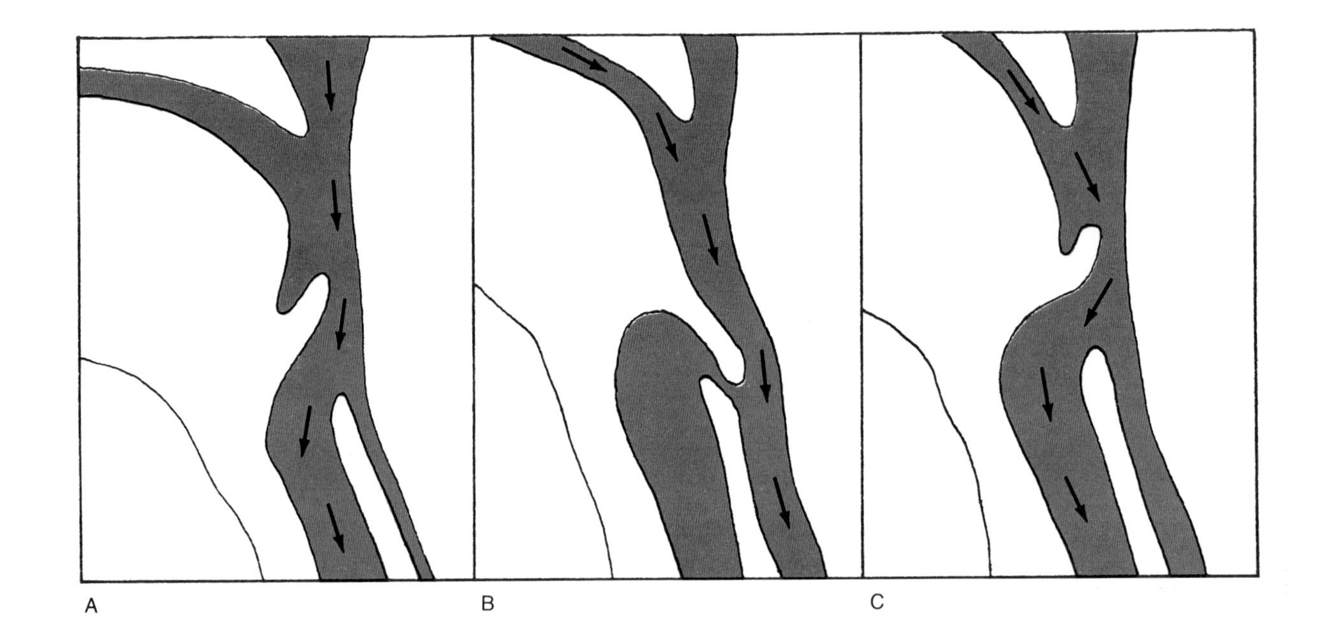

When a breath is taken, the air moves from the nasal passages, across the pharynx, and into the larynx at the top of the windpipe (A). When food is swallowed, the epiglottis snaps over the larynx so that food will not enter the windpipe (B). If food should "go down the wrong pipe" (C), it would have to be dislodged before the person choked.

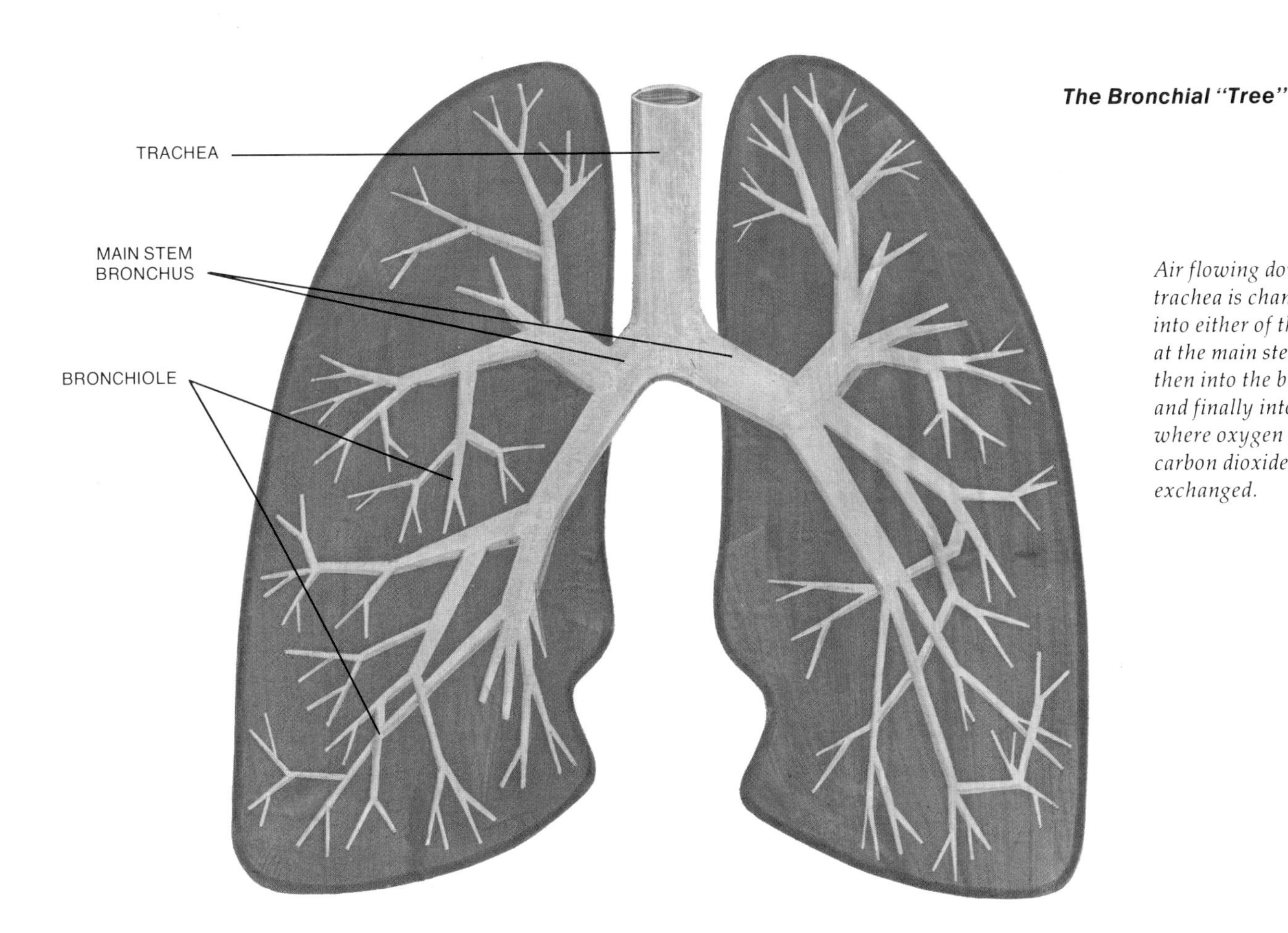

The Bronchial "Tree"

Air flowing down the trachea is channeled into either of the lungs at the main stem bronchi, then into the bronchioles, and finally into air sacs where oxygen and carbon dioxide are exchanged.

undersurface of the jaw vibrates the air within the oral cavity. By "mouthing" words and phrases, oral communication becomes possible.

The Windpipe (Trachea) and the Major Bronchi

The trachea connects the upper airway structures—nose, mouth, pharynx, larynx—with the lungs. It begins at the lower end of the larynx immediately in front of the esophagus from which it is completely separated. The trachea has multiple branchings, much as the trunk of a tree is subdivided. The major bronchi, or branches, send off progressively smaller branches eventually reaching to all subdivisions of the lungs.

Both the trachea and major bronchi consist of a series of cartilaginous rings embedded in a fibroelastic wall. The cartilage is absent posteriorly, or in the back, where it is replaced by muscle fibers. The right and left main bronchi are the first subdivisions and lead respectively to all successive generations, or series of bronchial branchings within the lung on each respective side. The cartilaginous rings are no longer intact after the second or third subdivision. Only irregular plates of cartilage can be found in these smaller bronchial walls. Thus, tertiary branches and beyond have much less stability than the larger bronchi with their intact cartilaginous rings. On the right side the main bronchus makes an angle of approximately twenty-five degrees with the vertical and is therefore almost a direct continuation of the trachea. On the left side the main stem bronchus makes an almost seventy-five-degree angle with the trachea. The right bronchus is also somewhat larger in diameter than the left. This size difference, plus the minimal angulation between it and the trachea, accounts for the fact that foreign bodies are more likely to enter the right lung than the left. Although primarily a passive conduit for movement of air into and out of the lungs, the smaller subdivisions of the bronchial tree and the blood vessels intimately related to them comprise the tissue of the lung. Their anatomical features and how these smaller airways work will be considered subsequently in connection with a review of lung structure and function.

Obstruction of the trachea, either by injury or foreign bodies, is an urgent medical emergency that requires immediate relief. Unless the obstruction can be removed, an opening, or tracheostomy, must be created below the obstruction. Despite reports of heroic lifesaving efforts made by using a pocketknife or nail file, the procedure is far from simple and requires considerable care to avoid excessive blood loss and permanent injury to adjacent nerves.

The Chest and Lungs

The lungs are located within the chest cavity, or thorax, which also contains the heart, major blood vessels, and the esophagus. In any review of how the lungs function, consideration must be given to the role of the bony and muscular components of the thorax—including the most important muscle related to breathing, the diaphragm. This dome-shaped muscle separates the thorax from the abdomen. Its rhythmic contraction accounts for the bulk of air movement into and out of the lungs. Each side of the diaphragm has separate nerve connections. Interruption of the nerves on one side has no dramatic effect on breathing. The motion of the remaining half of the diaphragm is sufficient to maintain adequate air movement in and out of the lungs. Paralysis of the diaphragm on both sides need not be fatal, as other muscles can provide minimal air exchange, but the patient will be unable to do anything other than merely subsist. The other important muscle groups are: (1) the intercostals—sheets of muscle tissue interconnecting the ribs; (2) the abdominal wall muscles, which, when contracted, transmit abdominal pressure toward the thorax and thereby displace the diaphragm; and (3) the neck muscles. These latter muscles are attached to the upper margin of the thorax and to various points on the head and neck. When contracted they stabilize or elevate the entire thorax. Following very strenuous exertion, or in persons with severe lung diseases causing shortness of breath, these muscles can be seen contracting when the individual tries to inhale. Paralysis or disease of the chest wall muscles can produce severe alterations in lung function and may necessitate frequent or continuous use of mechanical devices for breathing assistance.

The bony structure of the thorax is essential for normal breathing. As a result of muscle contraction, the thorax enlarges with each breath. A negative pressure is thereby created in the thorax, and air moves into the lungs. During expiration the process is reversed. If there is a loss of rigidity of the chest wall—as occurs with multiple rib fractures following automobile or other accidents—inspiratory effort produces an inward deviation of the chest wall with a much reduced inflow of air to the lungs. The motion of the injured area of the chest is paradoxical (in, when it should be out, etc.). In this situation it may be necessary to insert a tube into the trachea and breathe for the patient, blowing air into the lungs intermittently. This overcomes the paradoxical movement of the chest wall.

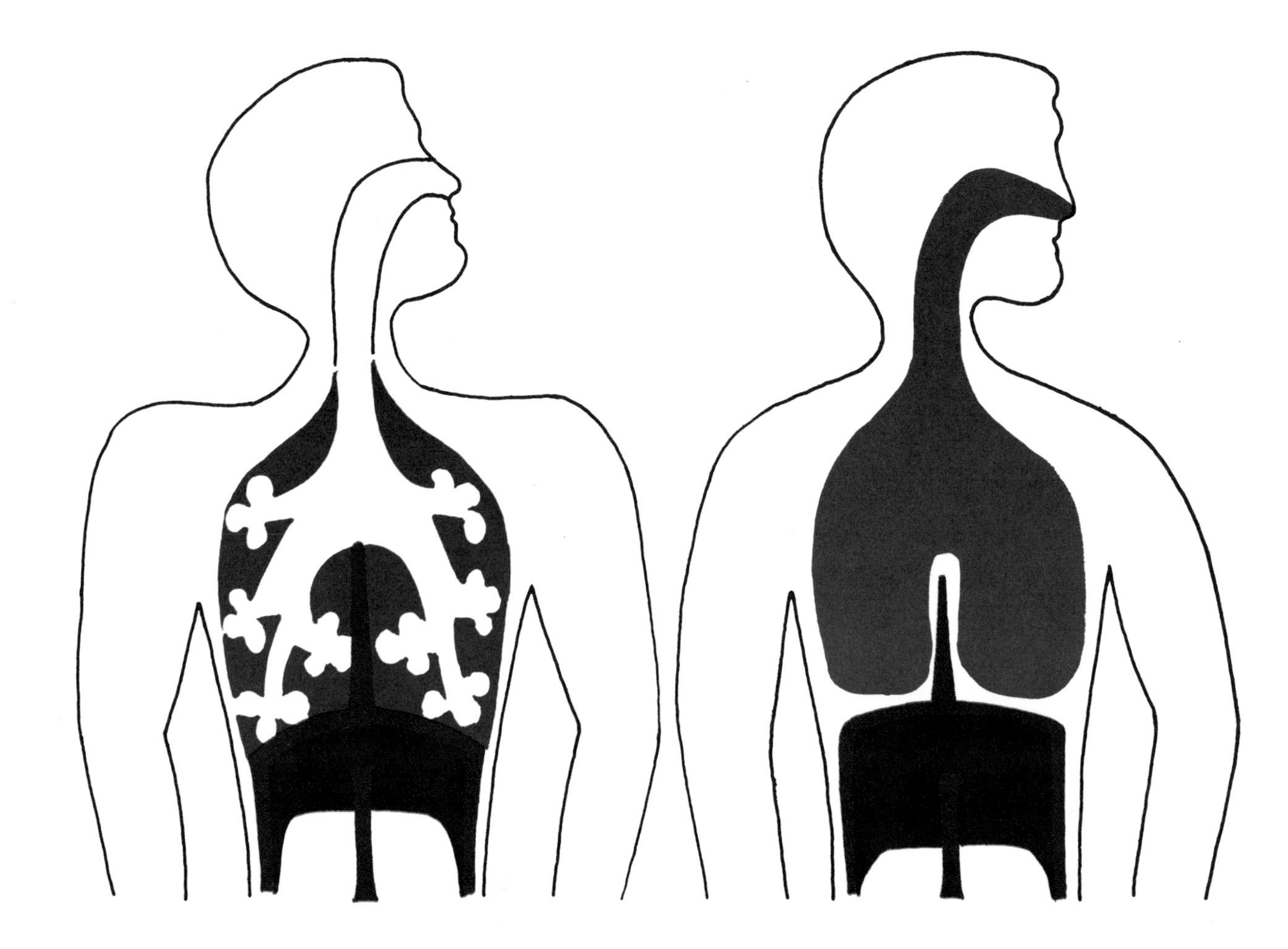

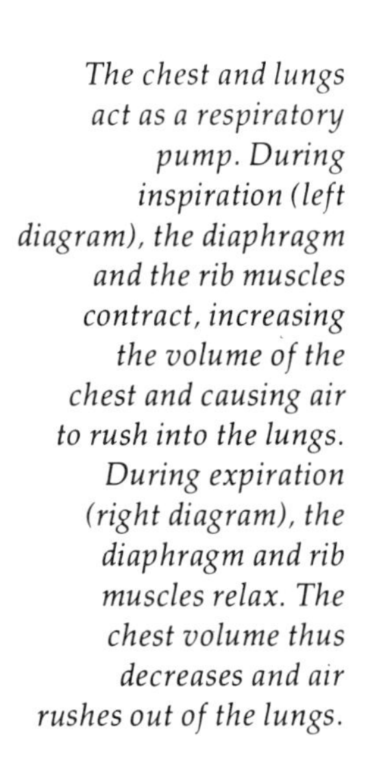

The chest and lungs act as a respiratory pump. During inspiration (left diagram), the diaphragm and the rib muscles contract, increasing the volume of the chest and causing air to rush into the lungs. During expiration (right diagram), the diaphragm and rib muscles relax. The chest volume thus decreases and air rushes out of the lungs.

Lung Structure

The lungs themselves are subdivided into five lobes, three on the right and two on the left. Each lobe is connected with a main bronchus by a lobar bronchus. As would be expected from the slightly larger right main bronchus, and because the heart is almost entirely within the left half of the thorax, the right lung is bigger than the left and accommodates about 55 percent of the total air capacity of the two lungs. A thin membrane, the pleura, covers the surface of the lungs, of each lobe individually, and of the inner wall of the chest. This is a glistening, moist membrane that is only a few cells thick in the average person. Inflammation of this membrane is called pleurisy and often causes severe pain with breathing. Sneezing and coughing are particularly painful for patients with acute pleurisy. In chronic long-standing inflammation of the pleura there may be large collections of infected material, old blood, or merely old fibrous tissue. Only minimal symptoms may be experienced. The pleural surface of the lung itself is called the visceral pleura, whereas the pleural surface lining the chest cavity is called the parietal pleura.

Shortly after entering a lobe, the bronchus runs adjacent to two blood vessels—the pulmonary artery and vein. The pulmonary artery carries blood from the heart to the lung. The pulmonary vein returns this blood to the heart. Additional subdivisions of the bronchi are accompanied by corresponding subdivisions of the blood vessels. There is progressive branching of the bronchi. Up to about the sixteenth generation, or series of separate branchings, they serve merely as passive conduits. From the seventeenth through the final or twenty-third generation the air passages participate in gas exchange, the major function of the lung. Beyond the sixteenth generation both the air passages and blood channels are sufficiently small that only microscopically thin layers of cells separate the two. Thus the number of channels is very great. It has been estimated that the area of the lung available for gas exchange approximates the size of a tennis court.

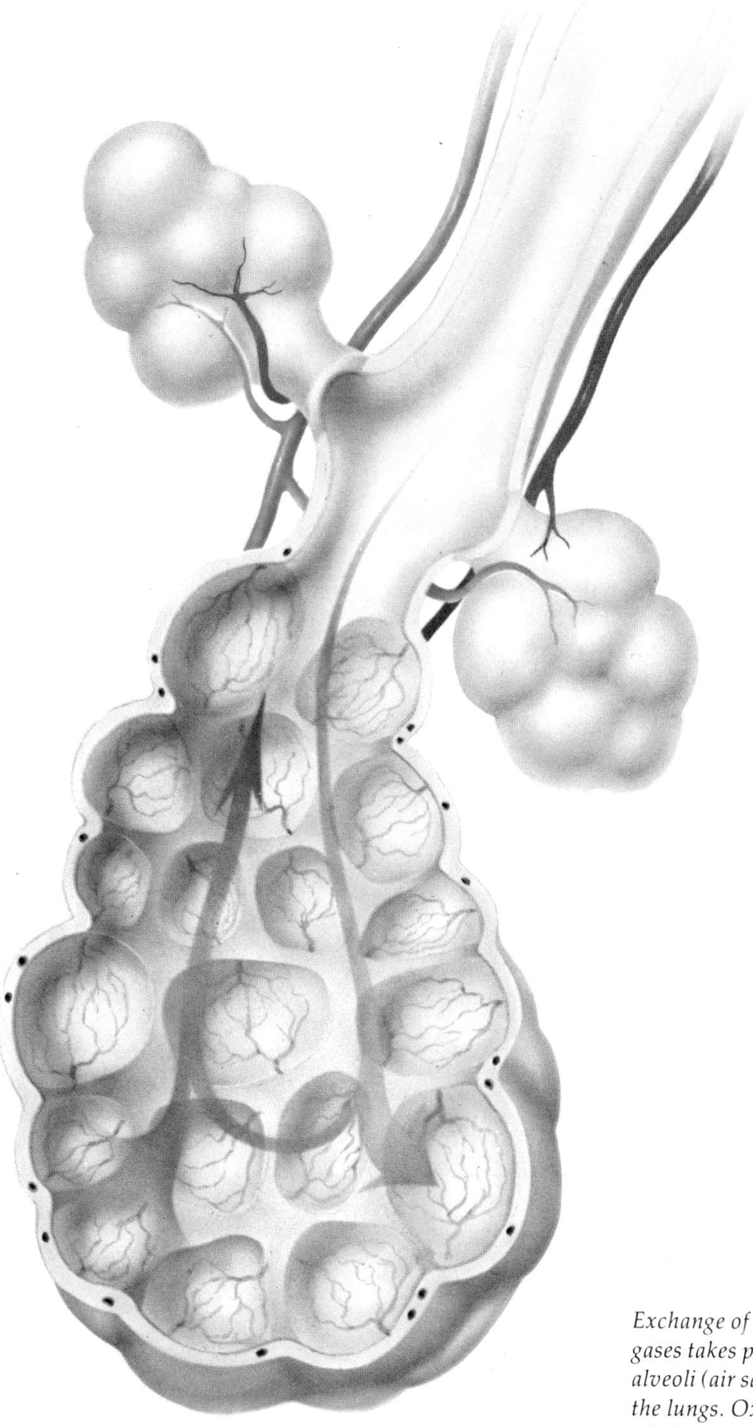

Exchange of respiratory gases takes place in the alveoli (air sacs) of the lungs. Oxygen inhaled (purple arrow) is exchanged in the capillaries around each alveolus with carbon dioxide (red arrow), which is then exhaled.

Gas Exchange in the Lungs

Blood entering the lung by way of the pulmonary arteries has a low oxygen content. Inspired, or inhaled, air has a high oxygen content. During a brief period of equilibration with structures suited for gas exchange the oxygen moves from the gas phase, where it exists at a high concentration, into the blood where the oxygen concentration is low. A similar but reversed exchange goes on simultaneously for carbon dioxide. Carbon dioxide moves from the blood into the air because its concentration gradient is in the opposite direction. The rhythmic contraction of the heart pumps blood into the lung and also distributes oxygenated blood throughout the body where it releases the newly acquired oxygen and accumulates carbon dioxide. This process of oxygen absorption and carbon dioxide excretion in the lung is dependent upon the maintenance of the various related functions. Unless adequate air flow into and out of the lungs is maintained, there will be reduced gas exchange with an oxygen deficit and carbon dioxide accumulation in the blood. Respiratory muscle paralysis, asphyxiation, rebreathing of expired air, chest injuries, and spasm of bronchial tubes are examples of conditions in which abnormal air flow can produce impaired gas exchange. Another cause for inadequate gas exchange is a reduced gas-carrying capacity of the blood. Hemoglobin—the pigment of the red blood cells—is a major carrier of both oxygen and carbon dioxide. An anemic individual or one in whom chemicals or drugs have diminished the gas-carrying capability of the blood may be unable to absorb and transport sufficient oxygen for essential needs.

A final category of causes for poor gas exchange is a mismatch in the distribution of air and blood within the lung. If a large blood vessel is plugged up, as occurs when a clot breaks loose and is swept into the general circulation, lodging eventually in the lung, then no gas exchange can take place in that portion of the lung which is fed from the obstructed vessel. Whatever air flows into and out of that portion of the lung which is deprived of circulation is "wasted" in that the air cannot participate in gas exchange. Similarly, if an airway is occluded, as may occur following inhalation of a peanut, for example, there can be no gas exchange in subsequent generations of the obstructed airway because they receive no air. The blood flowing into that lung segment with an obstructed airway is also "wasted"—no gas exchange can occur. Thus, a mismatch of distribution of air flow and of blood flow with "wasted" breathing or "wasted" blood flow can result in abnormal gas exchange.

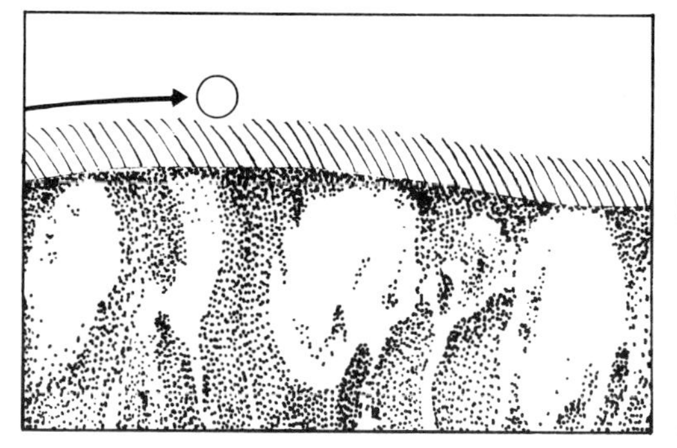

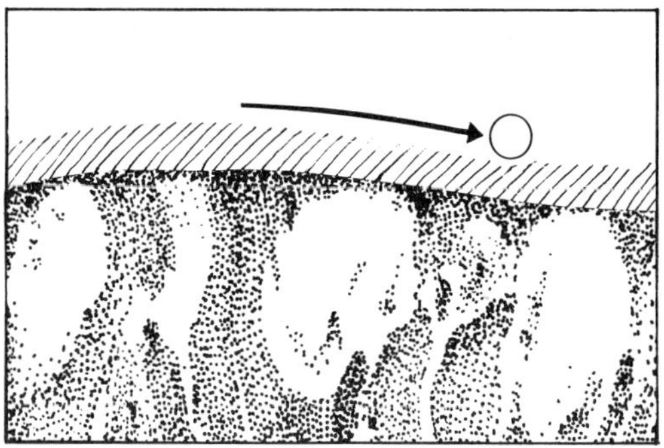

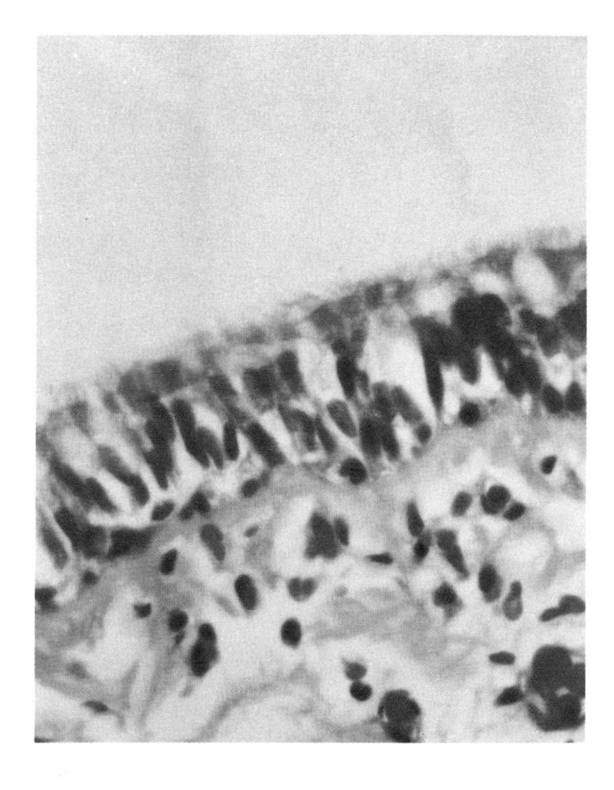

The sweeping motion of microscopically-small cilia moves particles (as shown in the illustration) that get into the respiratory tract back toward the nasal cavity for expulsion. A magnified view of cilia is below the illustration.

Cilia Protect the Lungs

Cilia, hairlike projections from the surface cells of the tracheobronchial tree and the nose, also exist in many organisms from the plant and animal kingdoms. Cilia are very small—about 0.0002 mm (millimeters) in diameter and 0.0005 to 0.015 mm long. All cilia, irrespective of their origin, have similar dimensions and comparable appearances when examined by electron microscopy. Each cilium consists of two centrally placed and nine peripherally arranged filaments. At their base the filaments extend into the cell body from which they originate, and the membrane covering the filaments becomes the cell membrane. The cilia have a coordinated, wavelike, continuous back and forth motion. They curl up and "relax" when moving backward, whereas they stand straighter when moving "forward." In the bronchopulmonary tree, "forward" is toward the outside of the body. The benefit of ciliary motion is the continuous transport of the mucus blanket that covers the nose, pharynx, larynx, and bronchial subdivisions. Cilia move mucus at a rate of 15 mm per minute. The body produces approximately 100 milliliters of mucus (not saliva) each twenty-four hours. Exposure to inhaled irritants—either in particulate or vapor form—may have a dramatic effect upon ciliary motion and upon mucus secretion. One whiff of cigarette smoke will paralyze the cilia so that all motion stops temporarily, thereby halting the otherwise continuous movement of the mucus blanket. Irritant fumes, foreign bodies, and all types of infections result in an increase in mucus production. Abnormalities of this process of clearing bronchial secretions may well be the initial change that occurs in patients destined to develop chronic disabling lung diseases many years later.

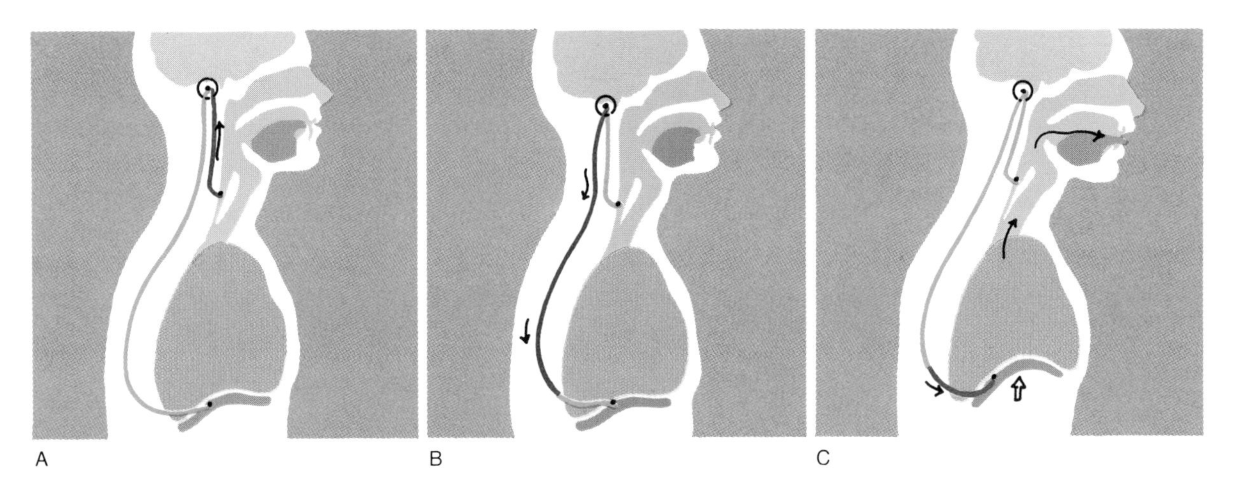

A particle lodged in the windpipe stimulates a nerve impulse to a cough center in the brain (A). The cough center then generates nerve impulses to the diaphragm and other muscles (B). When they contract, high pressure develops within the chest, and air is forcefully expelled.

Cough As a Lung Defense

The ability to generate a vigorous cough is essential for expulsion of foreign materials or excessive secretions. The involuntary, almost uncontrollable cough provoked by inadvertent aspiration of food or fluids is familiar to all. The cough reflex is particularly sensitive in the larynx, trachea, and major bronchi but is much less so in the lobar and smaller bronchial divisions. Thus, coughing is initiated primarily by foreign material in the larger air passages, but secretions from other regions may gravitate toward sites where the cough reflex is present.

The act of coughing requires a generalized contraction of the muscles of the abdomen, chest wall, and the diaphragm while the glottis remains closed, thereby creating a very high pressure inside the chest. Just as the maximum pressure is achieved the glottis opens, resulting in a rapid flow of air through the major air passages. The elevated intrathoracic pressure inverts the noncartilaginous portion of the trachea so that the trachea assumes a crescentic shape and is reduced to about 16 percent of its normal cross section area. The very rapid rush of air through the greatly narrowed airways dislodges foreign material or accumulated mucus. It has been calculated that peak air flow velocities during a vigorous cough may approach the speed of sound. Diseases which cause muscle weakness or paralysis reduce cough effectiveness. Pulmonary infection is a common complication following all types of surgery because the pain caused by the surgery inhibits the muscle contractions that are necessary to produce an effective cough.

The Macrophage in Lung Defense

Materials of 0.3 to 2.0 microns in diameter are so small that they can remain suspended in the inspired airstream until they reach the gas exchanging regions of the lung. There are no mucus secreting cells, no cilia, and the cough reflex is nonexistent in these locations. How do we dispose of the small inhaled particulates? An independent "cleaning" system exists for this purpose. Tissue cells, called macrophages, serve as scavengers and engulf the small particles. The macrophages then migrate and may move up to become embedded in the mucus blanket or they may stay put, with their engulfed foreign material. Not all particulate material that reaches regions where gas exchange occurs is removed. Silica particles, for example, may remain and provoke a reaction from adjacent lung tissue, eventually becoming the disease called silicosis. Other particulates may just sit there, such as coal dust, which can create considerable black deposits on the lungs. Bacteria, viruses, or other agents of disease may be neutralized by one or another of the various defense mechanisms, but, if not, they may cause a specific lung infection.

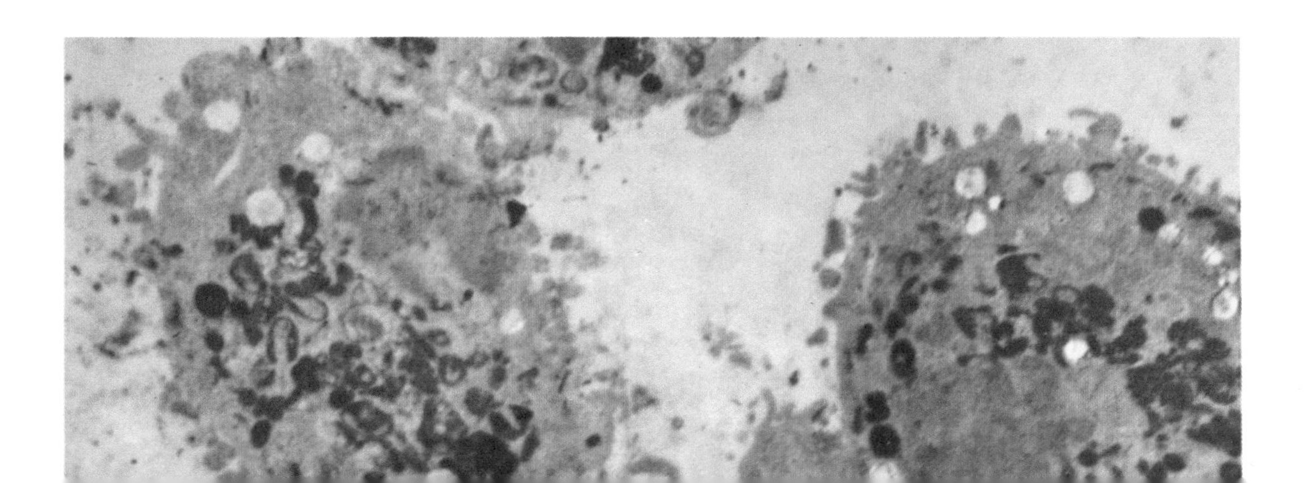

Macrophages are migrating body cells which are able to scavenge germs and debris in the respiratory tract.

Diseases of the Lungs

Because of their continuous exposure to all of the contaminants in our environment, the lungs are a frequent site of disease. These diseases can be classified according to several schemes: (1) the causative agent—pollen in hay fever for example; (2) the specific area in the lungs that is involved—pleura, bronchi, lung tissue; or (3) the duration or type of illness—whether it is brief or long-lasting.

Lung Infections

Infections in the lung tissue itself may be caused by bacteria, viruses, and other microorganisms. If confined to one region of the lung and if limited in duration, an illness is often designated as pneumonia. Generally, pneumonia, irrespective of cause, heals satisfactorily, although complications can occur, such as pleurisy, perforation of the lung, abscess formation, etc. Although now a general health problem of only minor magnitude, tuberculosis in the past was a major health hazard for all of mankind. The special type of microorganism responsible for this disease produces a different reaction in the tissues and thus tuberculosis is longer lasting and may not heal as completely as do lung infections caused by other types of organisms. Cough, sputum, fever, chest pain, and extreme sensation of a general sick feeling occur in all types of pneumonia. If the infection extends to the surface of the lung with inflammation of the pleural covering, the patient may also have the typical pain of pleurisy. Careful bacteriological examination of the sputum is essential for identification of the causative organism so that proper antibacterial medicine can be prescribed. Unfortunately, careful sputum examinations do not always reveal the organism. Virus identification is exceedingly difficult, and we have few if any effective anti-viral agents.

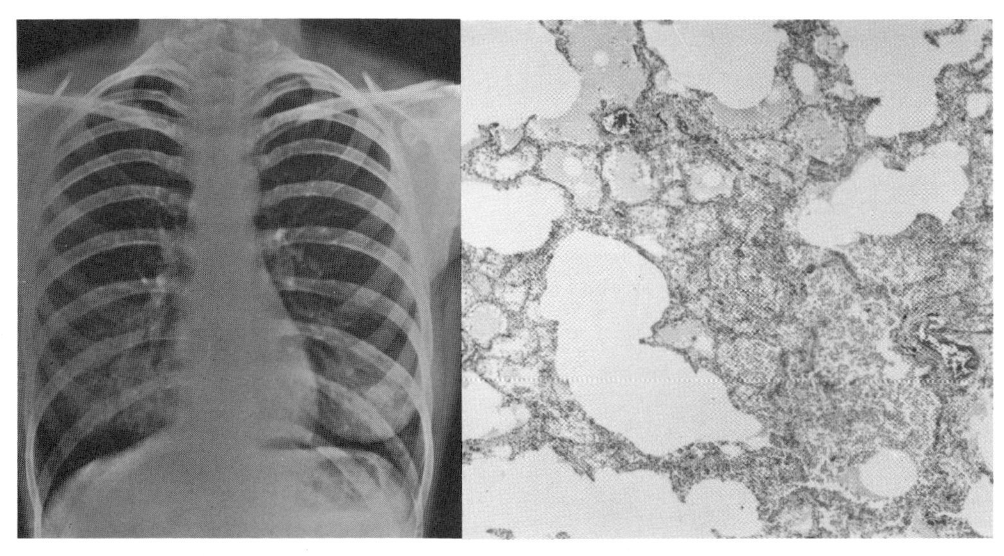

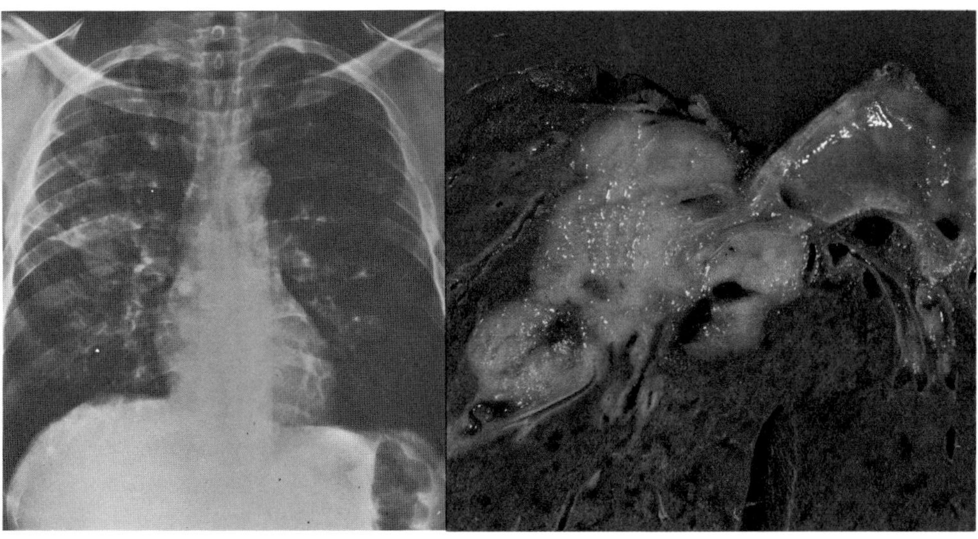

The upper photos show an X ray of a patient with pneumonia and the lung tissue affected by the disease. The lower photos show an X ray of lung cancer and a section of cancerous lung tissue.

Chronic Bronchitis and Emphysema

Some of the more common disabling and long-lasting lung diseases occur primarily in the larger bronchi and their subdivisions. Inflammation of the epithelial lining with excessive secretions is associated with chronic cough, sputum production, and shortness of breath. This constellation of symptoms, emerging gradually and increasing slowly but progressively over many years, is known as chronic bronchitis. Although mainly a nuisance at its outset, it may eventually lead to serious lung disease with interference in oxygen absorption and carbon dioxide excretion. It is often associated with widespread partial destruction of lung tissue—the hallmark of emphysema. These two diseases—chronic bronchitis and emphysema—which often cannot be clearly differentiated, are the third most frequent cause of disability payments (immediately after psychiatric and heart disease) in the United States. Unfortunately, we have no means of curing these diseases, and all available methods of treating them provide only modest, temporary benefit at best. The altered lung tissue cannot be restored to normal. There is a great deal of evidence that cigarette smoking is the single most important contributing factor to the development of these diseases. Although not everyone who smokes will develop them, nonsmokers are virtually certain to escape the horrors of these diseases. The "pleasure" of smoking should be thus viewed.

Asthma

Asthma describes a group of symptoms consisting of intermittent episodes of wheezing, coughing, and breathing difficulty. These symptoms occur whenever there is a decrease in the caliber of the bronchial tubes. In persons with long-standing bronchial diseases such as chronic bronchitis, symptoms of asthma may occur whenever there is extra bronchial infection, irritation, or other cause for an increase in bronchial secretions. Such patients may be relieved of their wheezing merely by vigorous coughing, but their basic disease and associated symptoms remain. On the other hand, persons with asthma caused by temporary spasm of the muscles in the bronchi will be relieved of all symptoms when the spasm subsides. Between spasms there may be no symptoms. Many agents can provoke bronchial spasm. These include molds, grasses, house pet dander, insect bites, and chemical fumes.

In general, bronchial asthma is primarily a childhood problem that is usually outgrown. Asthmatic symptoms in older persons, particularly if part of chronic bronchitis or emphysema, can be relieved but not eliminated.

Lung Cancer

Cancer of the lung was a most uncommon disease a generation ago, whereas it has become one of the most common cancers, particularly in men. Unfortunately, by the time these growths can be detected they are often so advanced that there is no hope for a cure. These tumors originate in the surface cells of the bronchial tree, and thus some early symptoms are: bloody sputum, wheezing, and recurrent pneumonias. Once the tumor has spread outside the chest many other symptoms may be present. Here, too, the evidence is virtually incontrovertible that cigarette smoking is directly responsible for the very high incidence of this disorder. All types of cancer can be classified on the basis of the microscopic appearance of the malignant tissue. One rather uncommon type of lung cancer is not related to cigarette consumption, but approximately 95 percent of all lung cancers in men are probably a consequence of cigarette smoking. The frequency with which this same type of cancer is observed in women has been increasing rapidly in parallel with the increase in their cigarette smoking.

Occupational Diseases

Another major category of lung disease is related to occupational exposure to suspended particulate materials or chemical vapors. The best known industrial lung disease is silicosis. Anyone exposed to finely powdered sand or dust with a high silica content is at risk. Workers in occupations that are particularly hazardous (sandblasters, hard-rock miners, foundry workers, etc.) must be protected by adequate ventilation, face masks, and hoods into which there is a high flow of filtered air. Silica particles that become deposited in the lung initiate a tissue reaction resulting in the formation of scar tissue, which interferes with movement of air into and out of the lungs and impairs gas exchange with eventual severe reductions in lung function. Diagnosis of industrial lung diseases can be exceedingly difficult. It is dependent upon the presence of characteristic alterations in the chest X-ray film, evidence of functional disturbance in the lung, and a very careful, detailed inquiry into the precise occupations of the patient throughout his working lifetime.

For further information on the subjects covered in this chapter, consult ***Encyclopædia Britannica:***

Articles in the Macropædia	*Topics in the Outline of Knowledge*
RESPIRATION, DISORDERS OF	424.E.6
RESPIRATION, HUMAN	422.D.3
RESPIRATORY SYSTEM, HUMAN	422.D.1 and 2
RESPIRATORY SYSTEM DISEASES	424.E.1,2,3,4, and 5
SINUS	422.I.6

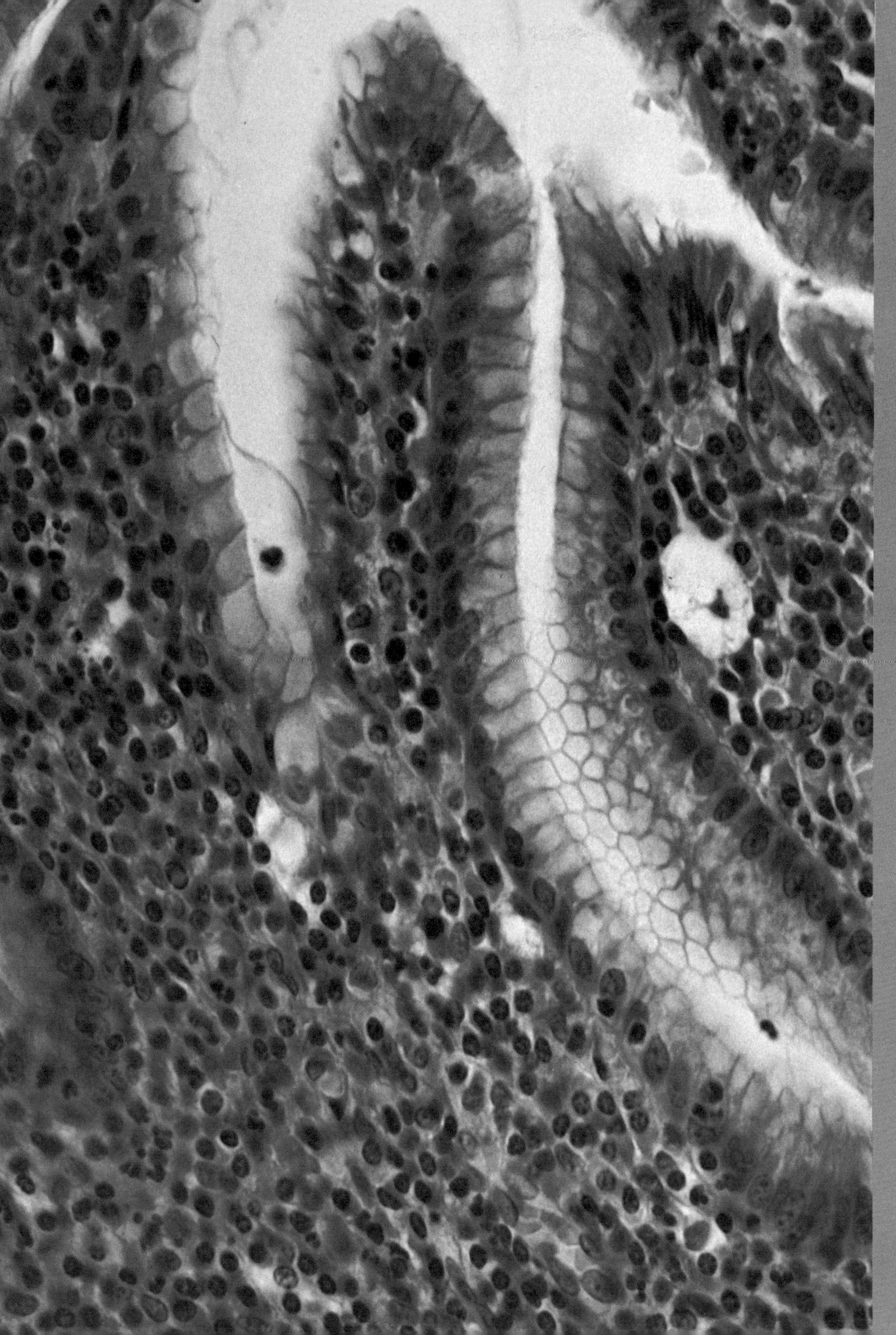

Secretions from stomach glands, such as this pyloric gland, provide a blanket of mucus that protects the stomach from the harsh substances other of its glands secrete to break down foods in more usable substances.
Photo, Joseph Delly, Walter C. McCrone Assoc., Inc.

Van Doren. *What a wonderful refinery I've got inside me!*

Kessler. *But more sensitive to attitudes than you'll ever see in an industrial plant.*

Van Doren. *Yes. My stomach and my emotions are awfully closely connected.*

Kessler. *There's a very great intimacy between the psyche and the whole digestive system—not just the stomach. The majority of patients who consult a physician about this part of their physical equipment have symptoms that can also be related to emotional problems. "I am what I eat"—that's another old saying with a good deal of truth to it—should be supplemented with "and how well I digest it depends on my frame of mind."*

6

Digestion

BY SUMNER C. KRAFT, M.D.

The digestive system is one of the most complicated systems of the body. Virtually everyone has had some kind of digestive problem at one time or another. It may be something as simple as "gas pains," diarrhea, heartburn, or indigestion. Fortunately, most of the "stomach-aches" people normally experience are relatively minor distress signals that something is wrong with the digestive process. These symptoms usually clear up in a matter of hours, or a few days at most. Other symptoms such as sudden altering of bowel habits or bloody stools cannot be considered lightly and should be brought to the attention of a physician.

Man's stomach has been called the mirror of his emotions, and there is no doubt that the stresses of daily living influence the digestive process. Knowing how the digestive system works can help one to keep it healthy, an important goal for maintaining good general health.

The digestive system is made up of the tubelike alimentary canal that extends from the mouth to the anus and of several accessory glandular organs. The main purposes of this system are to receive, move, break down, and absorb foods, and to eliminate waste products. It also serves as a

The Digestive System

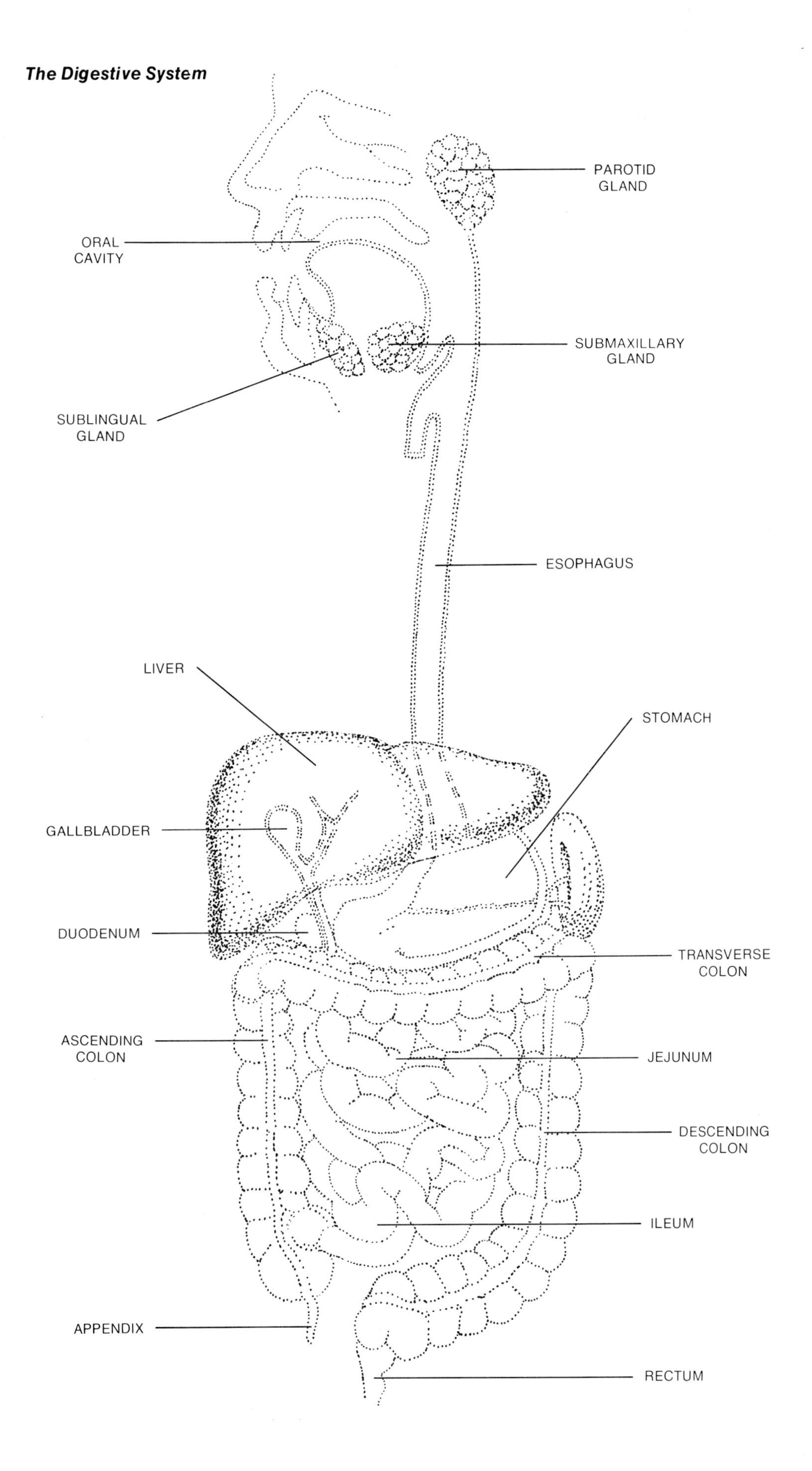

barrier to unwanted and unneeded substances that may enter and leave the body through it without really getting "inside." This chapter will show how the digestive system is able to do these jobs, which are critical to maintaining life and good health. In this regard anatomy and physiology will be discussed, as will the reasons that digestive disturbances lead to particular abnormal signs and symptoms. Furthermore, some behind-the-scenes insight will be provided into how a physician acts in response to a patient's complaints, in order to find out what has gone wrong and what corrective measures should be taken.

To place things in perspective, it has recently been estimated that half the population of the United States has digestive complaints, and that about 20 percent of all illnesses are in this category. In addition, digestive ailments lead all diseases in terms of time lost from work and money spent for hospitalization, and a third of all deaths from cancer involve tumors of the digestive organs. While the parts of the digestive system will be discussed separately, it will become clear that they have numerous functional and anatomical interrelationships, such as sharing common nerve and blood supplies, and also that the digestive tract works in close harmony with other systems of the body.

The Mouth and the Salivary Glands

The alimentary canal begins at the mouth, or oral cavity, the opening that starts at the lips and includes the teeth and the tongue. The teeth, gums, and adjacent structures have been discussed in great detail in the chapter on the dental system, but this area should also be looked at as the receptor site for food—the staging area for digestive processes. The tongue helps the teeth in the initial mixing and breaking up of food, or mastication, and the teeth help the tongue in modifying sounds in speaking. However, the tongue has two other important jobs—the reception of taste stimuli and involvement in the first step of the swallowing process, the movement of chewed food from the mouth to the pharynx.

The Tongue

The tongue is easy to examine, and for centuries physicians and patients alike have correlated the appearance of the tongue with the general state of health. While not completely true that it is the "window of the digestive system," a number of conditions may be associated with changes in the tongue, and the time-honored ritual of "stick out your tongue" is not without merit. Physicians are routinely taught to study its shape, color, projections, and movements—both voluntary and involuntary. Although tongue changes are not usually diagnostic of a specific condition, occasionally they may provide important clues, such as a suspicion of epilepsy when localized areas of scarring are present due to episodes of tongue biting. At the opposite end of the spectrum, a painful burning tongue may represent a difficult diagnostic challenge even after a thorough medical history and a complete physical examination.

Taste Bud Areas of the Tongue

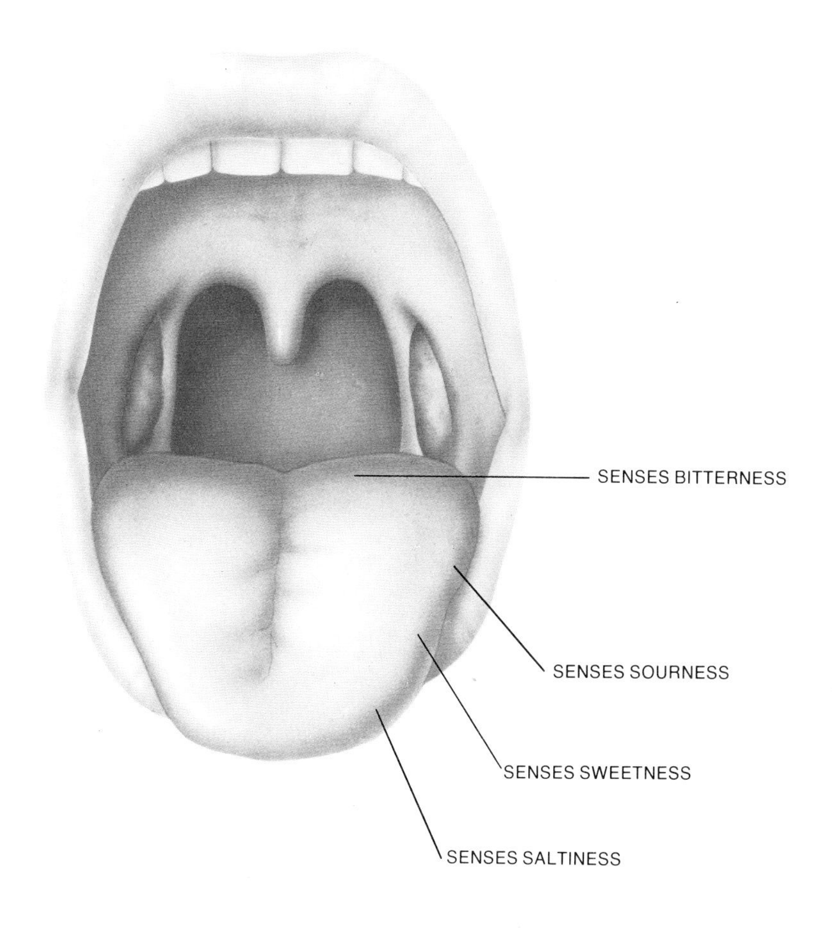

The Mucous Membrane and the Palate

The mucous membrane, or lining of the mouth, has several microscopic features that more closely resemble the surface of external skin and the lips than the lining of the gastrointestinal tract below. This light pink, smooth oral mucosa extends from the lips to the tonsillar area of the pharynx and provides mucus-rich lubricating secretions to aid in the softening and passage of ingested food. These secretions are derived from small buccal glands in the lining of the cheek and from the plasma in blood vessels that supply the mucous membrane. There are also small mucus-producing glands in the lips. Diseases of the oral mucosa can vary from small, virus-caused "cold sores" to more severe infections such as "trench mouth," and from relatively harmless swellings caused by certain medications to definite tumors. Mouth lesions may also be a manifestation of disease elsewhere in the body or of a generalized disease. Symptoms may include pain, discoloration, swelling, bleeding, and unpleasant odor. Halitosis, or bad breath, for example, may be caused by a wide variety of conditions, ranging from a dry mouth, dental caries, and other oral factors to abnormalities of the upper and lower respiratory tract, and finally to diseases such as diabetes, uremic poisoning, or liver failure. Appropriate mouth cultures, tissue examinations, X rays, and blood tests generally enable physicians and dentists to determine the cause of these symptoms.

The palate forms the roof of the oral cavity and separates it from the nasal cavity. The front part is hard and contains bone, whereas the back portion is made up of soft, fleshy tissue consisting mainly of muscle. Abnormalities of the palate, depending on their nature and severity, may interfere with chewing, mixing, and swallowing food; normal speech; and even normal breathing. The uvula projects down from the region of the soft palate in the middle of the roof of the mouth. It derives its name from the Latin meaning "little grape" and may become quite swollen in certain infectious diseases. The tonsils, also known as palatine tonsils, are small, almond-shaped structures located on both sides of the back of the throat between the base of the tongue and pharynx. Although not essential, they may help to destroy bacteria entering the mouth. The adenoids, also called pharyngeal tonsils, are very similar and are found in the nasal portion of the pharynx. The epiglottis is a thin, lidlike structure serving as a cartilage cover at the entrance of the larynx. It is important from a digestive standpoint because it acts as a valve to help ensure that food does not enter the respiratory tract.

The Salivary Glands

Three pairs of salivary glands make up the first of the accessory organs of the digestive system. They form the complex secretion called saliva, which passes into the mouth through small tubes, or ducts. The parotid glands, located at the side of the face below and in front of the ears, are the largest of the salivary glands, and are perhaps best known for their propensity to infection with the mumps virus. Secretions from the parotid glands enter the mouth through Stenson's ducts, located on the lining of the cheek opposite the crowns of the second upper molar teeth. The submaxillary, also called submandibular, glands lie on either side of the mandible or lower jaw. The third pair, the sublingual glands, are located under the tongue just below the mucous membrane on the floor of the mouth.

The flow of saliva starts when nerve endings in the mouth, stimulated by the presence of food, send signals to the salivary glands. Some types of food are better stimulants of salivation than are others. The sight, smell, and even the thought of food can be effective salivary stimulants, acting through the brain via highly specialized nerve pathways. A person's output of saliva is about 1 to 1.5 liters (a bit over a quart) per day, about two-thirds of it coming from the submaxillary glands.

One of the functions of saliva is to make chewing easier by moistening food in preparation for its passage down the alimentary canal. Two additional functions include cleansing the mouth by flushing out excessive bacteria and other foreign substances, and aiding in taste by permitting certain food substances to go into solution prior to stimulating the taste buds on the tongue. Saliva also aids in the breakdown of starch into sugars by the fermenting action of one of its components, amylase. Amylase is one of a class of body chemicals known as enzymes, which are important in accelerating the conversion of specific substances into simpler products. Other constituents in saliva are water, electrolytes, proteins, and some minor enzymes.

The Pharynx and the Esophagus

The pharynx is a muscular, saclike canal. Its upper part—the section continuous with the nasal passages and Eustachian tubes—is called the nasopharynx. The middle part is called the oropharynx and is that portion that lies between the soft palate and the upper edge of the epiglottis. Lower in the throat, the laryngopharynx joins with the larynx and esophagus. Clearly, the respiratory and digestive passages merge and cross in the pharynx. The opening to the oropharynx at the back of the mouth is surrounded by two pharyngopalatine arches separated by the uvula.

The wall of the pharynx is made up of three layers, an inner mucous lining continuous with those of the surrounding organs, a middle connective tissue layer, and an outer muscular coat. Its inner mucous membrane, or mucosa, is the portion that first comes into contact with food and other substances from outside the body and provides a mechanical and chemical barrier that prevents their gaining access to the body in general. The thin middle layer, the submucosa, serves as a fibrous coat upon which the fragile mucosa rests, while also providing a conduit for small blood vessels and nerves. The outer muscular coat is composed of so-called voluntary muscle which, as is discussed later, is important in the second stage of swallowing.

The pharynx is also called the throat, although commonly the outside of the front of the neck is referred to in this way. Probably the most common condition affecting the pharynx is acute pharyngitis, or a "sore throat," often caused by bacteria such as certain streptococci (e.g., strep throat) or a virus (e.g., the common cold). Because the base of the tongue may obscure a clear view of the oropharynx, a tongue blade is often useful in depressing the tongue to permit a better view of possible redness, pus pockets, enlarged lymph nodes, or other abnormalities on the surface of the pharynx or on the tonsils. Occasionally patients complain of the sensation of a lump in the throat, and careful local examination fails to reveal a cause. This condition, referred to for centuries as globus hystericus, was considered to have a neurotic basis. However, recent studies suggest that many such cases may be due to irritation of the lower esophagus by stomach acid, resulting in a secondary interference with the swallowing process.

The Swallowing Mechanism

Several steps are involved in swallowing, or deglutition. The first, or oral, stage is under voluntary control and involves the combined functions of the lips, teeth, palate, and tongue in receiving, chewing, hydrating, and partially homogenizing the

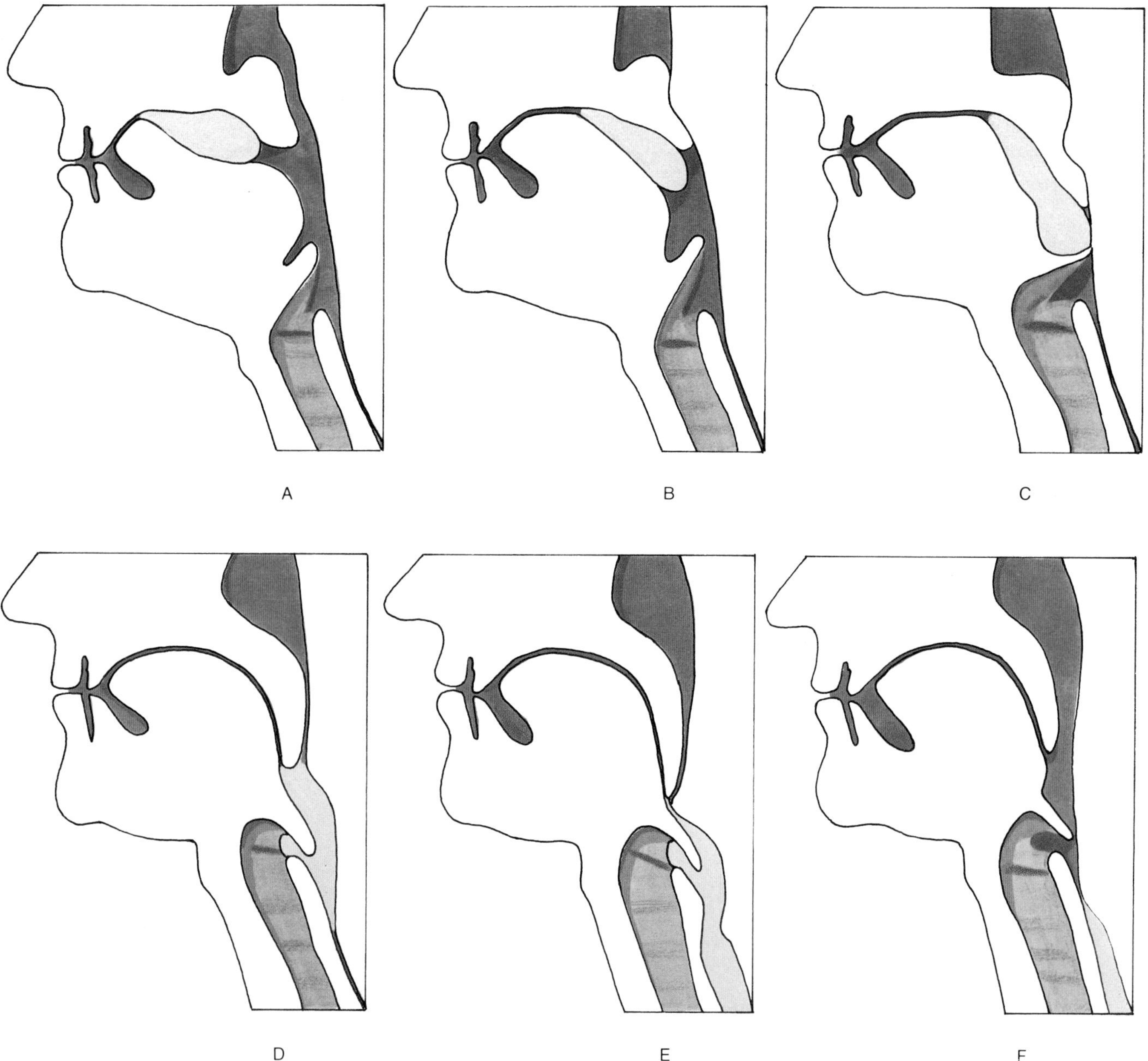

Swallowing is a reflex action; once started it cannot be stopped. The tongue forces the bolus, or chewed up food, back toward the pharynx (A, B). The bolus encounters the epiglottis (C) and pushes it downward. As the epiglottis closes the passageway to the trachea, food cannot "get down the wrong pipe" (D,E). With the bolus safely down the esophagus (F), the epiglottis can resume its usual position before the next swallow.

mass, or bolus, of food and forcing it back into the pharynx. The second, or pharyngeal, stage is necessary so that food does not "go down the wrong way." Coordinated activities of the palate, tongue, and uvula inhibit either the return of food into the mouth or its entry into the nose, while passage into the lower respiratory tract is blocked by actions of the larynx and epiglottis and the simultaneous transient restraint of breathing. Thus, this second stage involves rather complex reflexes that start as soon as the bolus touches the pharynx. It includes the interrelationships of local nerve endings with nerve connections that go to the brain and back to pharyngeal and nearby muscles. This is an involuntary part of swallowing and cannot be interrupted once initiated; it causes pressure to build up in the lower pharynx, forcing the bolus of food into the esophagus. The third, or esophageal, stage of swallowing begins after the food arrives at the first part of the esophagus. An upper esophageal sphincter, a ringlike band of muscle fibers, closes the opening at the top of the esophagus during rest.

The Esophagus

The esophagus is a muscular tube about ten inches long lined with a mucous membrane and leading from the pharynx through the chest cavity to the upper end of the stomach. The mucosal lining of the esophagus, like that in the mouth and pharynx, is made up mostly of squamous (scaly or platelike) cells, although its lowest part may microscopically resemble the mucosa of the stomach. This esophageal mucosa adds small amounts of mucus that serve as additional lubrication for the food bolus and aid in protecting the esophagus. The esophageal muscle is made up mostly of voluntary (skeletal) longitudinal and circular layers, but involuntary (smooth) muscle is found in the bottom one-third. The two types of muscle are intermixed toward the middle of the esophagus. An outer coat of thin connective and fatty tissue, the serosa, surrounds the esophagus and is also characteristic of lower parts of the alimentary canal.

The Esophageal Phase of Swallowing

The involuntary passage of material down the esophagus, the esophageal phase of swallowing, results from a process of wormlike movement known as peristalsis. This can be demonstrated by a specialized technique used to record esophageal pressure measurements after a person swallows a compact set of tubes. By this technique, the squirting of the bolus of food into the esophagus from the pharynx is seen to cause an initial positive pressure, which starts a primary peristaltic propulsion. This wavelike action, produced by contraction of the muscular esophageal wall, moves the food down toward the stomach. Secondary contraction waves develop as a result of distention of the upper esophagus and serve to empty the esophagus of residual food. Both the primary and secondary waves depend upon nerve connections with the "swallowing center" in the brain stem. There also may be a weak, nonperistaltic tertiary, or third, set of waves in some people.

The esophageal cavity is nearly collapsed in the resting stage, but there is a short high-pressure zone at the upper esophageal sphincter and another at the lower esophageal sphincter. The latter is located slightly higher than the actual junction of the stomach and esophagus. An integrated series of relaxations and contractions governs the activities of these two esophageal sphincters. After the upper esophageal sphincter muscles relax in response to the pharyngeal "push," primary peristalsis begins. By the time this propulsive wave reaches the lower esophageal sphincter its muscles have relaxed, permitting food to pass readily into the stomach. The contractions of the upper and lower esophageal sphincters also tend to prevent food from regurgitating into the pharynx or esophagus, respectively. The force of gravity is not a major factor in the esophageal phase of swallowing since deglutition can be accomplished when lying down or even when upside down.

A fiberoscope, which conducts light through flexible fibers, can be inserted into a person's digestive tract to enable a physician to look for signs of disease.

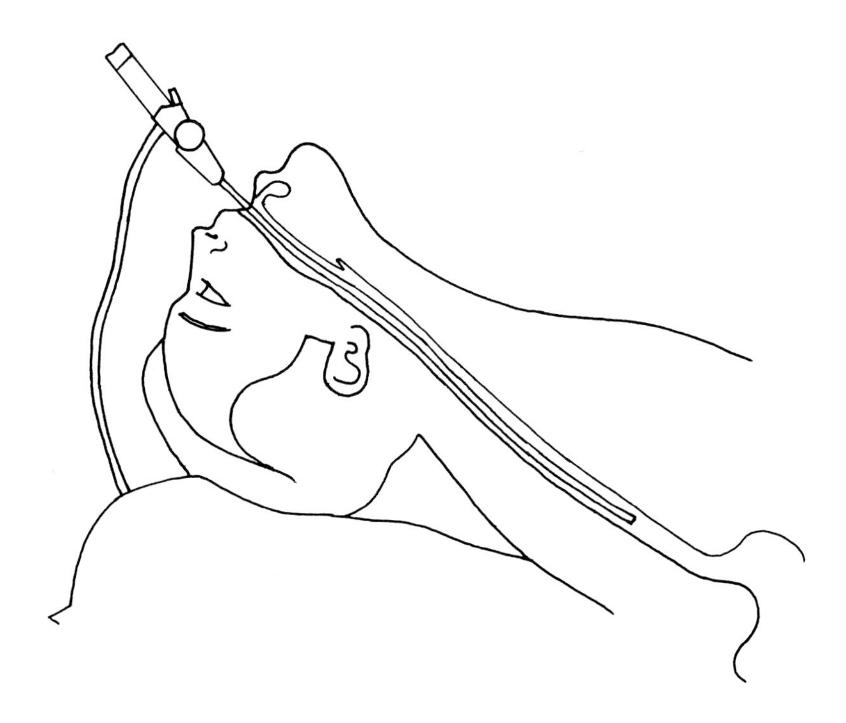

Swallowing and Other Esophageal Disorders

From the above information, it is not surprising that there are many disorders that can lead to difficulty in swallowing, or dysphagia. These vary from mechanical barriers to the passage of food down the esophagus, (e.g., inflammatory or caustic strictures, lodged foreign bodies, benign or malignant tumors), to the aftermaths of brain damage that might have been caused by strokes or poliomyelitis. Other conditions may directly affect the connective tissue or muscle of the esophageal wall; or there may be failure of the lower esophageal sphincter to relax so that the esophagus becomes greatly stretched, a condition known as achalasia.

Other esophageal symptoms include heartburn, which often occurs within an hour after eating and is felt as a burning in the center of the chest underneath the breastbone. It is caused by gastric juice backing up into the lower esophagus and can be a manifestation of esophageal irritation, or esophagitis, or even ulceration, which is the actual loss of a portion of the mucosal surface. The sensation of a lump in the throat sometimes may be caused by esophagitis, perhaps because of a secondary disturbance in esophageal muscle contraction. In some patients, stomach contents may actually back all the way up into the mouth, as many have experienced even in the absence of disease when bending over after eating a heavy meal. Belching involves the forceful passage of air from the stomach or esophagus. By itself, it rarely signifies organic or structural esophageal disease and usually is merely an annoyance. Frequently, more air is swallowed than is eliminated during belching, tending to promote a vicious cycle of belching and abdominal bloating.

When a person experiences such symptoms, an X ray of the esophagus, sometimes referred to as a barium swallow, is often indicated. Barium sulfate, a substance opaque to X-ray waves, is swallowed, and a physician-radiologist then follows its passage through the upper and lower esophagus with the use of a fluoroscopic screen. X-ray films are exposed to document the observations and to obtain a permanent record for more detailed study and comparisons. X-ray movies are occasionally made of the swallowing process, a technique called cineradiography. When these studies show a suspicious area, or when X ray findings prove inconclusive, direct vision of the entire length of the esophagus is possible through an esophagoscope—an instrument now generally composed of numerous fiberoptic light bundles encased in a highly flexible rubber tube. It is easily swallowed without appreciable discomfort. Through the esophagoscope, a trained physician can look directly at the mucosa of the esophagus, take still photographs or movies, or perform cell brushings or biopsies to look for microscopic evidence of cancer. Such an instrument may also be used to dislodge and remove foreign bodies. The esophageal pressure studies mentioned earlier can be quite useful in showing specific patterns of disturbed contraction waves in several conditions, such as achalasia and diffuse spasm. Other esophageal disorders that may be detected by one or more of these techniques include a diverticulum, a pouch or pocket leading off the esophageal wall, and esophageal varices, enlarged, twisted veins that are often secondary to increased pressure in the portal vein behind a diseased liver.

While it is not the aim of this chapter to get deeply involved in the area of treatment, the general principles of management of inflammatory conditions of the esophagus would include the use of multiple small feedings of liquid or mechanically soft foods of a bland nature, and measures to neutralize regurgitated stomach acid. In patients with an inflammatory narrowing, or esophageal stricture, it is sometimes possible to widen the involved area without surgery by using various stretching devices attached to tubes. A person with esophageal symptoms has no way of knowing whether they are functional or organic in nature, and medical advice should be sought as early as possible so that appropriate diagnostic and therapeutic measures can be taken.

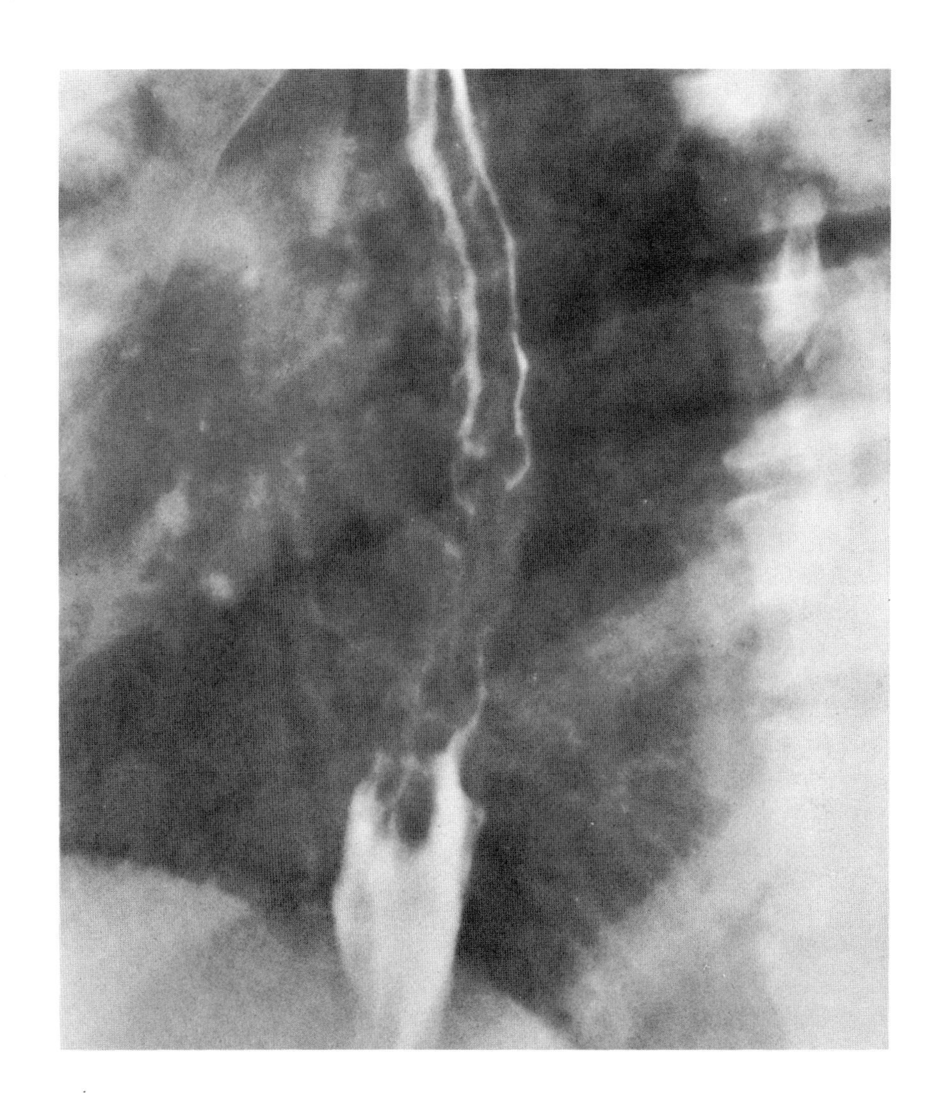

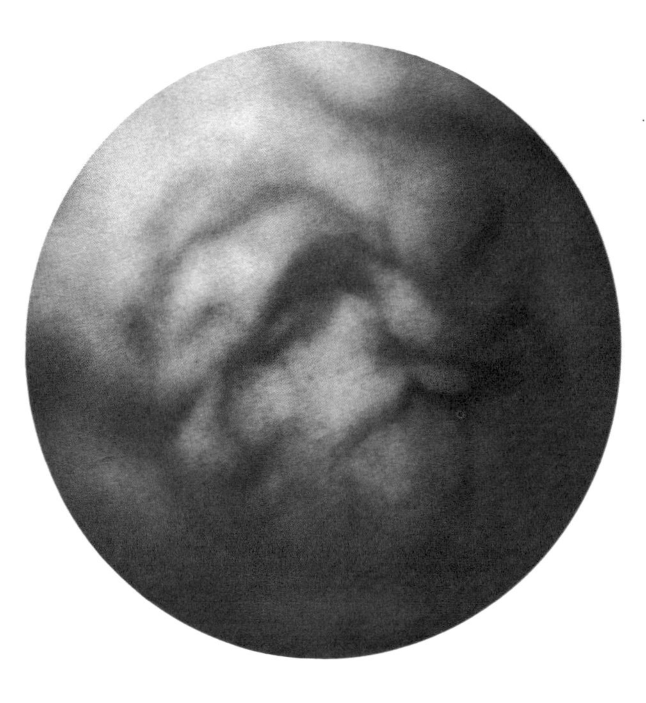

As barium sulfate, a substance that X rays cannot penetrate, moves through the digestive tract after being swallowed or given by enema, a radiologist can see the outlines (above) of the digestive organs and determine if their structure and movements are normal. By using the fiberoscope diagrammed on the opposite page, a physician can get a more vivid view of portions of the digestive tract (left).

The Abdominal Cavity

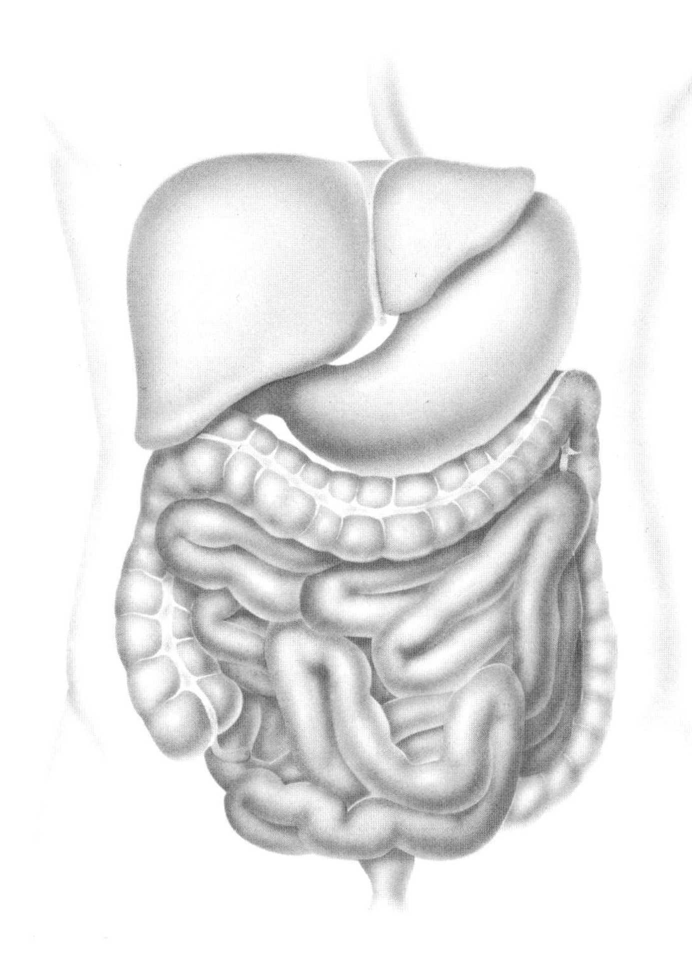

The abdominal cavity is separated from the thoracic, or chest, cavity by the dome-shaped muscular diaphragm and is continuous at its lower end with the pelvic cavity. Each of these cavities is lined by a smooth membrane that also covers the organs within them. The strong colorless membrane that lines the abdominal cavity is called the peritoneum. That portion which lines the abdominal and pelvic walls and the undersurface of the diaphragm is called the parietal peritoneum, while that which covers the stomach, liver, intestines, and most of the female pelvic organs is called visceral peritoneum. The peritoneum serves to hold the organs, or viscera, in position by extending its curtainlike folds, called the mesentery, to the body wall. Special mesenteric extensions join the stomach, liver, spleen, and colon to the body wall, and each bears a special name, such as the greater omentum extending from the stomach over the intestines and the lesser omentum between the stomach and liver. In addition, a number of peritoneal ligaments form membranes between the liver and diaphragm; between the stomach and colon, spleen, or liver; between the spleen and left kidney; and between the liver and duodenum. Abdominal organs that are located behind the peritoneum, such as the kidneys and the adrenal glands, are referred to as being retroperitoneal.

Disease processes involving organs within the peritoneal cavity may exhibit various signs and symptoms that are due to peritoneal inflammation, or peritonitis (e.g., perforated peptic ulcer, ruptured appendix). Peritonitis causes abdominal pain and tenderness, fever, nausea, and vomiting, and is a very serious situation. The process may remain localized or become generalized depending upon the extent of the inflammation and the body's ability to keep it in check before appropriate medical or surgical treatment is instituted. The pain of peritoneal irritation is always sharp and severe, in contrast to various types of dull abdominal discomfort that characterize many processes emanating from the digestive organs themselves. Pain arising from the peritoneum may also occur as referred pain, that is, pain felt at an area other than the site of its cause. For example, a perforated duodenal ulcer often causes pain at the top of the right shoulder.

Some of the other disease processes that may occur within the abdominal cavity include localized accumulations of pus, or abscesses, generally the result of perforation of one of the abdominal organs; cysts, or sacs containing various fluids, proteins, and sometimes red blood cells; and ascites, an accumulation of fluid in the peritoneal cavity that is most commonly seen in certain advanced types of cirrhosis of the liver. Ascites may also be caused by irritation of the peritoneal surface by chronic infection or tumor, or even by obstruction of the veins that return blood to the heart. Peritoneoscopy is a diagnostic technique whereby a telescopelike instrument is inserted through a small incision in the abdominal wall under local anesthesia. This permits examination of the peritoneal surface as well as the surface of the abdominal organs; it is also possible to perform limited procedures, such as biopsies, through the peritoneoscope.

A hernia is a rupture of the body wall with protrusion of part or all of an organ through the new or enlarged opening; among the most common are inguinal, femoral, and umbilical hernias. Postoperative hernias may occur along the length of the surgical excision. One of the more common types of internal herniation occurs when a portion of the stomach protrudes into the chest through an esophageal opening, or hiatus, in the diaphragm called an esophageal hiatus hernia.

The Stomach

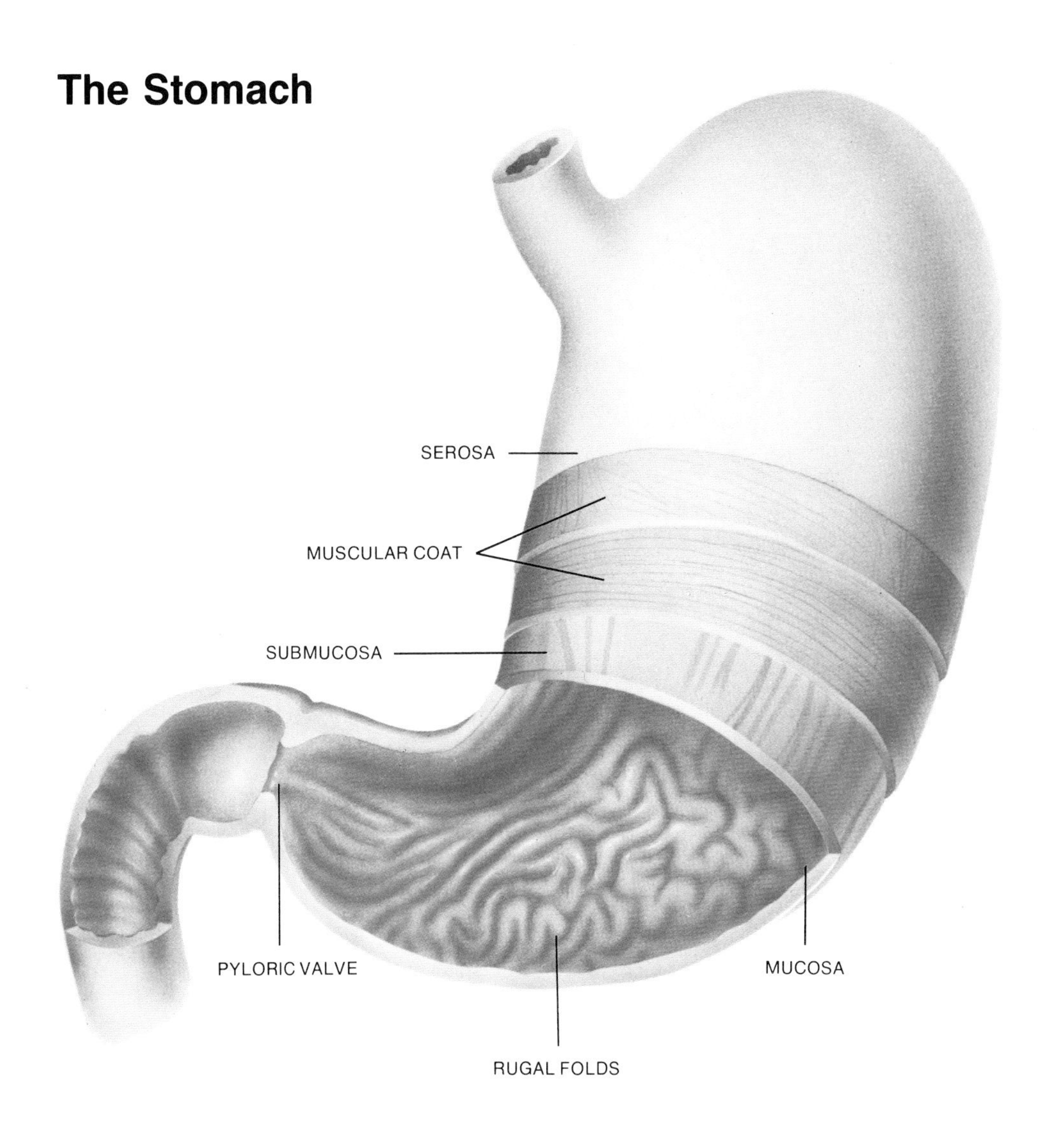

The stomach, a muscular pouch located just below the diaphragm in the left upper abdomen, connects the esophagus with the duodenum, or first portion of the small intestine. The stomach normally extends to the right of the midline of the body, and, depending upon body build and the tone, or state of tension, of the stomach musculature, it may actually reach into the pelvic cavity. The cardia is the upper portion of the stomach where the esophagus enters; the fundus is located immediately under the diaphragm but mostly above the cardia because of the stomach's curvature; and the body is the main portion of the stomach between the fundus and pyloric antrum, that portion just above the opening into the duodenum. The upper concave edge of the stomach is called the lesser curvature, and the lower convex edge is the greater curvature. The stomach—the widest part of the digestive tube—is shaped somewhat like the letter J.

The stomach wall is made of up four coats, the mucosa, submucosa, muscle, and serosa. The mucosa, or mucous lining, is thrown up into numerous rugal folds, or ridges, when the muscular coat contracts. The muscular coat consists of longitudinal, oblique, and circular involuntary muscle fibers, including a ringlike pyloric sphincter which acts as a valve to regulate the passage of food out of the stomach. The integrated actions of all these muscles permit expansion and contraction of the stomach so that it can receive and store relatively large quantities of food. This food is then released in small amounts when the rest of the digestive tract is ready to receive it.

The other major function of the stomach is to aid in the chemical breakdown of food. How is the stomach built to do this? Scattered over the surface of its mucosa are tiny pits, or foveolae, which mark the openings of different types of gastric glands; usually three or four gastric glands open into each foveola. Fundic glands are distributed throughout the fundus and body of the stomach, extending to the pyloric antrum, and are composed of three types of cells: parietal, chief, and mucous neck cells. The parietal cells make hydrochloric acid, while the chief cells produce pepsinogen. In the presence of hydrochloric acid, pepsinogen is changed into pepsin, an enzyme that splits the basic proteins in food into simpler peptides. Mucus is the primary product of the mucous neck cells, so-called because they are located predominantly in the neck of the fundic glands. The cardiac glands in the region of the esophagogastric junction also consist largely of mucus-producing cells; they contain very few parietal and chief cells. The pyloric glands of the gastric antrum secrete mucus and a number of biologically active substances—especially the hormone gastrin, an important regulator of gastric secretion. The surface epithelial cells throughout the stomach are another source of mucus; and some blood plasma leaves the rich network of gastric blood vessels and seeps between the surface cells into the stomach cavity.

Gastric Secretion

The adult human stomach daily secretes about 2 to 3 liters of gastric juice that contains a mixture of the above ingredients plus water, a number of electrolytes, and lipase, an enzyme that acts to start the breakdown of dietary fat. These various gastric secretions have a number of physiological functions which alter foods by mechanical and chemical means. They include dilution of the liquid mass, called chyme, into which food is converted by gastric digestion; splitting of proteins into small fragments; and production of a so-called intrinsic factor required for the intestinal absorption of dietary vitamin B_{12}. In addition to providing suitable acidity for converting inactive pepsinogen to pepsin, hydrochloric acid also has a bacteria-killing role but appears to be inhibited by gastric mucus from attacking the stomach mucosa itself. Mucus lubricates and protects the mucosa from mechanical abrasion and also can act to inhibit pepsin activity.

The regulation of gastric secretion involves a relatively complex balance of stimulatory and inhibitory processes. Gastric secretion may be considered to have cephalic (mental), gastric (stomach), and intestinal phases, but much overlap exists as a result of close interrelationships among nerve, endocrine, and other mechanisms. For example, circulating hormones from the pituitary and adrenal glands exert a continuous effect on gastric secretion.

The cephalic, or vagal, phase occurs when the brain centers are stimulated by the smell, sight, or taste of food so that gastric juice is secreted as a conditioned response. This type of secretion contributes heavily to ulcer formation because hydrochloric acid and pepsin are released into the stomach when there is little food present to digest. Since this cephalic phase is mediated mostly through the vagus nerve, the operation of cutting the vagus nerve, or vagotomy, is an important part of most peptic ulcer surgery.

The gastric phase of gastric secretion begins when food enters from the esophagus and continues until the chyme has been emptied from the stomach. It is triggered by dietary chemicals such as meat extracts and by food stretching the stomach walls, causing the release of the hormone gastrin from the pyloric glands. Gastrin goes directly into the bloodstream and circulates to stimulate the parietal cells of the fundus and body of the stomach to produce hydrochloric acid. Gastrin release is also triggered by vagal stimulation, illustrating the overlap of the gastric and cephalic phases. The presence of hydrochloric acid in the stomach tends to turn off the release of gastrin, either directly or indirectly. Gastrin actually occurs in different forms, and it is known to be capable of influencing esophageal and gallbladder muscle activity, as well as pancreatic and liver secretion.

The intestinal phase may result in the production of a comparatively low-acid-content gastric juice when the chyme enters the duodenum. The chyme is thought by some clinicians to stimulate the release of gastrin from the duodenum when the duodenum comes in contact with the digestive products of meat, or when the duodenum is distended by food. More importantly, gastric acid and pepsin secretion are inhibited by fats, carbohydrates, and acid in the duodenum as a result of the release from the duodenum of one or more unidentified hormones, or "enterogastrones," which circulate back to the stomach through the bloodstream. A specific inhibitory hormone released by acid from the mucosa of the duodenal bulb was described recently and called bulbogastrone. Other inhibitory mechanisms of gastric secretion include fear and fever (inhibition in the brain cortex), and vagal inhibitory reflexes of different types. The stomach has a rich blood supply, which greatly enhances its ability to be stimulated by circulating hormones. By the same token, in the presence of inflammation and ulceration, the stomach is predisposed to bleed.

Some of the biochemical activity in the stomach are outlined below. Secretions of the stomach glands are also indicated.

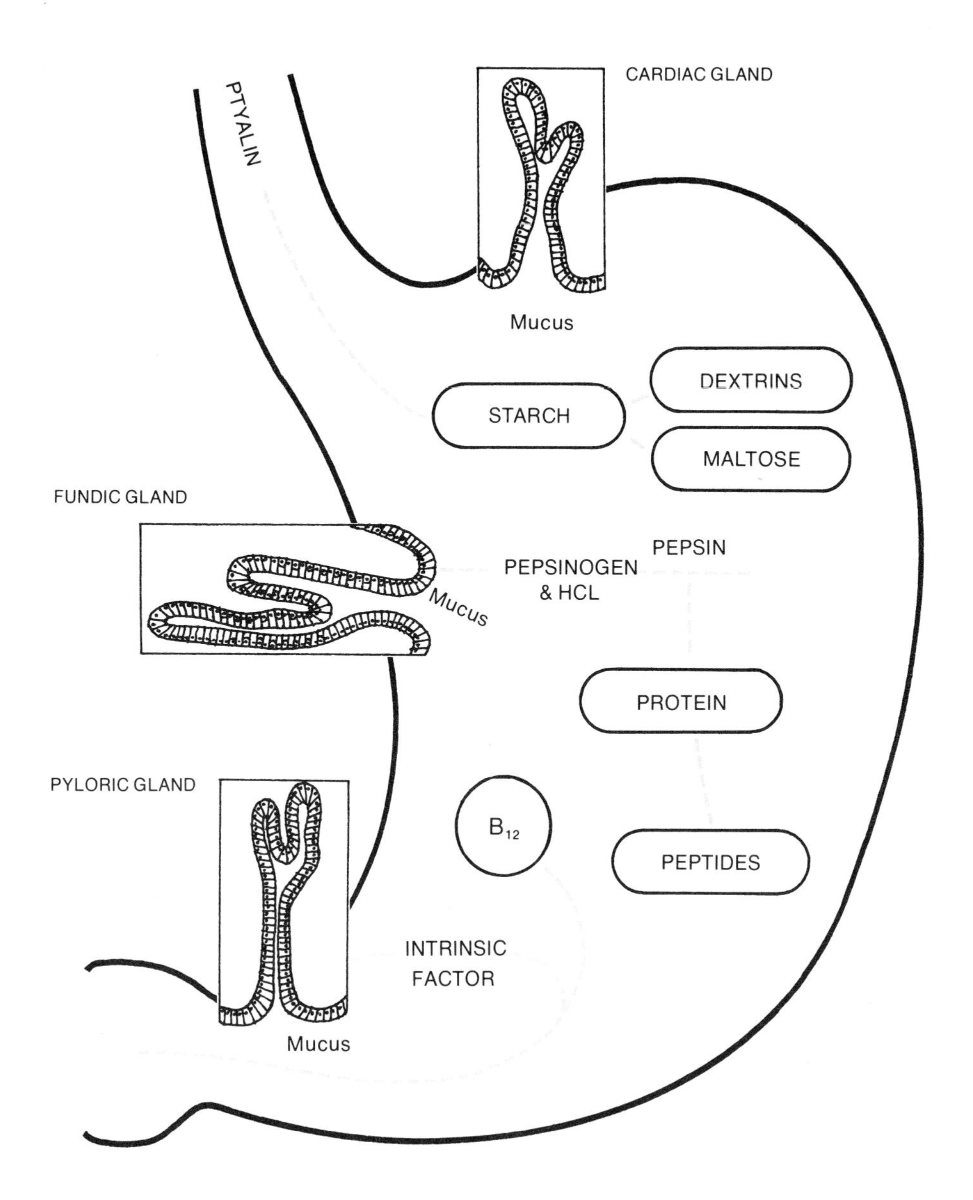

Stomach Movements

The motor functions of the stomach serve to help split, soften, and homogenize the food received from the esophagus and to move the chyme into the small intestine. These activities are accomplished by contraction waves that start high in the fundus of the stomach and increase in strength as they approach the pylorus. Throughout this process, the stomach is changing its size in response to other types of contraction waves. Additional but related factors that promote gastric emptying include the difference in pressure between the antrum and duodenum; the opening of the pyloric sphincter; the consistency and size of the bolus to be passed; and the volume, acidity, temperature, and chemical constitution of the gastric content. A gastric motor-activity-stimulating hormone, called motilin, was recently purified from the upper small intestine of the hog.

There are also a number of mechanisms that act to inhibit gastric motility. These include nausea and other reflexes that are stimulated from lower in the intestinal tract by the hormonal action of perhaps one or more enterogastrones released from the duodenum. The duodenal contraction waves that help carry the chyme are separate from those that originate in the stomach.

Stomach Disorders

Of the complaints frequently associated with disturbances of the stomach, nausea and vomiting are perhaps the best known. These are rather complex symptoms that may be due to problems both within and outside the digestive system, and that involve special centers in the central nervous system. In the case of vomiting, close coordination with the respiratory system is required to prevent the accidental aspiration of gastric contents into the bronchial tubes and lungs. Knowledge of the color, acidity, volume, particulate content, and forcefulness of the vomitus may be helpful to the physician faced with determining the cause of this most perplexing symptom. To gain such information is part of "taking a history" and is of great practical value. As regards abdominal pain, for example: When does it come? What time of day? Before or after meals? What relieves it? Milk? Food? What makes it worse? Did it follow the taking of aspirin or other medications? Does it go anywhere? To the shoulder? To the back? Upper abdominal pain, even when localized to the "pit of the stomach," may emanate from one of several organs in either the abdomen or lower chest.

As was the case for esophageal symptoms, a number of procedures are used to aid in reaching a diagnosis of stomach disease. These include an X-ray examination after a barium swallow, as well as direct vision of the inside of the stomach by gastroscopy. The latter provides the opportunity to obtain biopsies or cell brushings for microscopic analysis. Also, thin tubes may be passed into the stomach to collect specimens of gastric juice, used to measure the amount of hydrochloric acid produced in a given time period in response to specific stimuli.

Now that this discussion has considered what the stomach is made of, what it produces, how it moves, some of the distress symptoms that it may give rise to, and how physicians approach these complaints, it can be concluded by looking at some of the more common conditions involving this important digestive organ. Gastritis, or inflammation of the stomach mucosa, is probably the most common of these, although it is not diagnosed as frequently as it occurs. Often it may not cause symptoms, but it can result in lack of appetite, nausea, and upper abdominal pain, and may become severe enough to cause bleeding. The many causes of gastritis include taking aspirin and excessive alcohol consumption.

Gastric ulcers consist of a localized area in which the mucosal lining of the stomach has been destroyed, permitting the digestive juices to seep into the lower mucosa and submucosa and produce inflammation and often intense pain. These ulcers usually occur in the lower portion of the stomach and may be associated with marked gastritis; multiple ulcers may be present. The complications of gastric ulcers include bleeding and perforation; also, obstruction occurs when the swollen tissue is located so as to interfere with the passage of food from the stomach. The presence of a gastric ulcer always causes a physician to consider the possible presence of a tumor, and the diagnostic tests for stomach disease described above become especially important in helping to distinguish a benign ulcer from a malignant one.

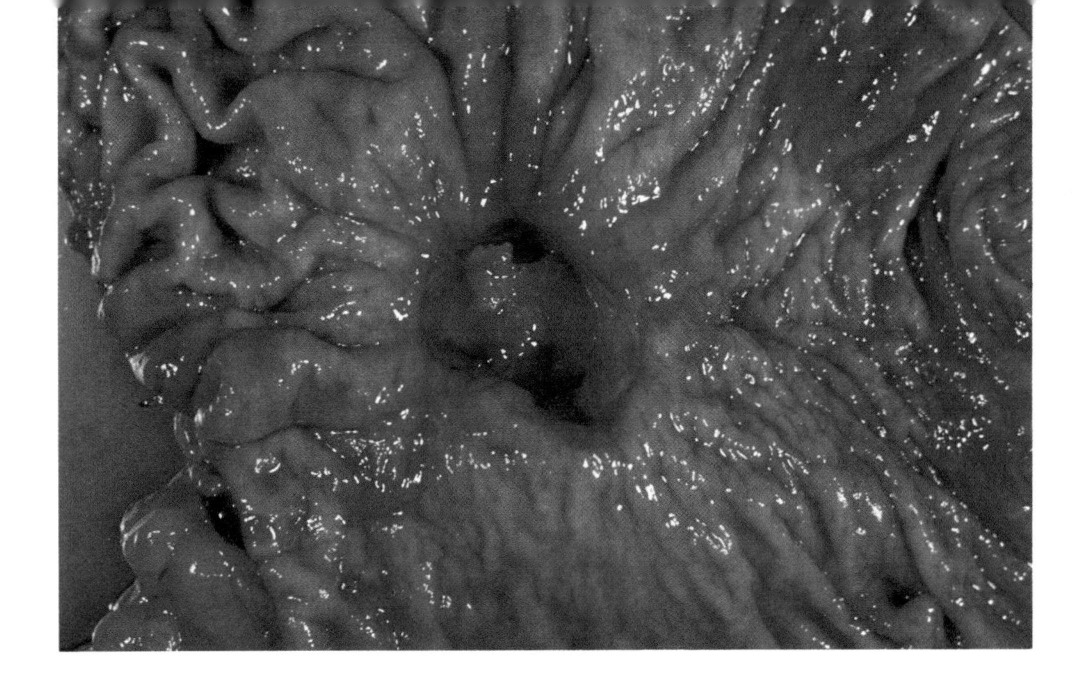

When the stomach lining becomes damaged, digestive juices seep into the resulting ulcer, inflame underlying tissue, and cause pain.

The Small Intestine

The duodenum, the first twelve inches of the small intestine, or small bowel as it is sometimes called, is a C-shaped tube which surrounds part of the pancreas and contains a small opening, called the ampulla of Vater, for receiving digestive juices from the pancreas, liver, and gallbladder. Therefore it is not surprising that the vital digestive process becomes accelerated as soon as the chyme passes into the duodenum. The glands of the duodenal mucosa produce mucus and other fluids that serve as protection against the corrosive action of gastric juice. A number of hormones are also produced in the duodenal wall, some of which—enterogastrone and bulbogastrone—have already been mentioned in regard to their suspected roles in the regulation of gastric secretion and motility. Others, including secretin and pancreozymin-cholecystokinin, which have been both physiologically and biochemically defined, will be discussed below.

Ulcers of the Duodenum

The most common abnormality involving the duodenum is duodenal ulcer disease. Approximately 80 percent of all peptic ulcers occur at this site, and at least 10 percent of the population of the United States is so afflicted at some time in their lives. Generally this is a disease of young people, although it may occur at any time between infancy and old age, and more often in men than in women.

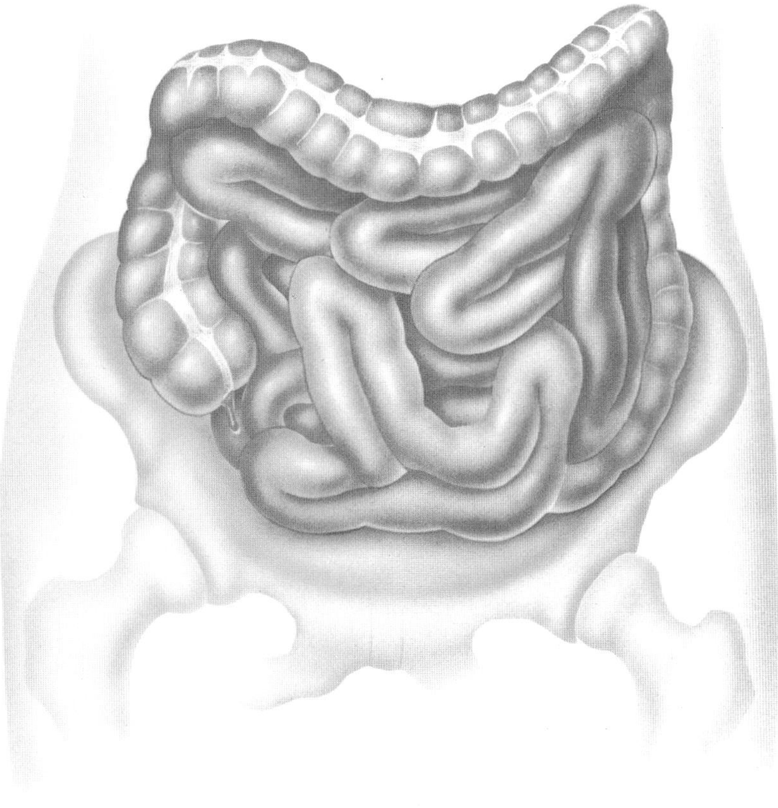

The exact cause of duodenal ulcer is still a source of controversy, but the overproduction of hydrochloric acid by the parietal cells of the stomach appears to be an important factor. The basic protective mechanisms of the duodenal mucosa seem to be inadequate to handle this increased acid production, resulting in a localized ulcer. These protective mechanisms involve the products of the local mucus-producing glands, plus the acid-neutralizing effect of alkaline secretions from the liver and pancreas.

Because ulcer pain is caused by excessive stomach acid that irritates exposed nerve endings in the ulcerated duodenal wall, the distress classically occurs in mid-morning, is relieved by eating lunch, recurs in mid-afternoon, and is relieved again by eating the evening meal. The pain may return at least once more before bedtime or a few hours after retiring; relief may be obtained by something as simple as drinking a glass of milk. Duodenal ulcer patients also often complain of a localized burning, cramping, stabbing, or gnawing pain in the upper abdomen. Treatment is directed to diminishing excess stomach acid. Since acid may be produced by a combination of factors, it is important to remove stimulants of the cephalic phase such as emotional stress, to avoid gastric irritants such as excessive alcohol and coffee, to eat frequent meals of bland foods, and to neutralize acid with substances such as milk, milk products, and antacid preparations.

The Jejunum and the Ileum

The remainder of the small intestine consists of the eight-foot jejunum and thirteen-foot ileum. The jejunum is the main site of homogenization and mixing of intestinal chyme and of the formation of intestinal enzymes. It is also an important absorptive site for most nutrients, including sugars, amino acids, and fats. The jejunal mucosa is thrown up into small folds or valves, which tend to slow down the passage of the chyme and, more importantly, greatly increase the surface area of the intestinal cavity. This surface contains about five million microscopic, fingerlike projections called villi. Each villus, in turn, has microvilli totaling in the billions for the entire small intestine. The villus contains networks of blood capillaries and lymphatic vessels, the routes by which the end products of digestion are absorbed into the bloodstream to be carried to all parts of the body.

A single layer of epithelial cells covers the villi and the intervening crypts. New cells constantly form in these crypts, move up to become part of the villi, and eventually are shed into the intestinal cavity from the villous tips. The intestinal surface layer is thus in a continual state of replacement.

In the jejunum simple water-soluble substances of low molecular weight are absorbed by diffusion, while more complex substances are absorbed by an active transport mechanism, that is, some type of carrier system through the cell membrane. These mechanisms are dependent on the provision of energy by the cell membrane across which the absorption is taking place, and they vary with the specific substance absorbed. Both diffusion and active transport mechanisms can be involved in the absorption of certain substances.

The ileum makes up the lower half of the small intestine. It is characterized by continuous wormlike movements, resulting in sequential narrowing and widening of its adjacent segments, and thereby significantly adding to the homogenization of the intestinal contents. The ileum also produces important digestive enzymes and is the site of absorption of vitamin B_{12}, bile acids, and remaining quantities of amino acids and fats. Further details concerning the digestion and absorption of some of these important nutrients are discussed below.

As in the esophagus and in the stomach, the entire small intestine has an autonomic nerve supply. The parasympathetic vagus nerve plays a stimulatory role, and the sympathetic intestinal nerve fibers have a restraining influence on intestinal movements and secretions.

Movements of the Small Intestine

The small intestine has four overlapping types of movements: rhythmic segmentation, peristaltic contractions, pendular movements, and contractions of the mucosal villi and smooth muscles. The rhythmic segmental contractions alternate with the relaxation of segments on each side. Peristalsis consists of those coordinated contractions of the longitudinal and circular muscle fibers by which the alimentary canal propels its contents. Pendular movements are probably of little importance in man, but in other animals they occur over small lengths of the small intestine as aids to both mixing and absorption. The intestinal villi move both by abrupt contractions and by swaying. These villous movements are thought to be caused by vagal stimulation and by an intestinal hormone and are primarily due to the contraction of smooth muscle in the mucosa.

Thus the quality and speed of intestinal movements are the end result of a complex series of neural, chemical, hormonal, and muscular phenomena. In this manner, the chyme becomes well mixed, and adequate time exists for its continued

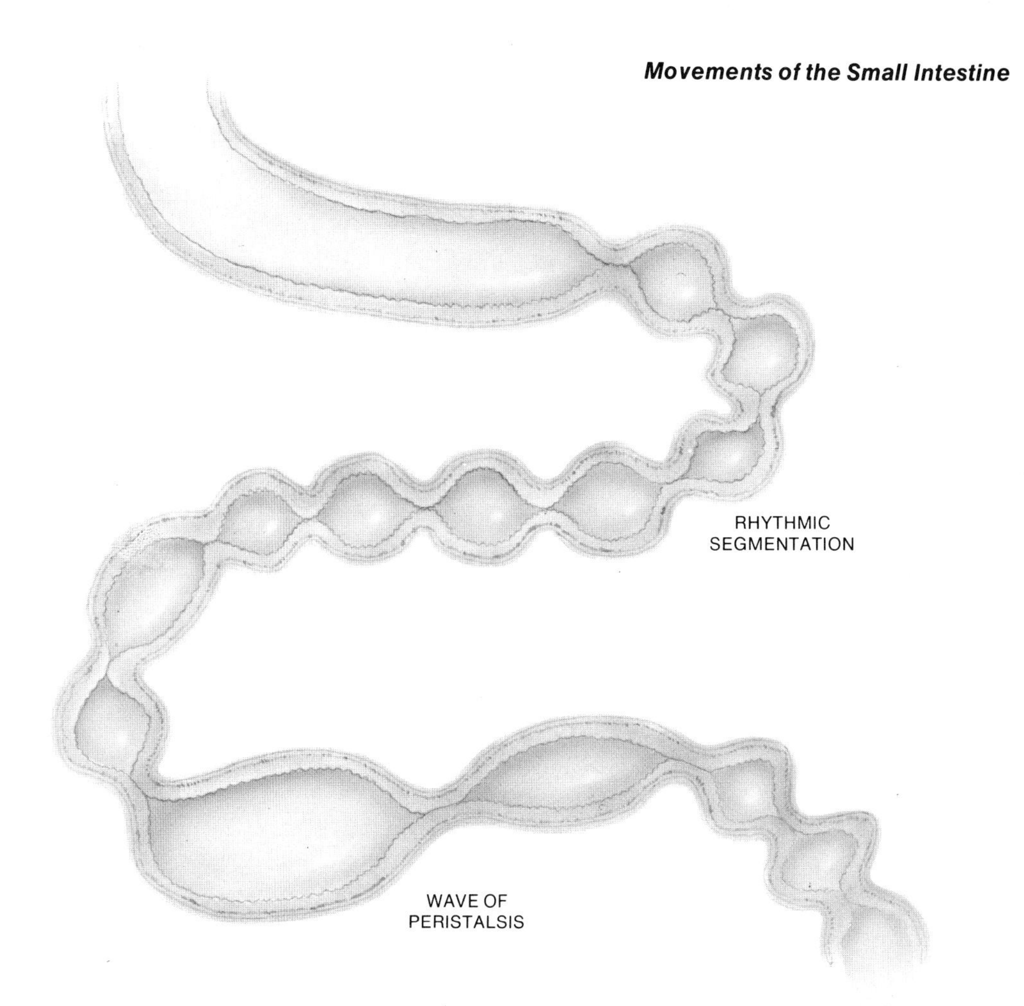

digestion and absorption. An ileocecal valve regulates the passage of material from the ileum into the cecum and, in a limited way, may prevent some backward flow of feces from the colon. The ileocecal valve opens when a peristaltic wave at the end of the ileum builds up enough pressure to overcome valve resistance, and then closes when the pressure in the cecum increases.

An interaction of topical, vagal, and hormonal stimuli results in the secretion of up to two liters of intestinal juice, or succus entericus, each twenty-four hours. This intestinal juice is made up of enzymes, intestinal mucus, water, and electrolytes. The enzymes are considered to be either extracellular, when secreted into the intestinal lumen (e.g., amylase; enterokinase), or intracellular, when contained in epithelial cells (e.g., peptidases, which break down proteins; intestinal lipases, which assist in breaking down long-chain fatty acids; and various disaccharidases, which split complex sugars into simple sugars). The functions of intestinal mucus appear to be both lubrication and protection of the surface lining. The electrolytes help to diminish the acidity of the intestinal contents. In cases of markedly excessive acid production, for example, the activity of certain enzymes can be reduced, leading to faulty digestion.

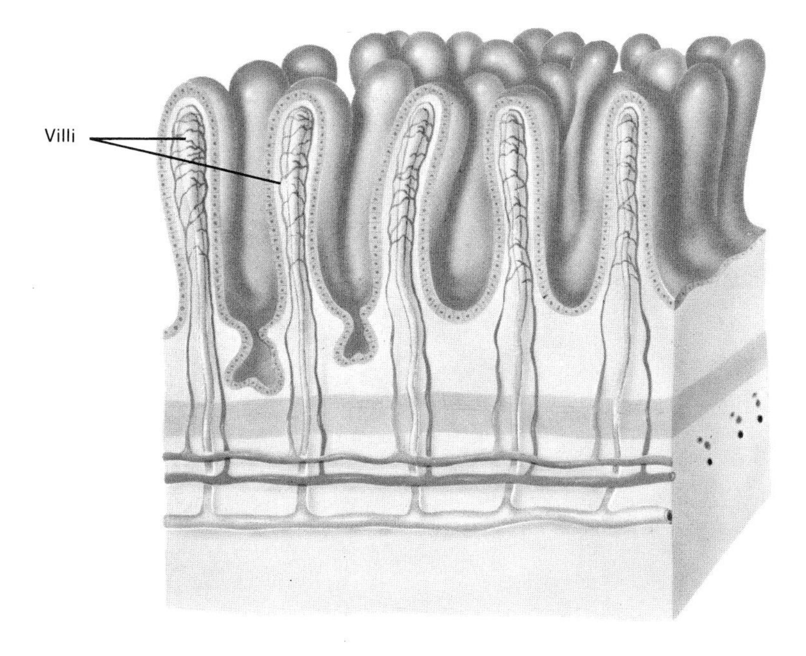

Most of the food in the daily diet is absorbed in the small intestine after being broken down by the secretions of glands in the stomach and small intestine. Absorption of the food components takes place through the many fingerlike villi in the small intestine.

Absorption by the Small Intestine

How are specific food substances absorbed through the intestinal wall? Carbohydrates, proteins, and fats ultimately are absorbed to some extent by active transport, a way of moving across cell membranes. Carbohydrates are split and absorbed as monosaccharides, or simple sugars, in the upper small intestine and pass via the blood capillaries into the portal vein. Proteins are broken down to peptides and amino acids, and are absorbed by both simple diffusion and active transport mechanisms. While most are transported by the portal vein to the liver, some resynthesis of these absorbed amino acids back into protein occurs within intestinal epithelial cells.

Fat absorption also involves active transport but is more highly specialized and merits closer inspection. Large insoluble fat globules are converted by the interaction of bile, pancreatic lipases, and intestinal juice into simpler water-soluble molecules—glycerol, monoglycerides, and fatty acids. The glycerol enters the mucosal cells as such, while the monoglycerides and fatty acids combine with bile salts in the intestinal cavity to form so-called fat micelles, which probably enter the intestinal epithelium by a passive process. Within the cells, short- and medium-chain fats are split to glycerol and short- or medium-chain fatty acids, and these enter blood capillaries in the digestive system. Larger or long-chain fats are resynthesized in the intestinal mucosal cells, exit from the base of these cells into the connective tissue of the mucous membrane as small fat droplets called chylomicrons, and are taken up into the lymphatic channels.

Disorders of the Small Intestine

As might be expected from the important functions of the small intestine, rather severe consequences may arise from abnormalities within it. Those suffering from such disorders may experience marked weight loss and diarrhea, for example, and may show widespread effects involving multiple body organs and systems, depending upon which required nutrients are missing. These patients may thus develop anemia, fragile bones, easy bruisability, skin rashes, altered skin pigmentation, a sore tongue, and swollen ankles, to mention just a few. These signs and symptoms are characteristic of the large and complex group of small intestine disorders collectively known as malabsorption syndromes. (A syndrome is a set of symptoms which occur together.) Malabsorption may be classified on the basis of inadequate digestion (e.g., disorders of the stomach, liver, biliary tract, or pancreas); inadequate absorptive surface (e.g., after excision of a part of the intestine); primary abnormalities of the mucosal cells (e.g., inflammatory, genetic, or endocrine disorders), or other problems (e.g., bacterial overgrowth in the small intestine, lymphatic obstruction, or vascular disorders).

One of the more common conditions affecting the small intestine, and undoubtedly present more often than diagnosed, is regional enteritis, or Crohn's disease. This is an inflammatory condition of unknown cause. It most frequently involves the lower portion of the ileum, but it may involve any part of the alimentary canal from the mouth to the anus. The bowel wall may become greatly thickened, causing narrowing of the intestinal cavity, and the nearby lymph nodes also enlarge. The inflammatory process may spread through the muscle and serosal layers of the intestinal wall, forming abnormal channels called fistulae that communicate to other loops of the small intestine or other organs. Perforation and abscesses also can develop as complications of the disease, but more common symptoms include weight loss, diarrhea, abdominal cramps, or fever. Thus, depending upon the extent and severity of the disease, patients with regional enteritis may have a malabsorption syndrome, primarily obstructive symptoms, or both.

Other mechanical-type small intestinal problems may be the result of obstructions due to foreign bodies, intestinal worms, or thin bands of fibrous scar tissue, or adhesions, that may surround and block the bowel. Symptoms of this type include intermittent abdominal colic, nausea, vomiting, abdominal distention, and constipation. Such obstructions are potentially hazardous and can lead to intestinal gangrene or perforation if not appropriately treated by the passage of a long, thin suction tube, or by surgery if necessary.

Until recent times, the twenty-two-foot-long small intestine was virtually "off limits" with respect to the physician's ability to diagnose some of its relatively subtle diseases and abnormalities. This problem has gradually been overcome. For example, X-ray techniques have been improved; intestinal biopsy tubes have been developed; and fiberoptic endoscopes are available to reach and photograph most of the small intestine, as well as to collect biopsy specimens and secretions under direct vision. It is even possible for experts to pass small tubes through the ampulla of Vater into the pancreatic and biliary ducts and to inject X-ray-detectable solutions. Also, biochemical tests can measure the ability of the small intestine to absorb fats, carbohydrates, and, to a lesser extent, proteins. These procedures involve various blood, stool, and urine collections, with or without the prior administration of test substances. Occasionally, however, none of these tests will solve a diagnostic mystery, and there is need for exploratory laparotomy, an operation in which a surgeon carefully examines all organs in the abdominal cavity.

The Liver, Gallbladder, and Pancreas

To really understand more of the actions making up digestion and absorption, it is necessary to take a side trip from the duodenum up through the ampulla of Vater. This permits examination of the accessory glands and organs that supply important fluids to the digestive tract at this location—the liver, gallbladder, and pancreas.

The Liver and Bile

The liver is the largest internal organ of the body. It is located just under the diaphragm in the upper part of the abdomen, with the largest of its four lobes to the right of the midline. Its surface is smooth and its texture rubbery; the gallbladder is located along its bottom edge. The portal vein collects blood from the capillaries and veins of the stomach, pancreas, spleen, gallbladder, and small and large intestines and delivers this blood directly to blood cavities situated between the individual liver, or hepatic, cells. These cells are arranged around the central veins of the hepatic lobules and are constructed so that each corner of the polygonal hepatic lobule is adjacent to a triad of vessels—a venule, an arteriole, and a bile ductule.

Via the portal circulation, the liver receives nutrients absorbed from the gastrointestinal tract before they get into the general circulation; the liver determines what to do with these end products of digestion. Thus, sugar can be released into the circulation for use as fuel by body cells or be stored in the liver; fats similarly may be burned for energy or stored; and amino acids may be used or rebuilt into new proteins. For example, the liver is

Vessels of the Biliary System (Rear View)

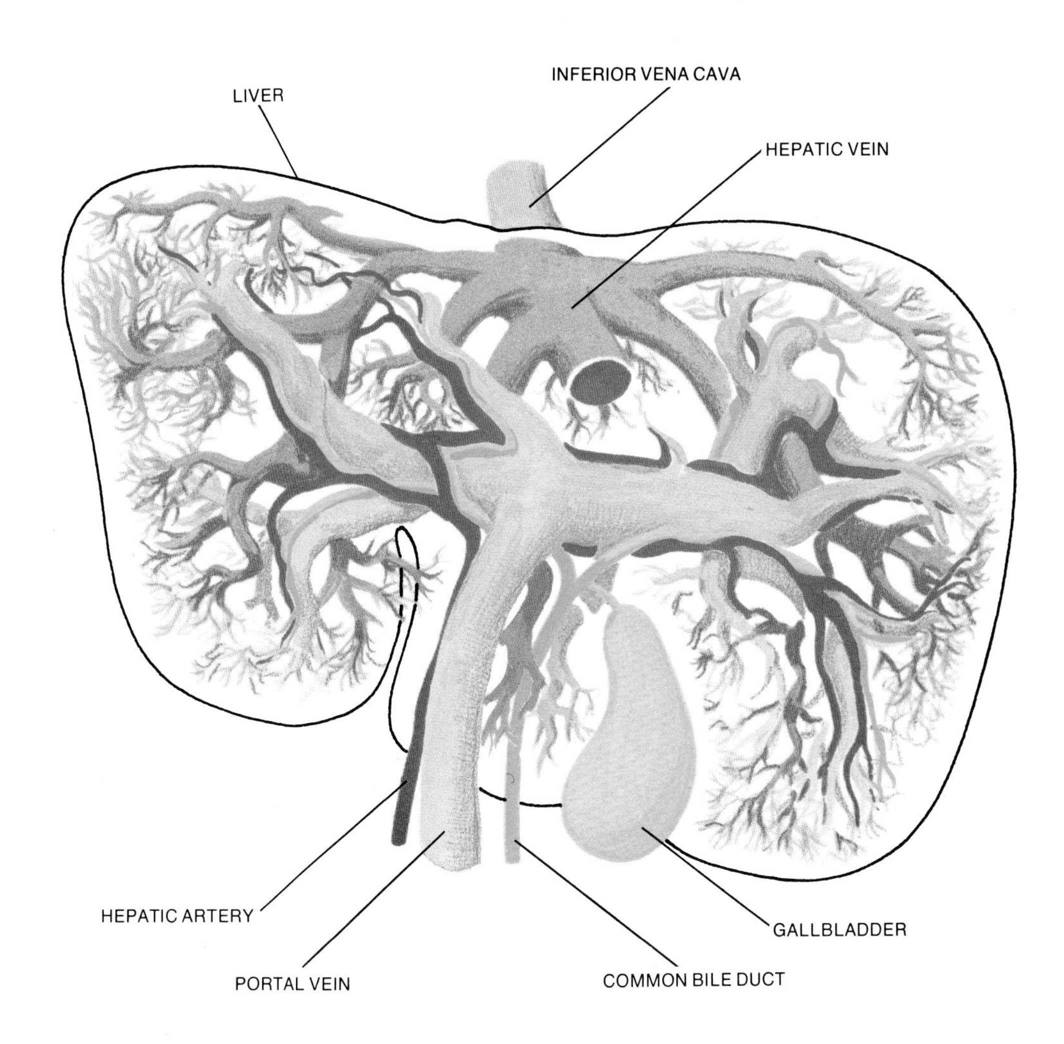

an important production site for the proteins of blood plasma, with the exception of the gamma globulins. The liver cells also separate toxic materials out of the portal and general circulations and break them down for inclusion in its secretory product, bile.

Bile is golden brown to yellowish green and contains mainly water, bilirubin, and bile salts. Bilirubin is the principal bile pigment; and the normal brown color of feces is due primarily to the presence of high concentrations of breakdown products of bilirubin—itself the result of the breakdown of the hemoglobin in red blood cells. Several types of bile salts are produced in the liver cells from cholesterol. They aid in the breakdown of fats, acting like detergents to split up the insoluble fats into tiny water-suspendible particles. The bile salts play a dual role by emulsifying fat within the intestine and stimulating the resynthesis of fat within the intestinal mucosal cell.

The Gallbladder

The gallbladder is a dark green sac, shaped like a blackjack and lodged in the hollow in the undersurface of the liver. The main function of the gallbladder is the storage and concentration of bile. Its duct, the cystic duct, joins the common hepatic duct to form the common bile duct. This, in turn, meets the main duct from the pancreas to drain into the ampulla of Vater in the duodenal wall. Thus, materials from the liver, gallbladder, and pancreas can reach the duodenum simultaneously, permitting the body to coordinate their important digestive activities.

The liver secretes bile continuously, but bile is needed in the duodenum only during periods of digestion. The rest of the time, the bile flows up the cystic duct for storage in the gallbladder. When fats and other foods are in the duodenum, the duodenal mucosal epithelial cells are stimulated to release pancreozymin-cholecystokinin. This hormone circulates in the bloodstream, stimulates the smooth muscle of the gallbladder to contract, causes relaxation of the muscular sphincter surrounding the ampulla of Vater, and permits bile to enter the duodenum. A less important mechanism for gallbladder emptying is mediated by the vagus nerve when products of protein digestion contact the duodenal mucosa. The hepatic and common bile ducts are able to concentrate and store small but adequate quantities of bile when the gallbladder is either nonfunctional or removed. The normal adult human secretes almost one liter of bile per day. This is not all excreted from the body, however, because a so-called enterohepatic circulation results in much bile salt absorption from the terminal ileum with its return to the liver via the portal circulation.

Jaundice and Other Liver-Related Disorders

Jaundice, or icterus, is a syndrome characterized by increased amounts of bilirubin in the blood, skin, mucous membranes, and other body tissues. This causes the patient to take on a yellow color, as best seen by examining the sclerae, or white parts of the eyes. The many causes of jaundice include blockage of the bile ducts with backing up of bile into the blood, failure of the chemical and enzymatic processes by which bilirubin is handled and excreted by liver cells, excessive destruction of red blood cells so that the liver has difficulty keeping up with the elimination of bilirubin, poorly functioning liver cells due to toxic or viral inflammation of the liver, and the presence of primary or secondary liver tumors. Aside from jaundice and the itching it causes, patients with liver abnormalities may have symptoms varying from mildly diminished appetite and nausea to severe debilitation and coma. In cases where liver damage causes an increased pressure in the portal vein, additional problems may include the accumulation of fluid within the abdominal cavity, enlargement of the spleen, esophageal varicose veins, and the appearance of increased collateral blood vessels on the surface of the abdominal wall.

One of the more common afflictions of the accessory organs is the formation of stones in the gallbladder due to incompletely understood changes in the solubility of certain bile components. Gallstones may block the cystic duct or otherwise cause inflammation and thickening of the gallbladder wall, or smaller stones may become lodged in the common bile duct, leading to jaundice. An inflamed gallbladder may result in nausea, vomiting, fever, or pain in the right upper abdomen. This pain may awaken the patient at about five or six o'clock in the morning, may radiate through to the back in the region of the right shoulder blade, and is commonly referred to as a gallbladder attack. It may often follow a fatty meal, a normal stimulant to gallbladder contraction. While surgical removal of the gallbladder generally is needed at some time for patients with a history of repeated gallbladder attacks, newer information about the interrelationships between bile salts and other bile constituents may lead to medical ways to dissolve gallstones.

In attempting to learn the cause of liver dysfunction, with or without jaundice, a physician may perform a number of diagnostic studies. Many of these are biochemical tests that attempt to measure the health of the liver cell, the status of blood circulation through the liver, and other functions. In cases of jaundice, it is important to differentiate between causes originating inside or outside the liver. In some situations, a needle biopsy (obtaining a specimen of tissue through a large-bore hypodermic needle) of the liver may be easily obtained through the skin using a local anesthetic. The radiologist may be helpful in diagnosing liver disease by injecting trace amounts of radioactive materials and showing their distribution within the liver. This technique is called a liver scan. Also, patients may be asked to swallow a tube that passes through the stomach into the duodenum, coming to rest near the ampulla of Vater. This procedure makes it possible to collect secretions leaving the common bile duct and to examine them microscopically for evidence of gallstones or malignant cells.

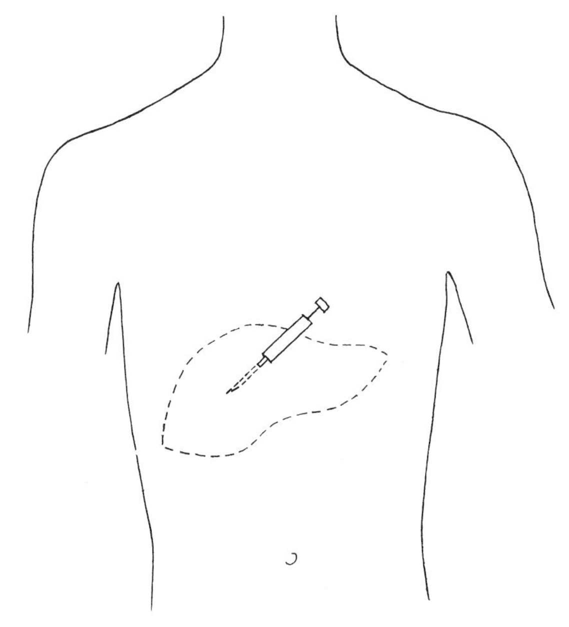

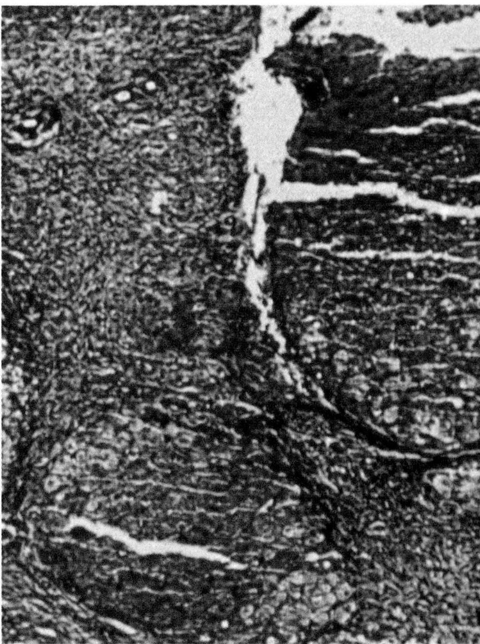

A physician performs a liver biopsy by inserting a large-bore needle attached to a syringe through the skin and into the liver, shown in the illustration at the left. The small bit of tissue extracted is examined microscopically. The photo above shows what the liver tissue of a person suffering from cirrhosis looks like under a microscope.

The Pancreas

The pancreas is a long tapering organ lying behind and below the stomach and, in part, adjacent to both the duodenum and spleen. It is pinkish white and anatomically consists of a head resting in the concavity of the duodenum, a central body, and a tail that extends to the left toward the spleen. The pancreas has both endocrine and exocrine functions. The endocrine portion consists of the accumulated islets of Langerhans, which produce insulin and other hormones concerned with carbohydrate metabolism. The exocrine portion of the pancreas has an elaborate ductal system that produces and secretes pancreatic juice, a watery alkaline fluid rich in such a variety of potent enzymes that it is capable of digesting almost all foods even in the absence of other digestive secretions. The most active of these enzymes are trypsin and chymotrypsin, which split large protein molecules; pancreatic lipase, which is greatly responsible for the digestion of fats; and pancreatic amylase, active in the breakdown of starches.

The pancreatic juice also has a high bicarbonate content, which serves to neutralize the strongly acidic gastric juice entering the duodenum. This provides a proper chemical environment for the optimum action of pancreatic enzymes and also may be useful in the prevention of acid-induced duodenal ulcers. In an adult human, the pancreas produces and secretes a total of about 700 ml (milliliters) of fluid daily in response to food stimuli. The rate of pancreatic secretion is slow until chyme enters the duodenum. Increased production is then caused largely by two hormones, secretin and pancreozymin-cholecystokinin, which are produced in and released from the duodenal mucosa into the bloodstream and which circulate to stimulate the pancreas.

Three of the more common pancreatic disorders are pancreatitis, cystic fibrosis, and pancreatic cancer. Pancreatitis may be associated with gallstones blocking the pancreatic duct or with excessive alcoholic intake. It is an acute inflammatory process leading to the escape of the highly irritating pancreatic enzymes into the surrounding tissues. These enzymes actually begin to digest the substance of the pancreas, leading to increased amounts of pancreatic amylase and lipase in the blood and urine. The symptoms of pancreatitis include nausea, vomiting, upper abdominal pain and tenderness, and fever. Such patients may be extremely ill and actually be in a state of shock, partly due to the loss of enzyme-rich bloody fluid into the abdominal cavity. In cystic fibrosis, a generalized hereditary disease of children and young adults, the glands of the pancreas become clogged with a very thick mucus; the resulting diminished supply of pancreatic enzymes to the duodenum may contribute to severe nutritional problems.

The techniques used by physicians to diagnose pancreatic disorders are relatively new and are constantly being improved upon. Thus it is possible not only to do blood and urine tests of pancreatic function, but also to position the duodenal drainage tube opposite the ampulla of Vater and collect and study pancreatic secretions after the intravenous injection of a small amount of secretin.

The Large Intestine

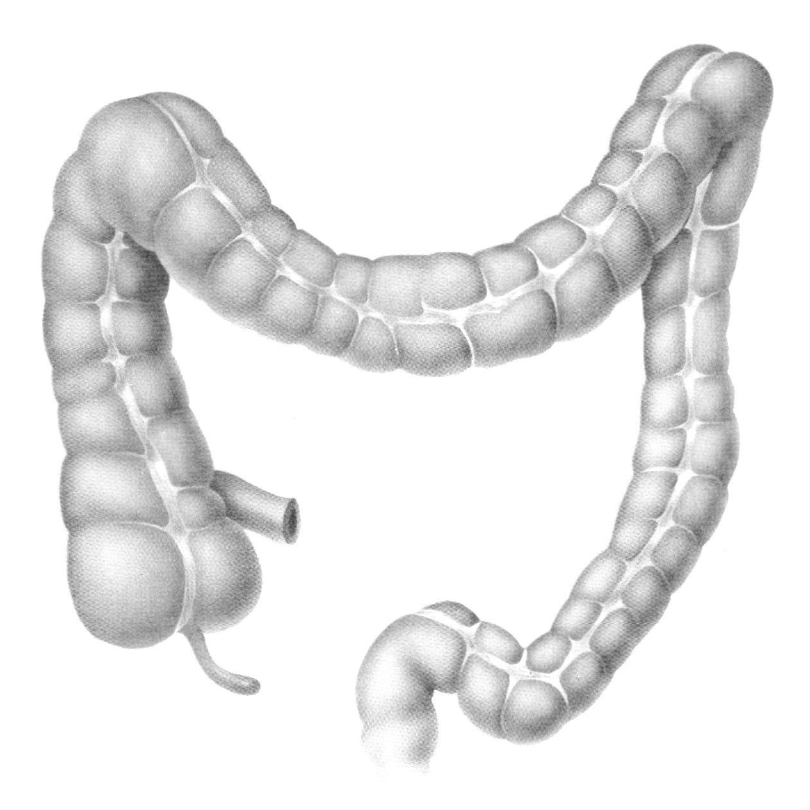

The large intestine consists of the colon, cecum, rectum, and anal canal. The colon is about five feet long, extending from the cecum, a blind sac located in the right lower abdomen, up the right side (ascending colon), across the upper abdomen (transverse colon), and down the left side of the abdomen (descending colon). The connecting sigmoid colon makes an S-curve toward the center and rear of the abdomen and ends in the rectum. The rectum is about five inches long and follows the curve of the sacrum and coccyx (tailbone) until it bends back into the short anal canal. Surrounding the anal canal is a strong sphincter muscle that is normally under voluntary control so as to close off the external opening at the lower end of the digestive system, the anus. The vermiform appendix is a wormlike blind sac arising from the cecum just below the ileocecal valve. Acute appendicitis may develop when the cavity of the appendix becomes blocked and inflamed.

The wall of the colon is made up of the four coats that are common to all parts of the alimentary canal from the esophagus on down—the mucosa, submucosa, muscle, and serosa. The colonic mucosa does not have villi such as are found in the small intestine, but numerous half-moon-shaped folds project from the surfaces that enclose the colon, and the epithelium contains a large number of mucus-producing goblet cells. The colon receives a rich blood supply via the inferior mesenteric artery, and there is a single marginal arterial arch in the mesentery sending off short arteries called endarteries to the colon. As with the upper parts of the digestive system, the colon is innervated by both parasympathetic and sympathetic nerve fibers. Stimulation of the parasympathetic system increases the motor activity of the colon, while the sympathetic action results in inhibition or slowing. Again, as in higher portions of the gastrointestinal tract, clusters of nerve cells are located within both the submucosa and the muscle layer.

The colonic muscle differs from muscle elsewhere in the gastrointestinal tube in that the outer longitudinal layer does not wrap itself around the colon but is composed of three narrow longitudinal strips. The spastic contractions of the circular muscle and the shortening of the longitudinal muscle cause division of the colon into sausagelike outpouchings separated by recesses known as colonic haustrations. This haustral segmentation results in kneading and rolling of the feces to retard flow and to aid in the absorption of water for conservation within the body. Colonic peristaltic waves also serve to move the colonic contents toward the anus. Also, three or four times daily, propulsive mass movements of the colon occur in which an eight- to ten-inch segment contracts simultaneously. These mass movements may be aided by reflexes set up by the distention caused by food in the stomach or duodenum.

Expulsion of Feces

As the fecal mass is pushed through the colon, water and certain electrolytes are absorbed; mucus is made; a number of gases are formed; and the bacteria in the cavity produce certain vitamins. The viscous and alkaline mucus lubricates and aids the passage of feces, helps the fecal matter hold together, and serves to neutralize irritating acids produced by bacterial fermentation. Water forms about 75 percent of the feces; and solids about 25 percent; while bacteria compose about one-third of the dry weight of the feces under normal conditions. Diets rich in vegetables increase the quantity of the stool, but a large amount of the fecal mass may be of nondietary origin (e.g., mucus, bacteria, and cells shed from the mucosa) since appreciable amounts of feces continue to be passed during prolonged fasting and starvation. Of the gas in the colon, a portion is swallowed air, and the remainder is produced in the small and large intestines.

These gases are nitrogen, oxygen, carbon dioxide, trace atmospheric gases, methane, hydrogen, hydrogen sulfide, and ammonia. The total amount of gas expelled as flatus in man may be as much as half a liter daily. While about eight liters of water are absorbed by the small intestine each day, another 300 to 500 ml are absorbed from the liquid chyme during its transit through the colon. The rectum, therefore, receives semisolid feces with which to form relatively firm stools.

When feces are pushed into the rectum, it becomes distended, a pressure is created, an internal anal sphincter relaxes, there is an urge to defecate, and the external anal sphincter may be either opened through conscious control or kept closed by voluntary contraction. When a person wishes to defecate, a sitting or squatting position is assumed; and an increased intra-abdominal pressure is brought about by fixation of the diaphragm, closure of the glottis, and contraction of the muscles of the abdominal wall. When coupled with the contraction of the distended rectum and relaxation of the external anal sphincter, this usually results in ejection of the stool, or defecation.

Food Intake–Defecation Time Span

It now becomes possible to consider how long it normally takes for ingested food to be digested and eliminated. Within a few minutes after a meal is swallowed, it begins to pass through the lower sphincter of the stomach. After the first hour the stomach is half empty; at the end of six hours none of the meal should be present in the normal stomach. The chyme goes through the small intestine, and the first part of it reaches the cecum in from twenty minutes to two hours. At the end of the sixth hour most of it should have passed into the colon, and by twelve hours after initial intake all should be in the colon. The transit time for the passage of feces from the ileocecal valve to the rectum is between eighteen and twenty-four hours. However, only part may be defecated and the rest retained for varying periods.

Colonic Disorders

Signs and symptoms suggesting colonic disease include a change in bowel habits leading to constipation, diarrhea, or both; diffuse or localized crampy abdominal pain; blood on the stools or toilet tissue; anemia and weakness; and perhaps even nausea, vomiting, or fever. The term constipation, commonly misused by persons, should refer to the infrequent and often difficult passage of hard and dry stools. Generally, the stools should be of regular consistency and be passed with ease, although not necessarily daily. Transient constipation may occur when traveling, with interruptions of the daily routine, or with dietary changes, but it is up to a physician to make the important distinction between functional causes of this type and the presence of organic disease, e.g., blockage secondary to inflammation or tumor. Diarrhea, the abnormally frequent passage of loose stools, may even be a consequence of a partial blockage and resultant mucosal irritation.

Unlike many of the other organs of the digestive system, the colon and rectum are not essential in order to maintain an excellent state of health. When the colon must be partially removed and continuity with the rectum cannot be reestablished, a colostomy, or opening in the colon through the skin, is surgically produced. The feces are then collected in an odor-proof bag. Similarly, when the entire colon and rectum must be removed, an ileostomy (opening in the ileum) is produced through the
of the right lower abdomen.

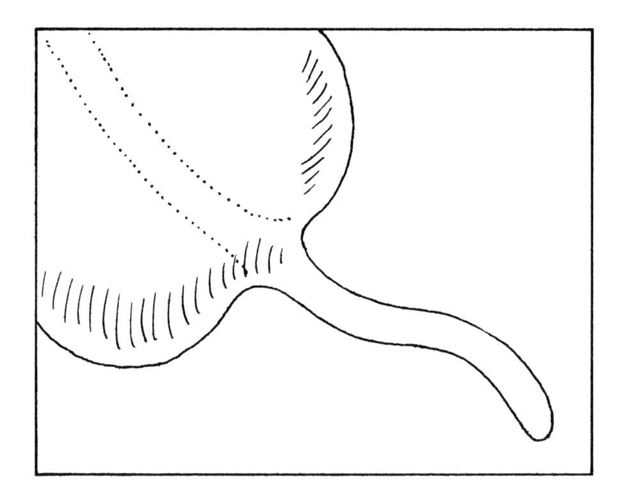

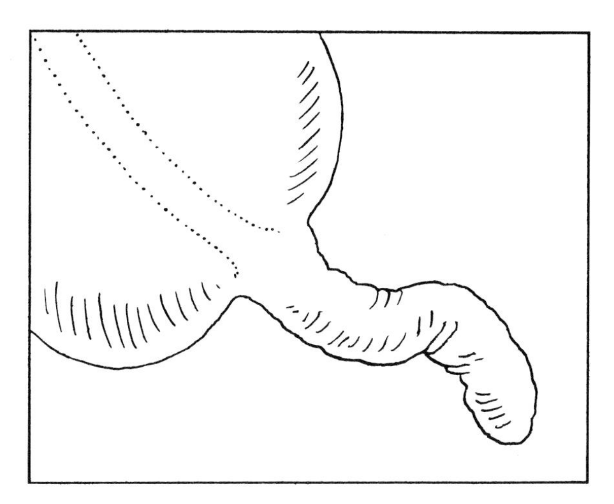

Appendicitis, or inflammation of the vermiform appendix, can lead to serious infection if the organ should burst. An inflamed appendix (right) is contrasted with a normal one (left) in the illustrations.

Diagnosis and Treatment of Colonic Diseases

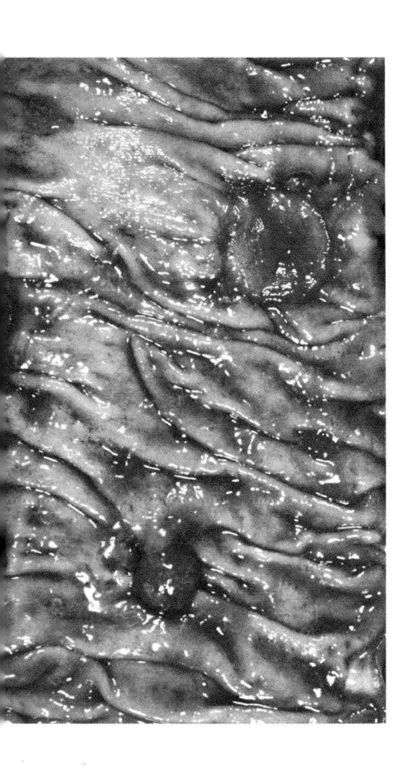

Cancer of the colon is a leading cause of death in the United States. The above photo shows a colonic cancer above and slightly to the right of a benign polyp.

What is a physician's approach to a story of a change in bowel habits? The basic procedures which may be performed include X rays after barium enema; endoscopy with the flexible fiber-optic colonoscope that can reach the full length of the colon; and stool examinations for occult blood ("hidden blood"), parasites and their eggs, and disease-producing bacteria. Since diseases of the digestive tract can involve other organs, especially those within the abdominal cavity, other tests such as kidney X rays are sometimes done. Some blood tests are also used in characterizing an anemia or suggesting the possibility of a colon cancer. Above all, however, a careful history is an important aspect of the physician's approach and of great help in formulating a list of the possible diagnoses. He must be told details such as the onset, duration, and severity of any change in bowel habits; of any recent need to use laxatives for the first time; and the diameter and color of the stools. A careful inquiry regarding the type, location, and radiation of abdominal pain is also of great use.

Constipation and Anal Discomforts

The most important principle in prevention or relief of chronic constipation is to maintain an adequate fluid intake--to drink at least a quart of water daily in addition to ample liquid consumption with meals. This need will vary according to climate and the body temperature and physical activity of the individual. The person's diet should include adequate bulk and fruits. Exercise is important, as are sleep, relaxation, and avoidance of harsh medication for treatment of constipation. Non-irritating bulk producers are available for those who cannot obtain adequate bulk from the diet.

One of the complications of chronic functional constipation is the development of lesions about the anus, such as anal fissures and hemorrhoids. A fissure is a small superficial crack in the lining of the anal canal and can be quite painful. The terms hemorrhoids or "piles" refer to swelling, protrusion, and possibly a localized thrombosis or blood clot in the hemorrhoidal veins which are normally present. Very often such local anal problems respond to a conservative medical approach involving low tub baths of warm water (sitz baths) several times daily, alleviating constipation by dietary means (e.g., extra cooked and canned fruits and vegetables; extra water), and avoiding further abrasion of the area (e.g., switching from toilet tissue to the use of cotton balls dipped in mineral oil; avoiding prolonged sitting and straining at stool). Occasionally, a small amount of petroleum jelly inserted with a finger cot or the use of a lubricating suppository may facilitate the passage of particularly hard stools.

As regards diarrhea, medications to slow down the movement of feces through the small and large intestine are sometimes useful, but they should be used only as temporary measures while a physician searches for infectious, inflammatory, or other causes of the symptom. Itching around the anus, or pruritus ani, may be a chronic, intermittent, low-grade problem associated with poor personal hygiene, pinworms, chronic irritation from locally applied medications, or even with a generalized skin or metabolic disorder. The determination and correction of the underlying cause is the first task for the physician when he is consulted. He always keeps in mind certain internal diseases which may first be manifested by this and other symptoms in the anal region.

Irritable Colon Syndrome, Inflamed Colon, and Other Colon Disorders

The person with "irritable colon syndrome" may have frequent abdominal noises, become bloated, and pass excessive amounts of flatus. The physician in examining is likely to observe the drumlike sound of gas as he percusses the abdomen and to note that the patient is tender as pressure is applied over the colon. He attempts to explain the symptoms in terms that can be easily understood and to outline briefly the anatomy and physiology of the colon in lay terms, as well as the distinction between functional and organic disease. Medication may be prescribed, but the individual afflicted with this syndrome should not attempt self-treatment, particularly if someone who "had the same symptoms" was relieved with a particular medication. The patient may be told, for example, that the bowel is made up of muscles, glands, and nerves that must function together as a team. Should the muscles or nerves overreact to improper foods and nervous tension, the result may be bowel spasm, delayed colonic emptying, or abdominal cramps. Although each component of the bowel wall is structurally normal, "teamwork is missing."

Among the organic diseases known to involve the colon, some of the more common are due to inflammation or tumors. The inflammatory processes may be caused by infectious microorganisms such as amoebae, other parasites, bacteria, or viruses. Such severe reactions as perforation of the bowel and large abscesses leading to obstruction and bleeding also may result from inflammation within pockets that sometimes develop as outpouchings from the colonic wall—a situation known as colonic diverticulosis complicated by diverticulitis. Another category of inflammatory bowel disease has no known cause. Regional enteritis, or Crohn's disease, in the small intestine has already been mentioned in this category; a closely related condition involving the large intestine is known as ulcerative colitis. This is an inflammation of the lining of the colon that leads to tiny mucosal ulcerations, bleeding, and secondary infections. As the inflammation subsides, either with appropriate treatment or by itself, scarring with contraction of the bowel may result. This process may involve the entire colon or be limited to the rectum, in which case it is called ulcerative proctitis. Also, the same changes as described in regional enteritis may sometimes be limited to the colon and referred to as Crohn's disease of the colon.

Cancer of the colon ranks second to lung cancer as the leading cause of death from cancer in the United States. These tumors most frequently develop in the rectum and sigmoid colon, but they can occur in any part of the bowel. The symptoms and signs depend upon factors such as the location, size, shape, and microscopic appearance of the tumor. For example, cancer of the cecum and ascending colon may produce loss of appetite, right-sided abdominal pain, or gradual blood loss with anemia and its associated symptoms of weakness and dizziness. Patients with cancer in the rectum and the sigmoid colon may complain of a change in bowel habits, rectal bleeding, and a feeling of incomplete stool evacuation; ultimately, weakness and weight loss may develop. These colon cancers can be completely cured if diagnosed early enough; thus it is critical for a person to inform a physician about these signs and symptoms promptly.

A polyp in the colon, or elsewhere for that matter, is nothing more than tissue which projects into the space within a hollow organ. Although polyps may be inflammatory or developmental, many turn out to be tumors—either benign or malignant. There is much controversy among physicians as to whether a benign polyp may become cancerous. Nevertheless, it is clear that a colonic mucosa that has once developed a benign polypoid neoplasm is predisposed to develop other tumors in the future. Such a "colon polyp patient" should be sure to see his physician as often as requested.

Digestive Problems Being Solved

In 1837 the Rev. Sydney Smith (1771–1845), a brilliant theologian, writer, and wit, wrote: "I am convinced digestion is the great secret of life." After reading this chapter one might not be willing to go as far as the Rev. Mr. Smith but is certain to realize that a healthy digestive system is vital to general health and well-being.

Although much is known about the digestive system, much also remains a mystery. The processes of metabolism, the specific mode of action of digestive enzymes, and the role of various hormones in aiding digestion are only now being fully understood.

Diseases of the digestive system are many and varied. Virtually every one has had some sort of gastrointestinal complaint at some time in his life. However, as is true of all the body systems, medicine is finding the answers to digestive problems.

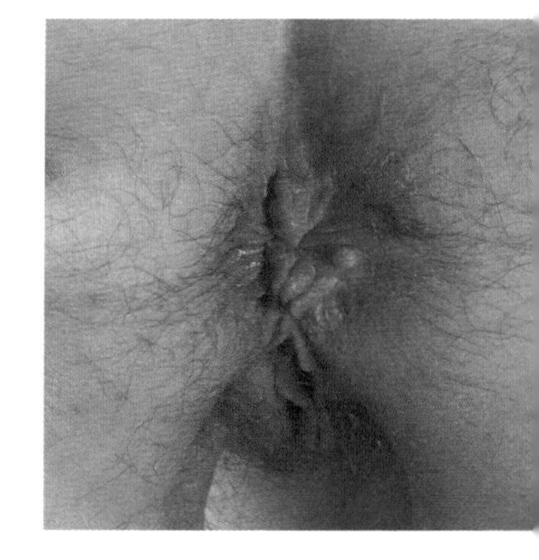

Hemorrhoids, or "piles," are enlarged blood vessels in the wall of the rectum or the anus.

For further information on the subjects covered in this chapter, consult ***Encyclopædia Britannica:***

Articles in the Macropædia	Topics in the Outline of Knowledge
DIGESTION, DISORDERS OF	424.F.7
DIGESTION, HUMAN	422.E.2
DIGESTIVE SYSTEM, HUMAN	422.E
DIGESTIVE SYSTEM DISEASES	424.F.1,2,3,4,5, and 6
LIVER, HUMAN	422.E.1.f.ii

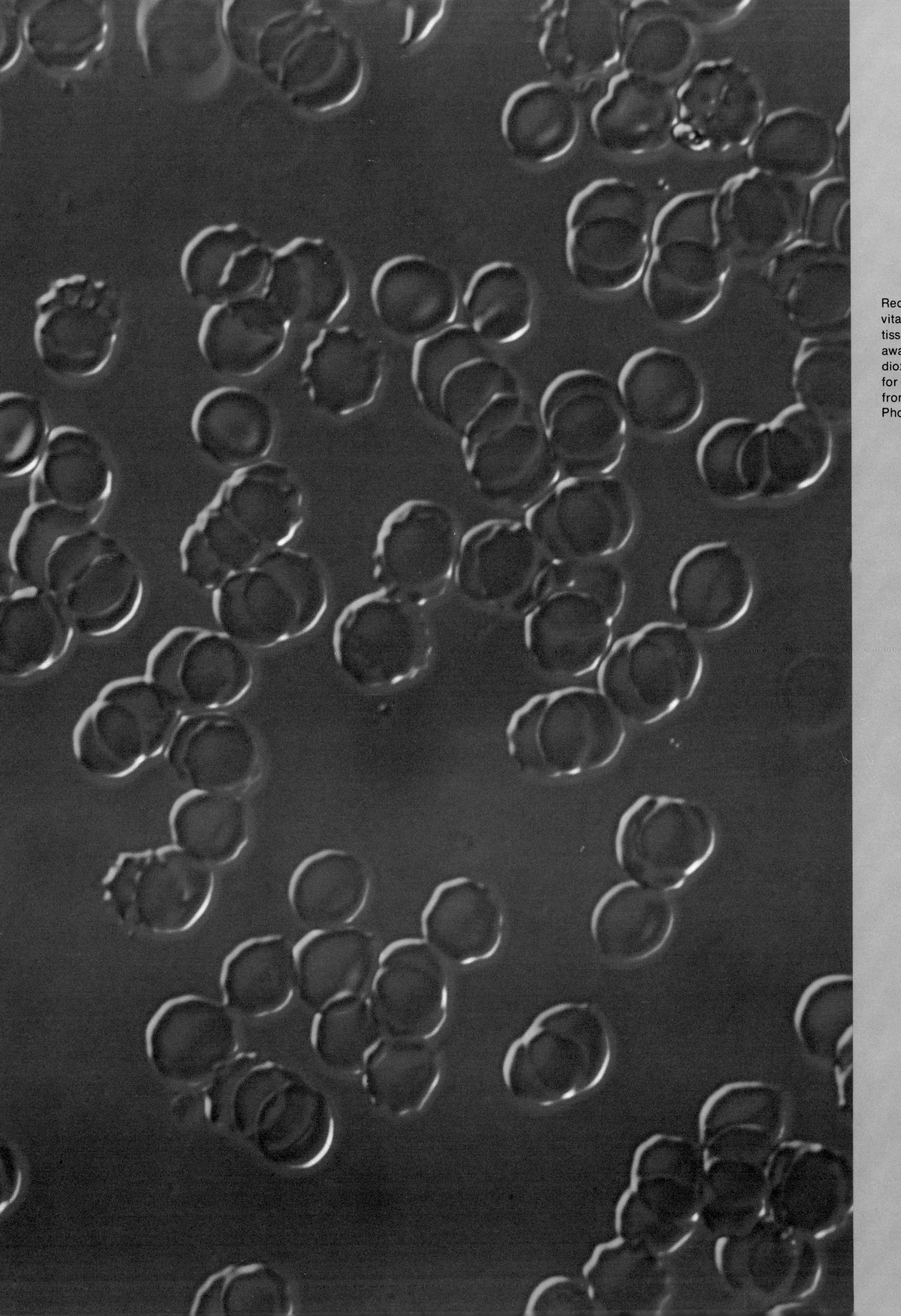

Red blood cells carry vital oxygen to the tissues. They also carry away needless carbon dioxide from the tissues for eventual removal from the body.
Photo, Camera MD

Van Doren. *I knew the human heart was a marvelous organ. But I'm surprised to learn how simple it is.*

Kessler. *In one sense, it's really the simplest major organ in the body. It's nothing but a pump. The remarkable thing is that it keeps on pumping so long and faithfully.*

Van Doren. *What is it—about 40,000,000 times a year for 70 years?*

Kessler. *Equally impressive to me is that it circulates 190 gallons of blood every day. That's some pump, man!*

7

Circulation

BY MORTON D. BOGDONOFF, M.D.

The circulatory system is responsible for moving supplies for energy to all organs of the body and also for taking away the waste products that result from organ activity. It constitutes the major transit network of the body. The medium in which the products are transported is the blood, discussed in detail in another chapter.

At the center of this transit network is the heart, which provides the force needed to move the blood through the system. Leading out from this pump are the arteries, which are large-bore, blood-carrying tubes that become smaller in diameter and in lumen size (the space within them) as they approach the organs. At the individual organs, or tissues, these blood vessels divide many times, each time becoming smaller, and branch out extensively through the substance of an organ, becoming first arterioles and then, finally, smaller vessels, the capillaries. The arrangement of blood vessels in an organ is very much like the elaborate branching of a tree. At the capillaries, the oxygen that is carried in the red blood cells within the arteries is released to the cells of the organ.

The capillaries gradually begin to join together after a distance and to lead into the smallest of the

Some Major Blood Vessels

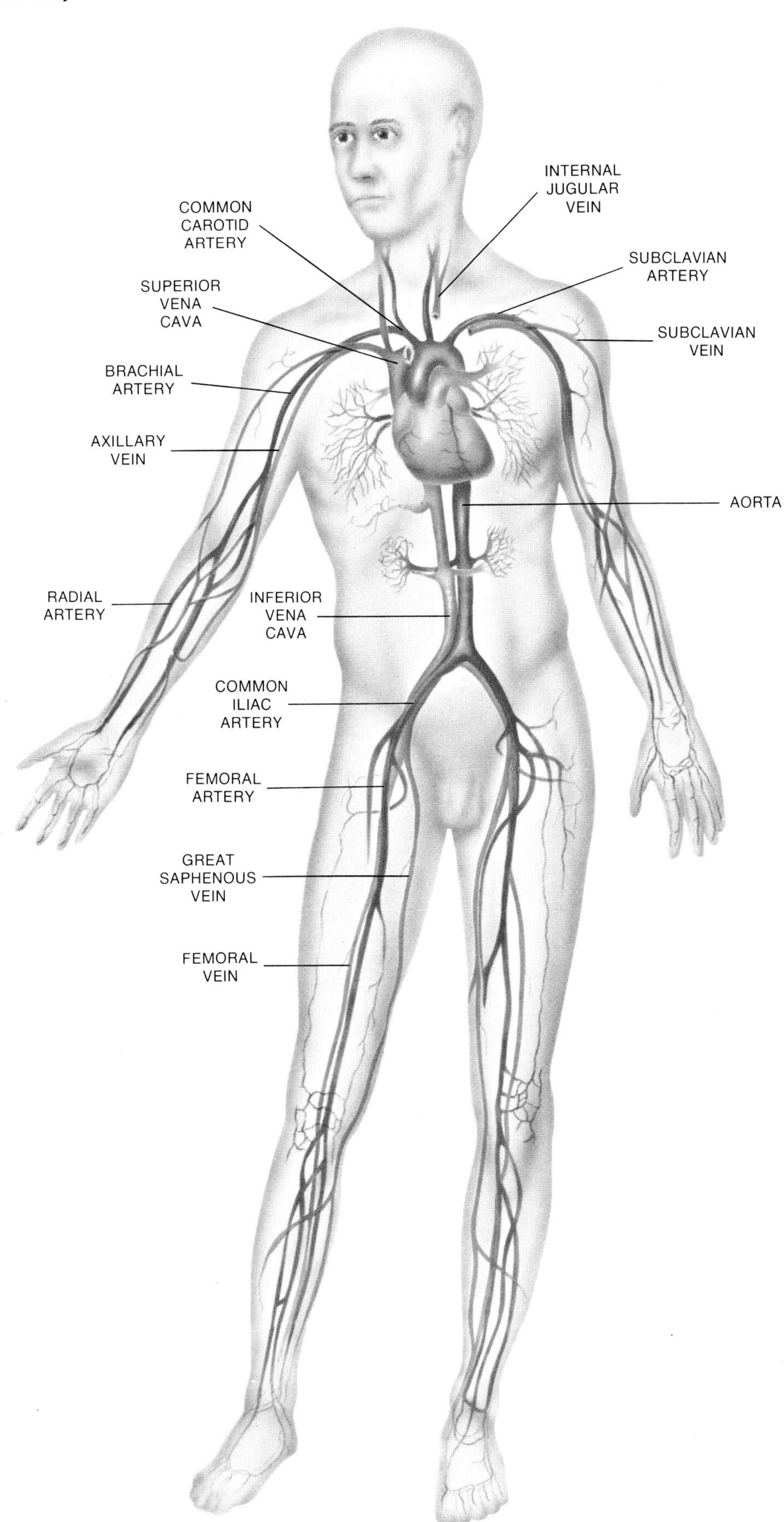

veins, or venules. These veins come together to form larger veins until finally a very large vein is formed. This vein leaves the organ and leads back toward the heart. At the capillary level the carbon dioxide and other products of organ activity cross into the blood and begin their course back to the heart. The arrangement of veins is also much like that of the arteries, a branching, treelike array, except that while the branching moves outward into the fine radicals and tips in the arteries, it moves in toward the trunk from the smaller tributaries into veins. Thus, the heart is a central point in the vascular distribution system, pumping out into the arteries and receiving blood back from the veins. The blood, in effect, makes a circle, or is circulated through the body, hence the term circulatory system.

The Heart

The heart is located in the mid-portion of the chest, just beneath and slightly to the left of the breastbone—within a part of the chest called the mediastinum (middle area). The heart conventionally is described as being shaped like a pear or apple and has a series of designations: the lower point is the apex; the upper part is the base. The large blood vessels that lead out of and into the heart are attached at the base.

Actually, the circulation has two divisions: the greater circulation, which delivers and returns blood from the organs of the body, and the lesser circulation, which delivers and returns blood to and from the lungs. The reason for this arrangement should be clear; it is through the lungs that the carbon dioxide is blown and dispersed, and that the oxygen is breathed in and absorbed into the blood. Thus, the major site of this gas exchange is isolated as a separate circuit from the remainder of the circulation.

Heart Chambers

The two circuits operate from one pump, and, therefore, the pump, the heart, is divided into two sets of chambers, right and left, separated by a structure called the septum. On each side at the site where the blood flows into the heart is a

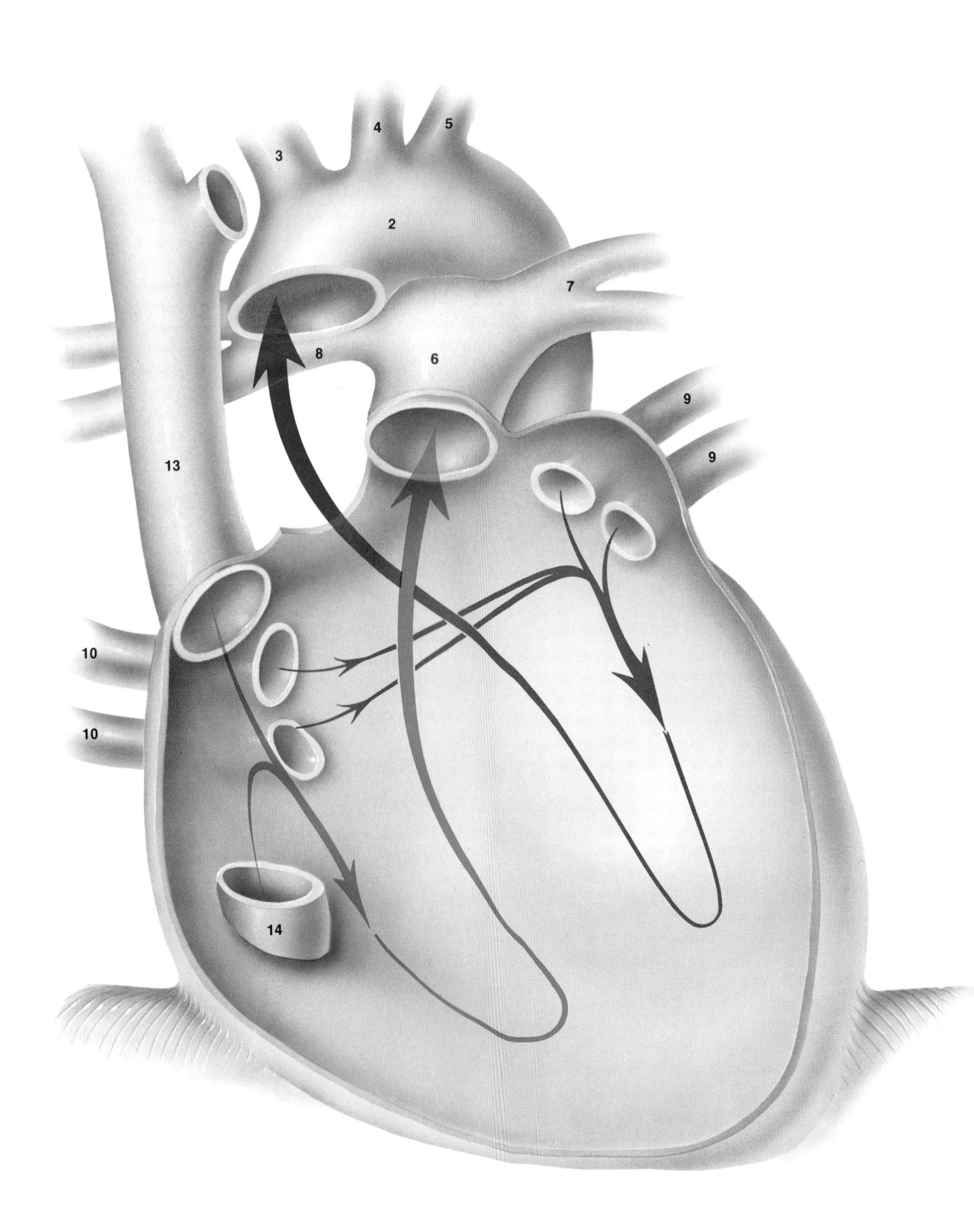
4
5
3
2
7
8
6
9
9
13
10
10
14

The Heart

1. Ascending aorta, 2
2. Aortic arch, 3
3. Brachiocephalic trunk, 3
4. Left common carotid artery, 3
5. Left subclavian artery, 3
6. Pulmonary trunk, 3
7. Left pulmonary artery, 3
8. Right pulmonary artery, 3
9. Left pulmonary veins, 3
10. Right pulmonary veins, 3
11. Left coronary artery, 1
12. Right coronary artery, 1, 2
13. Superior vena cava, 3
14. Inferior vena cava, 3
15. Left atrium, 1
16. Right atrium, 1
17. Left ventricle, 1
18. Right ventricle, 1
19. Bicuspid valve, 2
20. Tricuspid valve, 2
21. Papillary muscle, 2
22. Aortic valve, 2
23. Pulmonary valve, 2
24. Trabeculae carneae, 2
25. Interventricular septum, 2
26. Site of S-A node (where heartbeat impulse starts and spreads to A-V node), 2
27. Site of A-V node (which triggers pumping action of heart), 2

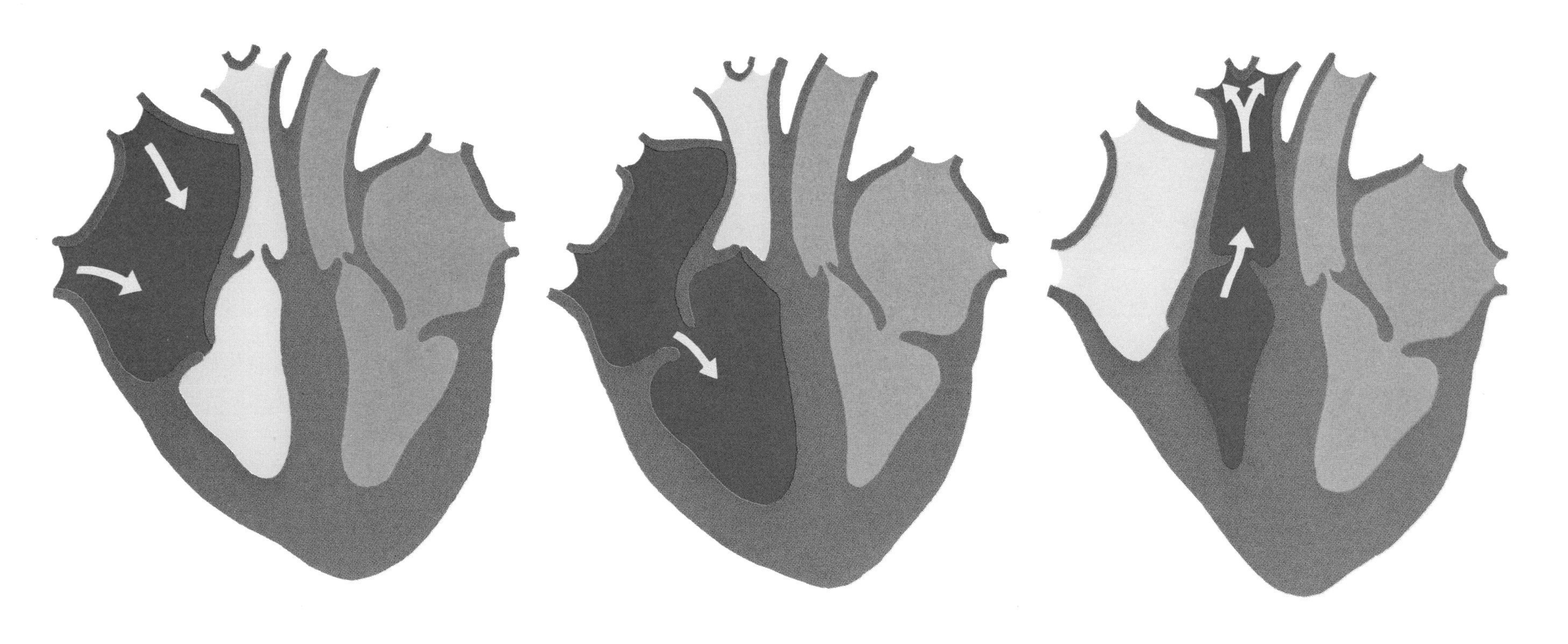

During diastole (relaxation) in the right side of the heart, deoxygenated venous blood enters the right atrium from the superior vena cava and the inferior vena cava.

The tricuspid valve between the right atrium and the right ventricle opens during diastole, and the oxygenated blood enters the right ventricle.

During systole (contraction of the ventricles), blood in the right ventricle is forced through the pulmonary valve into the pulmonary artery for oxygenation in the lungs.

chamber, the atrium. Blood from all the organs of the body (except the lungs) returns in the large veins (superior vena cava for the upper portion of the body; inferior vena cava for the lower part of the body) to the right atrium. Blood then flows from the right atrium into a second chamber on the right side of the system, the right ventricle. The ventricle is the chamber that does most of the work of pumping, although the atrium pumps to a slight degree. Blood in the right ventricle is pumped into the lungs by way of the pulmonary artery (*pulmone* = lung in Latin). After passing through the lungs for gas exchange the blood returns to the heart, this time to the first chamber on the left side, the left atrium, by way of the pulmonary veins. It then flows into the left ventricle, the second chamber on the left side, from which it is pumped out through the aorta and the other great vessels: through the carotids to the brain, the subclavians to the upper extremities, the mesenteric vessels to the intestinal organs, the renal vessels to the kidneys, and the iliac vessels to the lower extremities.

The heart is primarily a large array of muscle fibers surrounding the four chambers. The chamber that has the greatest amount of muscle fiber is the left ventricle, the one that pumps blood to the farthest reaches of the body. When these muscle fibers are stimulated by an electrical impulse, contraction takes place, and the blood within the chambers is subjected to a squeezing force that pushes the blood.

Heart Valves

To maintain the proper direction of flow, that is, from atrium to ventricle and from ventricle out to the large arteries and not back into the ventricles, a series of valves exists in the heart and at the origins of the great vessels leading out from the heart. The valve between the right atrium and ventricle has three parts, thus it is called the tricuspid valve. The valve between the left atrium and ventricle has two parts and looks like a bishop's mitered hat, thus the term mitral valve. When the ventricles contract, the valves between the ventricles and the atria close and blood flows out toward the lungs and the remainder of the body. When the heart muscles relax, blood flows into the atria from the veins. Then the valves between the atria and the ventricles

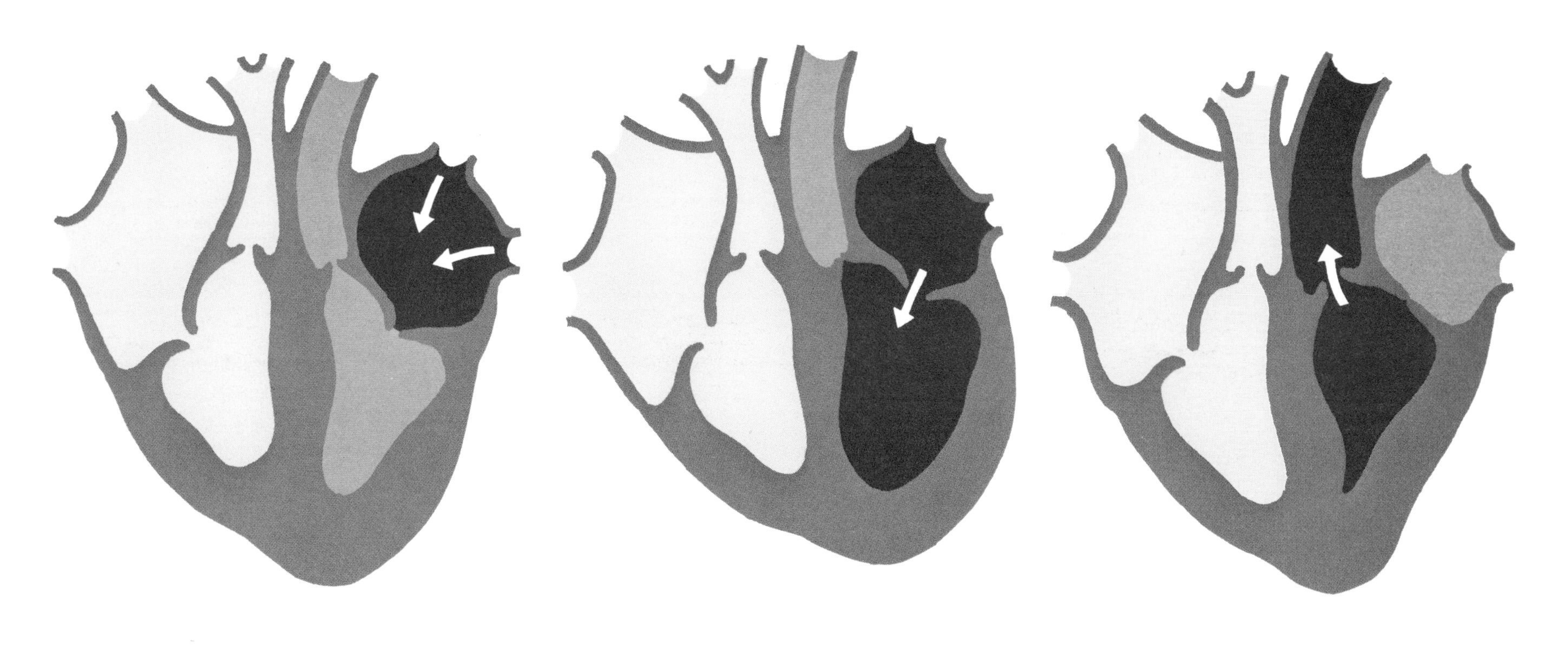

During diastole in the left side of the heart, oxygenated blood from the lungs enters the left atrium from the pulmonary vein.

The bicuspid valve between the left atrium and the left ventricle opens during diastole, allowing oxygenated blood to enter the left ventricle.

Oxygen-rich blood is pumped from the left ventricle during systole into the aorta for distribution throughout the body.

float open and the ventricles fill. The period during which the heart muscles contract is systole; the period during which they relax is diastole. Additional valves at the root of the aorta and the pulmonary artery close shut during diastole, preventing the blood that has gone into the major arteries from flowing back into the relaxed ventricles. The elastic arteries distended by blood from the heart gradually and constantly push the blood toward the capillaries until the next heartbeat. This wave of distention is called the pulse.

A small nest of special cells in the muscle of the right atrium has the remarkable property of periodically firing off electrical stimuli that spread throughout the muscles of the heart by means of another set of specialized cells, the conduction system, causing the muscles to contract. The heart, therefore, has its own pacemaker, which synchronizes the contractions. This stimulating network is remarkably responsive to the energy needs of the body. It is subject to stimuli from other parts of the body that help to govern the rate at which the pacemaker fires off (faster during exercise; slower during sleep).

Coronary Blood Vessels

The heart itself, being a very active array of muscle fibers, requires blood like any other organ. It has a full and elaborate set of blood vessels to carry oxygen to the muscle and to take the carbon dioxide away. The arteries supplying the oxygen to the heart are the coronary arteries. These arteries start from the aorta just at the beginning of that blood vessel and spread around and across the muscle, encircling it like fingers. A right coronary artery goes to the right side of the heart; a left coronary artery goes to the left side of the heart. The left coronary artery divides sharply after its origin into two branches, an anterior descending branch that sits on the front of the heart and courses downward, and a circumflex branch that makes a turn around toward the back and down the side of the heart. These vessels continue branching similar to the tree arrangement in other organs. At the capillary level the veins begin to form, and a comparable venous system leads to the main coronary veins, which deliver the blood that has had its oxygen removed by the heart directly to the right atrium.

Blood Pressure and Blood Supplies

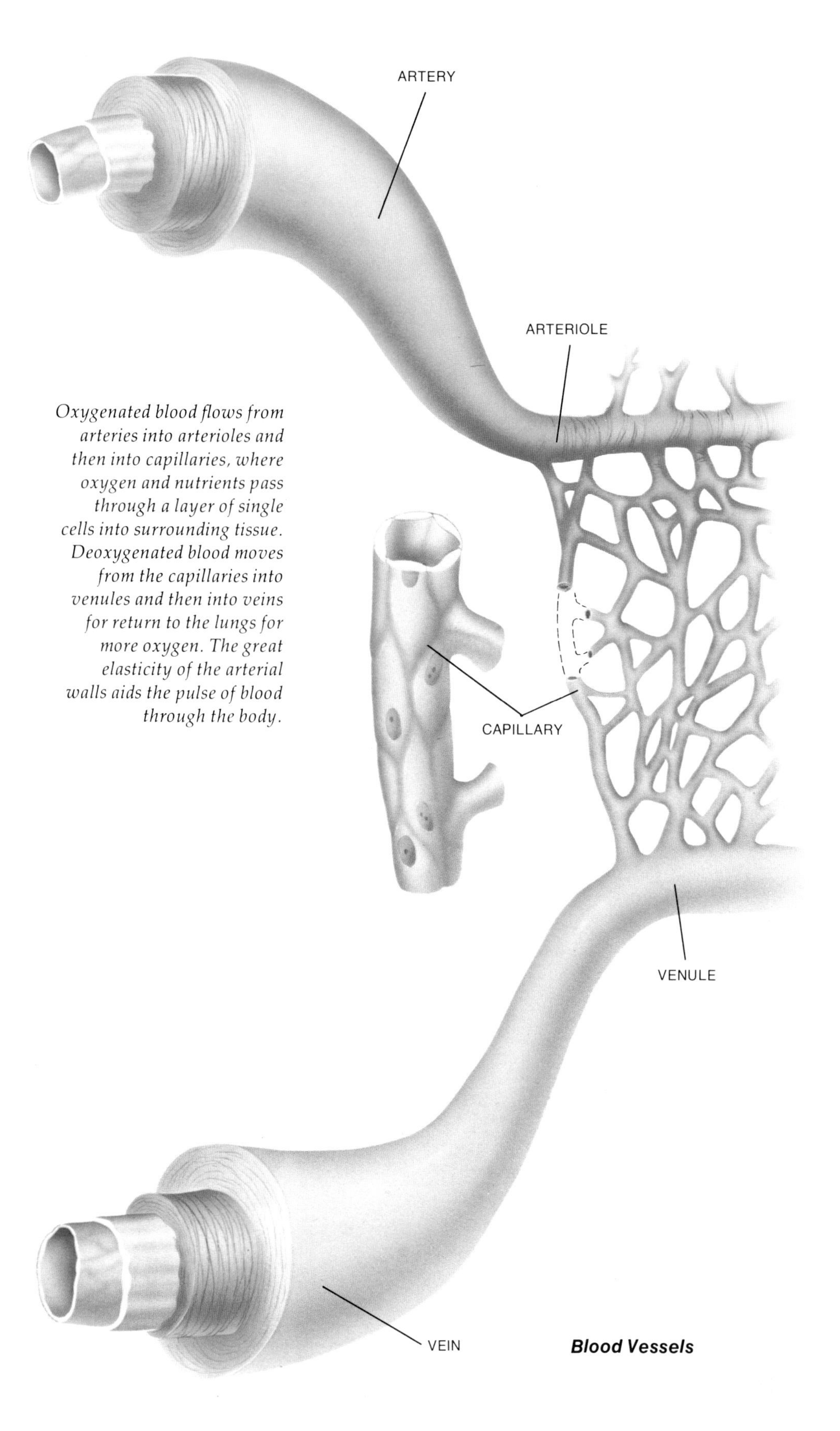

Oxygenated blood flows from arteries into arterioles and then into capillaries, where oxygen and nutrients pass through a layer of single cells into surrounding tissue. Deoxygenated blood moves from the capillaries into venules and then into veins for return to the lungs for more oxygen. The great elasticity of the arterial walls aids the pulse of blood through the body.

Blood Vessels

The circulatory system of arteries and veins is not merely a passive series of pipes. These vessels also actively contribute to the effective circulation of the blood by maintaining a certain diameter and consistent wall structure. This is accomplished by muscle fibers that make up a large part of the walls of the arteries and veins. These fibers are arranged in a circular fashion around the blood vessels and are maintained by the nervous system in a degree of contraction at all times. In general, the walls of the arteries consist of three layers of muscle or elastic fibers that run at right angles to each other. The inner layers run longitudinally along the length of the artery. The middle layers extend around the artery. The outer layers run longitudinally. The veins are not so complex and have no significant role in the pumping. One can, therefore, visualize the heart, the arteries, and the veins as being a large system of pumps and pipes, the walls of which are predominantly muscle, having an extensive network of capillaries (which do not have muscle fibers in their walls) arranged between the arteries and the veins.

Clearly, these muscles are very important: they augment the pump action of the heart and maintain the integrity and strength of the walls of the arteries and the veins. The energy or force maintained within this pump-and-pipe system creates a level of pressure in the system, the blood pressure. When the heart contracts, the pressure goes to its highest point, the systolic pressure. When the heart is relaxed, the pressure falls to its lower level, the diastolic pressure. The pressure within the heart and arteries of the greater circulation is higher than the pressure in the lesser circulation. As the blood courses through the capillary bed, the pressure falls markedly and the actual level of pressure in the veins is much lower than in the arteries.

Measuring Blood Pressure

Since the measurement of the blood pressure, or more properly, the pressure inside the blood vessels, is so common, an explanation of how blood pressure is usually measured is appropriate. A cuff of inflatable material is placed around an extremity, usually the arm, and this cuff is inflated so that the outside pressure is exerted on the limb. As this external pressure rises it compresses the artery in that limb. When the compression is great enough to reduce the inside diameter of the artery, the blood flowing through the artery becomes turbulent and a very faint sound is created at the site of the compression. This sound can be heard downstream, thus the use of the stethoscope below the cuff. As more and more external pressure is ap-

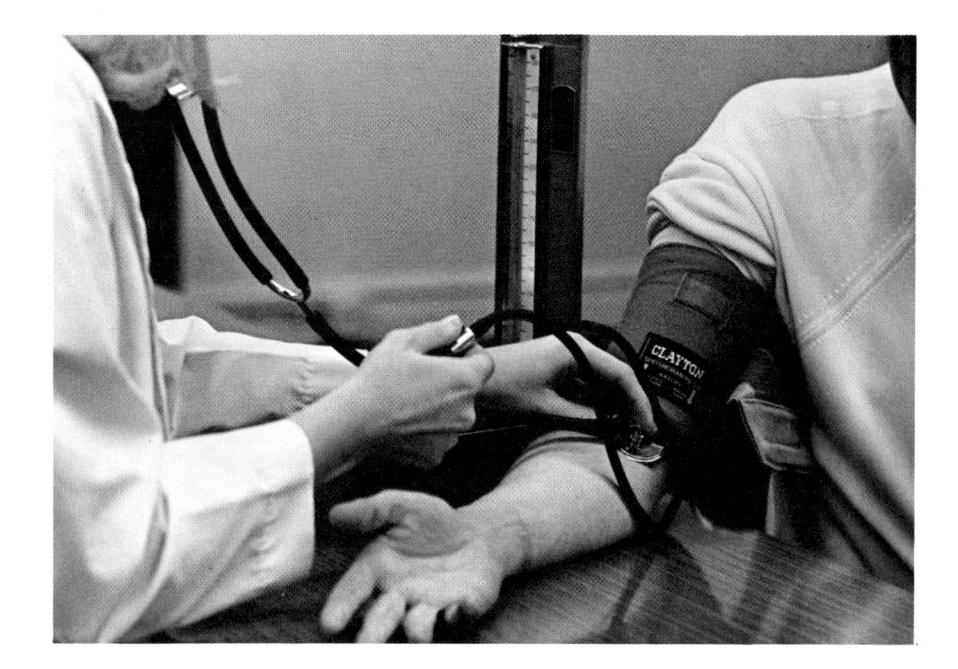

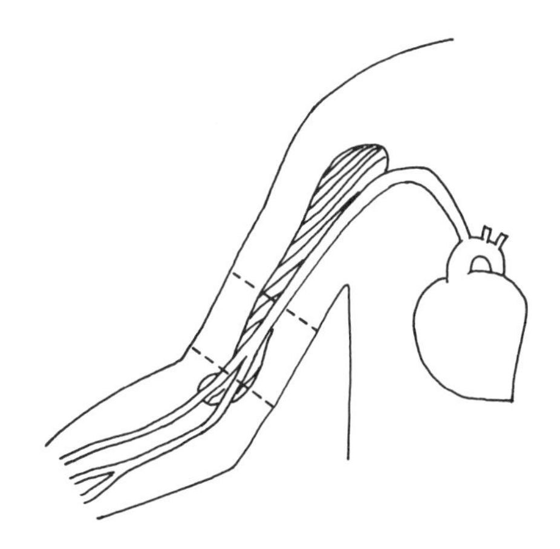

Blood pressure is measured by means of a pressure cuff with an attached gauge (upper left). When inflated, the cuff forces compression of an arm artery (upper right). While deflating the cuff, a doctor listens for the sound of the systolic pulse as it overcomes the cuff pressure. The gauge reading is then noted. The cuff is reinflated for similar detection of diastolic pressure (lower than systolic pressure).

plied, the artery is further compressed until the external pressure exceeds the pressure inside the artery. The artery then is fully collapsed; no blood flow occurs and no sound is heard. The cuff is then slowly deflated, and the first beat is heard. When that occurs, it is at the point of greatest pressure within the arterial system, or when the heart is contracting (systole), and is called the systolic pressure. With minimal compression by the external pressure, the lowest pressure inside the arteries is just exceeded. This level is attained when the heart is relaxed (diastole) and is called the diastolic pressure. This lower limit of pressure is greatly affected by the strength of contraction of the muscles in the walls of the arteries, the "tone" of the vessels, and by the flexibility and "give" of these walls.

Hypertension

If the muscles are "tight," or if the walls become relatively stiff and rigid, then the pressure inside these vessels will rise. This change from the normal represents an elevated blood pressure or hypertension. The usual levels of pressure in the arteries are recorded as 120 mm (millimeters) of mercury (Hg) at systole and 80 mm Hg during diastole. These levels mean that enough force develops in the blood vessels to raise a one-quarter inch column of mercury 120 mm (approximately five inches) up a tube at systole and 80 mm at diastole.

Blood Supplies

Normal function of the circulatory system exists when the heart and vessels pump and carry enough blood to provide adequate oxygen for the functions of all the organs of the body. When the individual is at rest, the demand upon the cardiovascular system is minimal; during vigorous exercise (running uphill), the demand may well be maximal. Intermediate conditions occur, for example, during moderate physical exercise (slowly jogging), in the state of the body following a large meal, or under the effects that follow exposure to high environmental temperatures. All of these situations call for an increase in the amount of oxygen that must be delivered to the tissues: to the muscles for exercise, to the intestinal organs for digestion, to the skin and sweat glands upon exposure to heat.

Adjusting the Blood Supply

In several ways the cardiovascular system increases the total blood supply to the tissues and thereby provides the necessary increased oxygen supply. The heart may squeeze harder at each beat and increase stroke volume. Also, the heart may beat many more times per minute. Or, the blood vessels may open up to the organs that need the increased blood supply and close down to organs that require less. In so doing, the muscles around the arteries going to the organ that needs the increased blood supply relax, with the result that the vessel diameter becomes larger and more blood flows through the vessels. Such changes occur during physical exercise, resulting in an increased amount of blood flow to the muscles. Conversely, the arteries may narrow down and decrease the blood supply to organs when the need passes. The nervous system controls the shifting of blood to areas as needs change. Blood volume may not be adequate to meet demands. When much of the blood supply to the intestines and liver is increased following a heavy meal, diversion of the blood supply from the limbs to the intestines increases the chances of developing muscular "cramps" while swimming after a heavy meal.

What makes the heart beat faster, or stronger, or what makes the blood vessels open or close at times of need has been of great interest. It is believed that the waste products of organ activity—carbon dioxide and the various products of tissue metabolism—influence the nervous system and the internal secretion organs that directly affect the pacemaker of the heart, the strength of contraction of the heart muscle, and the state of contraction of the muscles of the arteries. Thus, the brain sends out signals by way of the nerves to the heart, which then can increase or decrease the rate of electrical activity of the heart pacemaker. Epinephrine is one of the internal secretions that may be affected and that directly affects the rate of the heart. Certain other specific products of organ function are also believed to make the heart contract faster and stronger. It is the appropriate balance of all these elements that constitutes the normal function of the circulatory system. When any of these component parts is unable to function properly, the likelihood of some disease being present is very great.

Malfunctions of the Circulatory System

Many processes that can take place in the body alter the structure and function of the heart and the blood vessels. Some have become well understood because of extensive research; others are only well recognized, with the reasons for their occurrence still needing study.

Atherosclerosis

One of the most common alterations that creates a malfunctioning of the cardiovascular system is atherosclerosis. This is a disease in which the inner lining of the arteries, or the intima, becomes filled with fatty material. These collections of fat swell up within the lining and extend out into the lumen of the vessel. The actual amount of space through which the blood can flow is reduced. In fact, the artery may become blocked, or occluded. When the blood that is necessary for an organ cannot reach it because of narrowed or blocked arteries, the organ may not function well, and certain symptoms will develop.

Angina Pectoris and "Coronaries." Any of the arteries of the greater circulation may be affected by atherosclerosis, although certain ones are more likely to be than others. Among these are the coronary arteries to the heart. As a result of a severe narrowing, or occlusion, of the coronary arteries by atherosclerosis, spasm, or other causes, little blood gets to the heart muscle. With exercise such as climbing stairs or walking fast, a sensation of pressure and pain develops in the front of the breast and may include the neck and left arm. If it occurs with exercise and is relieved by resting, it is called angina pectoris. When the blood supply is so severely reduced that the muscle cannot survive, a part of the muscle may actually

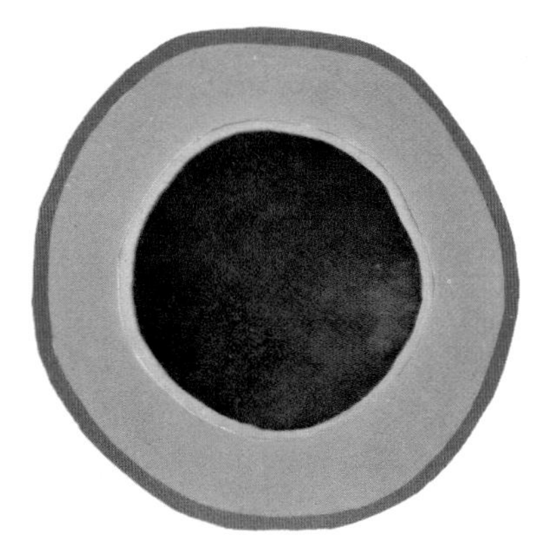

NORMAL ARTERY

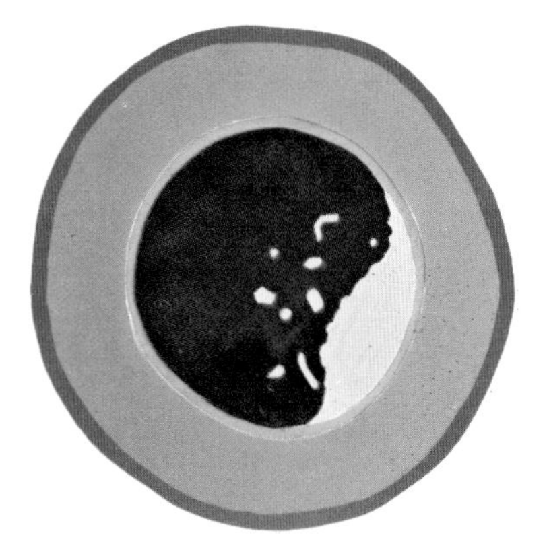

ATHEROMA FORMING IN ARTERY

ARTERY NEARLY BLOCKED BY ATHEROMA

die, resulting in a myocardial infarction from coronary thrombosis. The myocardium is the muscle of the heart, and infarction means death of tissue caused by too little blood supply. If the impairment is in the coronary artery, the condition is called a "coronary."

Arrhythmia and Cardiac Pacemakers. Most people with atherosclerosis of the coronary arteries continue to function quite well, though perhaps at a reduced level of physical and emotional activity. If a portion of the heart muscle dies, usually some healing follows, and the remainder of the muscle continues to work adequately. Occasionally, the blood vessel that supplies the pacemaker and the electrical conduction system of the heart may become involved. When this happens, the normal, steady, constant beating of the heart (the normal rhythm) may become disorganized or irregular. This irregularity is called an arrhythmia, which can often be treated successfully with certain drugs. If the area of the heart that controls the impulse to beat is damaged (thus causing a heart block), an external cardiac pacemaker can be applied. Cardiac pacemakers are small battery- or nuclear-powered devices that send out steady, regular electrical stimuli. Wires from a pacemaker are placed on or within the heart, and the stimuli from the pacemaker determine the rate of beating of the heart. If the heart's own pacemaker can perform normally at times, then the external pacemaker is superseded; if the heart's pacemaker fails temporarily, the external pacemaker will take over. This type of arrangement is called a demand pacemaker because it becomes active only when the heart needs or demands it.

The external pacemaker is usually inserted beneath the skin of the chest, and the wires are run through the veins to the heart. Most batteries last about two years, but researchers are constantly developing sources of longer lasting power to make possible less frequent changing of the pacemaker. A recently developed pacemaker permits recharging the battery through the skin, thus making surgical replacement unnecessary. A nuclear-powered pacemaker is also being used.

Cardiac Arrest. Impairment or stoppage of blood flow through the coronary arteries can also cause a sudden stoppage of all cardiac activity. When this happens, the heart abruptly stops pumping, resulting in cardiac arrest. Usually the individual also simultaneously stops breathing. During recent years, techniques to restart the heart (cardiac resuscitation) have been developed. It must be instituted immediately to be successful because the brain cannot be without oxygen for longer than

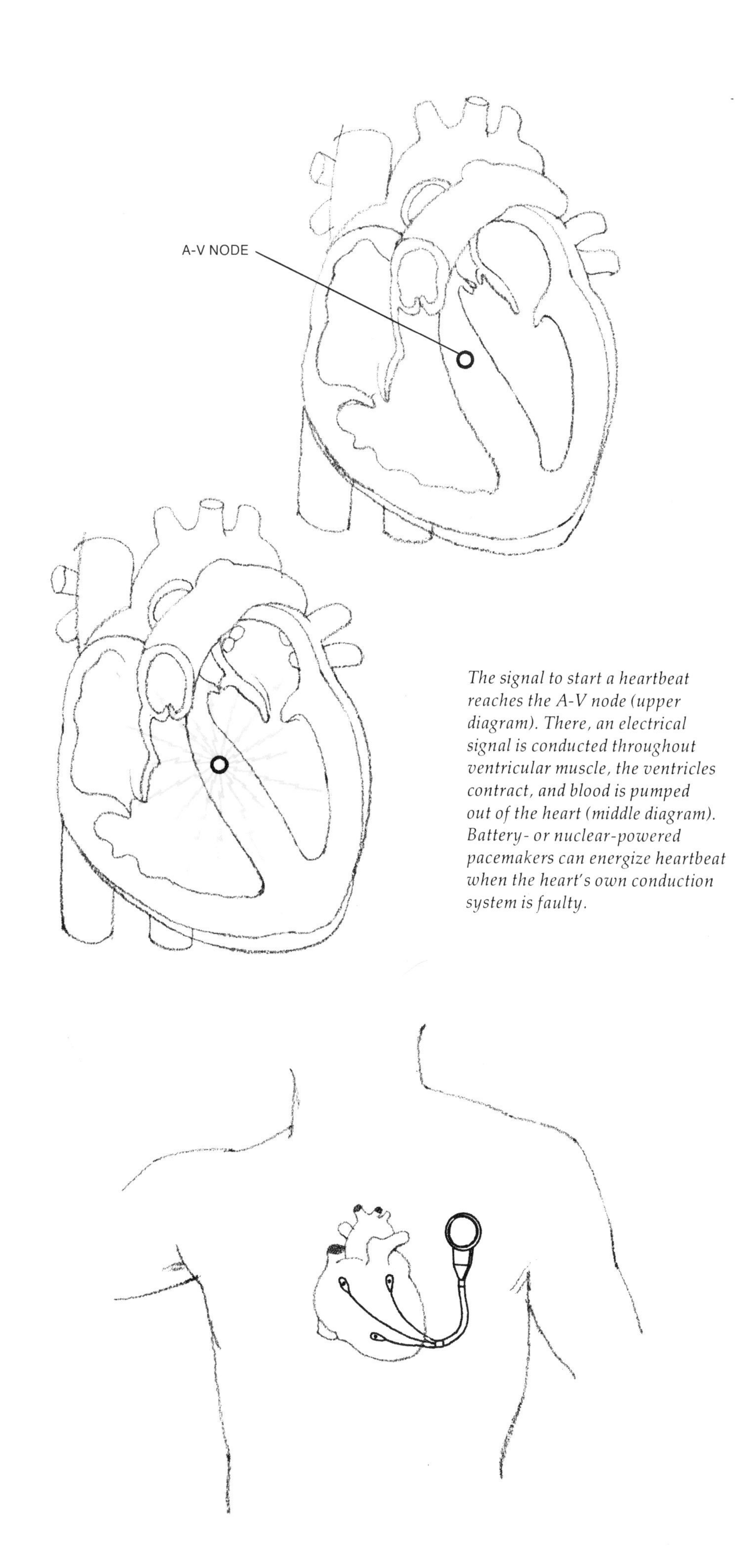

The signal to start a heartbeat reaches the A-V node (upper diagram). There, an electrical signal is conducted throughout ventricular muscle, the ventricles contract, and blood is pumped out of the heart (middle diagram). Battery- or nuclear-powered pacemakers can energize heartbeat when the heart's own conduction system is faulty.

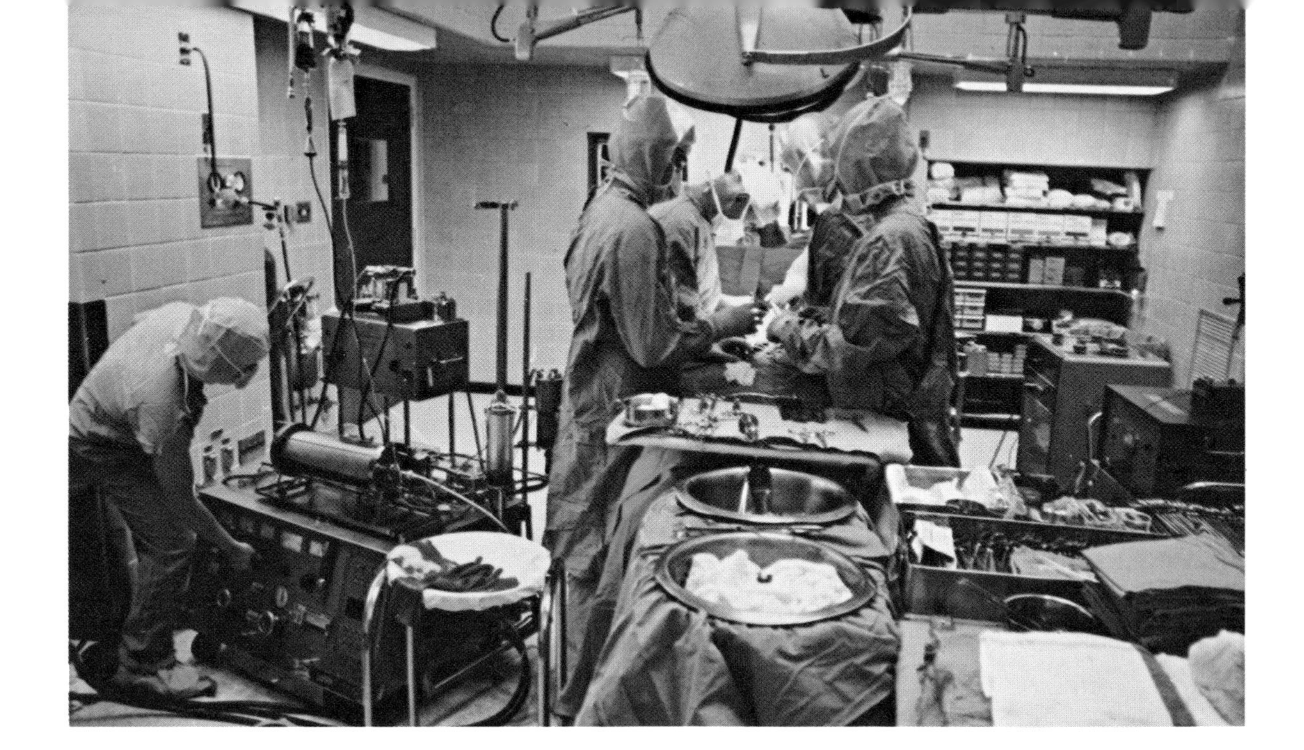

During open-heart surgery, a special pump (at the left in the photo) keeps the patient's blood flowing and supplies oxygen to his bloodstream.

three minutes. With proper training, many persons other than physicians, such as nurses, aides, firemen, and policemen, can recognize cardiac arrest and provide the help needed to restart the heart and to reestablish breathing.

Treating Atherosclerosis. Although the cause of atherosclerosis is not completely known, important studies suggest that many factors contribute to its development. These include cigarette smoking, excessive coffee drinking, and the eating of a large amount of animal fat. Therefore, physicians recommend that atherosclerotic patients avoid these contributing factors to try to avoid further disease or to reduce it. In addition, it is surgically possible to remove some of the large arteries that become clogged and replace them with grafts—not unlike inserting a new length of pipe in a clogged water system. These grafts consist of other unclogged blood vessels or of synthetic tubes. If the clogged vessel cannot be completely removed, the blocked area may be bypassed. A section of vein from the patient's thigh is used to make a detour channel around the block in the coronary artery. For patients with coronary artery disease, coronary bypass surgery has become a commonly used treatment. In some cases, instead of removing or bypassing a blocked vessel, some of the fatty deposits inside the lining of the blood vessels may be scraped out to reduce the narrowing.

Stroke. One of the most important organs to which blood supply may be affected is the brain. The blood flow may be impaired slightly or severely. If the impairment is in an artery in the neck, a bypass may restore the circulation and avoid the clogging of an artery in the brain, which results in a cerebral thrombosis, one cause of stroke. Another type of stroke is a rupturing in the brain of an artery wall either from a weakness in the vessel, high blood pressure, or both.

Preventing Atherosclerosis. Because atherosclerosis is relatively difficult to reverse or remove once it has developed, many physicians strongly urge an intensive program of prevention. It usually includes the elimination of cigarette smoking, the maintenance of a consistent level of moderate physical exercise, and the eating of a diet low in animal fat.

Heart Diseases

Although atherosclerosis affects the lining of the arteries of the body, the disease does not usually involve the valves of the heart. One of the diseases that does affect the heart valves is rheumatic fever.

Rheumatic Fever most often occurs in children and young adults and is related to repeated sore throats caused by streptococcus. Presumably in response to a streptococcal infection, the body makes a substance that not only damages the bacteria but, strangely enough, also causes an inflammation in the joints and sometimes in the valves of the heart. The valves swell and eventually become thick and scarred. The acute inflammation, called rheumatic fever, subsides; but the scarring process in the heart valves continues, and soon the valves do not open or close properly. The malfunctioning of these valves will place a greater load upon the pumping action of the heart, and if the heart cannot then meet the demands of the rest of the body, it is said to be "in failure." When that happens, excess fluid begins to collect in the body, in a condition called edema, or what was once popularly termed dropsy. When the excess fluid collects in the lungs, it is difficult for the patient to breathe properly, and he complains of shortness of breath. When this congestion in the lungs occurs the condition is known as congestive heart failure.

Congestive Heart Failure treatment includes the use of drugs that get rid of excess fluid from the body by increasing urine output. This process is called diuresis. The force of the pumping action of the failing heart can also be increased by administration of a drug called digitalis, a remarkable product of the foxglove plant. Finally, if the scarred and thickened valves are the real reason for the heart being overloaded, it may be possible by open-heart surgery to relieve the scarring or replace the diseased valves surgically with recently developed plastic and steel devices. While the surgeon is operating on the heart, the patient's circulatory system is connected to an artificial heart device, a cardiac bypass pump oxygenator. Such surgery has been successful in many patients. Before the surgery is performed, it is usual to measure directly the amount of blood flowing across the various diseased valves. This is accomplished by inserting a long tube, or catheter, into the heart by way of the veins and arteries. This procedure is called cardiac catheterization.

Since rheumatic fever presumably derives from infections with the streptococcus, physicians treat such infections from the very beginning with antibiotics to prevent the actual occurrence of rheumatic heart disease. Accordingly, sufficient antibiotic treatment for the streptococcal sore throats when they occur is considered to be one of the most important ways of eliminating the actual development of rheumatic heart disease. Many patients, once they have been identified as having some rheumatic heart disease, are placed on antibiotics for the remainder of their lives in an effort to prevent a recurrence of the fever.

Congenital Heart Disease. One important type of cardiovascular disease is the alteration of structure and function that occurs as a result of some faulty development of the heart and great vessels during fetal life. A child with congenital heart disease may have a defect in the wall separating the chambers of the right and left sides of the heart—a narrowing or pinching of the aorta or pulmonary artery as it arises from the heart, or a misconnection between the right and left sides of the circulation so that some of the blood in the veins goes out through the aorta before it goes to the lungs to receive more oxygen. This defect leads to dark venous blood mixing with bright arterial blood, causing cyanosis, or a "blue baby." Some of these congenital heart disorders can be surgically repaired at a very early age.

Diseases of the Veins

While the heart and arteries are the sites for many disease processes, the veins are less frequently involved. The most common disease of the veins involves those of the lower limbs. The walls of these veins may become weak and balloon out, resulting in varicose veins in which the blood pools and slows its rate of return toward the heart. The walls of the veins become chemically inflamed by the blood stagnation. Also, the vein may become injured by crossing and other use of the legs. This combination of pooling and inflammation leads to clotting and stoppage of the vein, or venous thrombosis. Surgical removal of these enlarged veins, some of which may be thrombosed, is often required. When these veins are removed, the blood returns satisfactorily through the deeper veins.

For further information on the subjects covered in this chapter, consult ***Encyclopædia Britannica:***

Articles in the Macropædia	*Topics in the Outline of Knowledge*
BLOOD CIRCULATION, HUMAN	422.A.4
CARDIOVASCULAR SYSTEM, HUMAN	422.A.1 and 2
CARDIOVASCULAR SYSTEM DISEASES AND DISORDERS	424.C.1,2, and 3

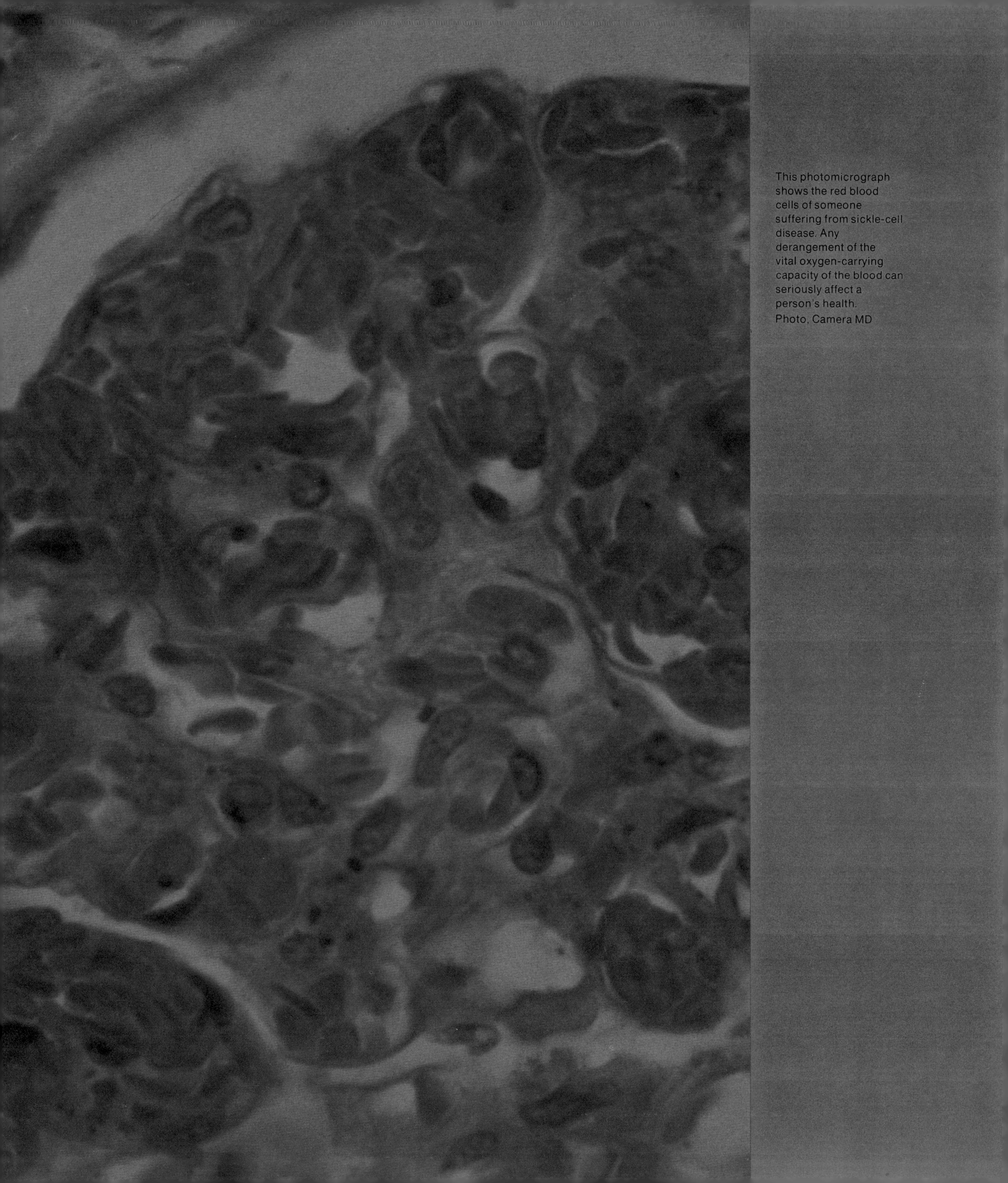

This photomicrograph shows the red blood cells of someone suffering from sickle-cell disease. Any derangement of the vital oxygen-carrying capacity of the blood can seriously affect a person's health.
Photo, Camera MD

Van Doren. *This is a wonderful chapter—so full of information, and excitement, and mystery.*

Kessler. *I agree. We might therefore just let it speak for itself. Except that I'd like to point out that the blood and lymph, like the skin, is a single organ. That's rather remarkable, I think.*

Van Doren. *Because it's a group of cells having a single overall function?*

Kessler. *Exactly.*

8

Blood and Lymph

BY HAU C. KWAAN, M.D.

During evolution, as an organism develops from a more primitive to a higher biologic order, it moves from the sea to the land. Once it becomes a land animal it no longer derives nutrition from its former aqueous environment. Nor can its body eliminate wastes in simple relation to pure water but must begin to relate to a new internal fluid that it acquires, namely the blood, which brings the organism life-supporting oxygen and food and removes waste products from its body tissues. In man the blood is a highly complex substance composed of a fluid part and a solid part. The fluid part is a mixture of protein, fat, and carbohydrates suspended in an aqueous solution called plasma. The solid part is composed of three main groups of blood cells, red cells (erythrocytes), white cells (leukocytes), and platelets (thrombocytes).

Approximately 8 percent of the human body weight is blood. In an adult this amount is the equivalent of 6,000 cc (cubic centimeters), or roughly 12 pints. The entire amount circulates through the body every minute, carrying not only food and oxygen but also, in cooperation with the lymphatic system, a number of infection-fighting devices and highly efficient means for eliminating

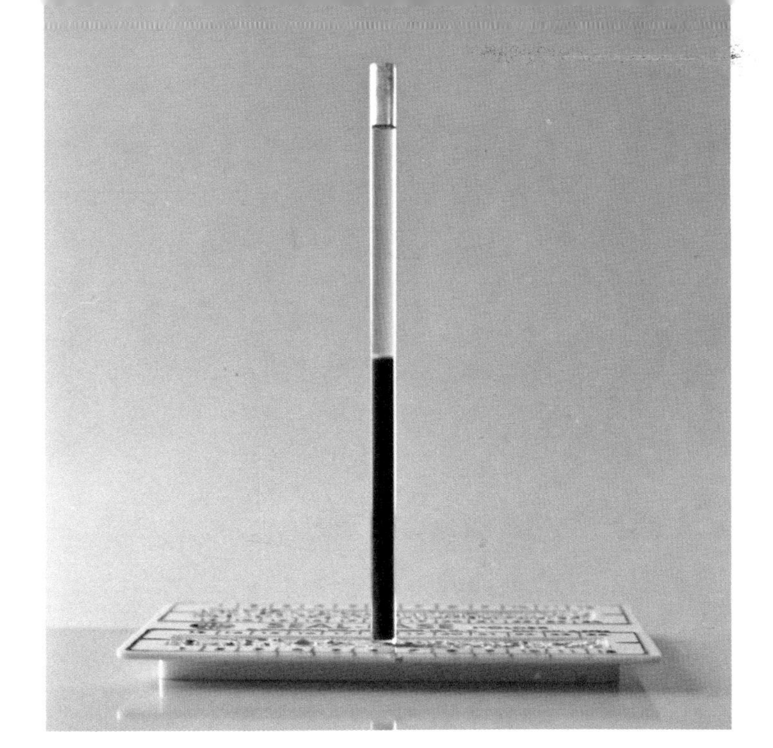

After being spun very quickly, the components of blood separate into several layers. Red blood cells are at the bottom, plasma at the top.

unwanted materials. By looking separately at each of the components of the blood supply, beginning with the solid ones, the role that each plays in the functioning of a healthy body can be assessed.

Red Blood Cells

Erythrocytes are small, flat, disclike cells measuring 7.6 microns (three ten-thousandths of an inch) in diameter. Approximately 30 trillion of these red blood cells are circulating in the blood of an adult. Normally, some 21 billion are formed each day in the marrow, the soft tissue that fills the hollow centers of bones; six to eight times that amount, however, can be produced if necessary. After being released into the circulation, erythrocytes live for about 120 days. When they are aged, they break down easily and are removed, primarily by the spleen, an organ usually regarded as part of the lymphatic system.

The red color of the erythrocytes is caused by a protein pigment called hemoglobin, which is contained within the cell membrane, or envelope. The hemoglobin is important to the role of the cell in transporting oxygen. In order to maintain life, 250 cc of oxygen must be delivered to the tissues every minute. The plasma, or fluid part of the blood, can carry only one one-hundredth of this amount; the remainder is transported by a more efficient oxygen carrier—hemoglobin.

This red-colored protein has the ability to absorb from the lungs as much as 21 cc of oxygen for every 100 cc of blood. The intricate way by which oxygen leaves the air in the lungs and diffuses into the bloodstream is described in the chapter on the respiratory system. Blood leaving the lungs is saturated with oxygen to 95 percent of its capacity, much more than is needed by a person at rest. Blood returning to the lungs from the veins may bring back 14–15 cc of oxygen per 100 cc of blood, having released only 6–7 cc to the tissues. This tremendous reserve enables the blood to provide sufficient oxygen to tissues under stressful situations, such as during strenuous exercise, fever, or failure of the blood pump system (the heart), when a greater demand for oxygen arises. In addition, this reserve permits the maintenance of life during diseases associated with anemia. Even when the normal blood content of 15 g (grams) of hemoglobin for every 100 cc of blood is halved by anemia, a person can function and carry out a sedentary life

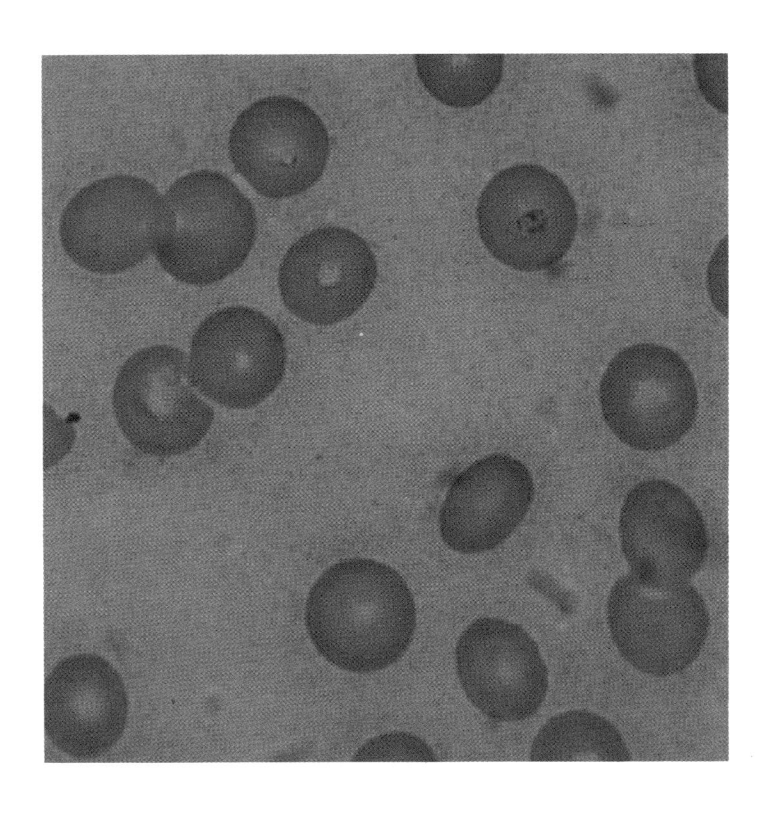

Red blood cells in their normal state are disklike with flat centers.

HEMOGLOBIN LEVEL OF BLOOD (in grams per cubic centimeter)	PHYSICAL STATE OF A PERSON
.15 G/CC	Healthy
.075 G/CC	Distress going up a hill
.05 G/CC	Distress while reclining
.03 G/CC	Near death

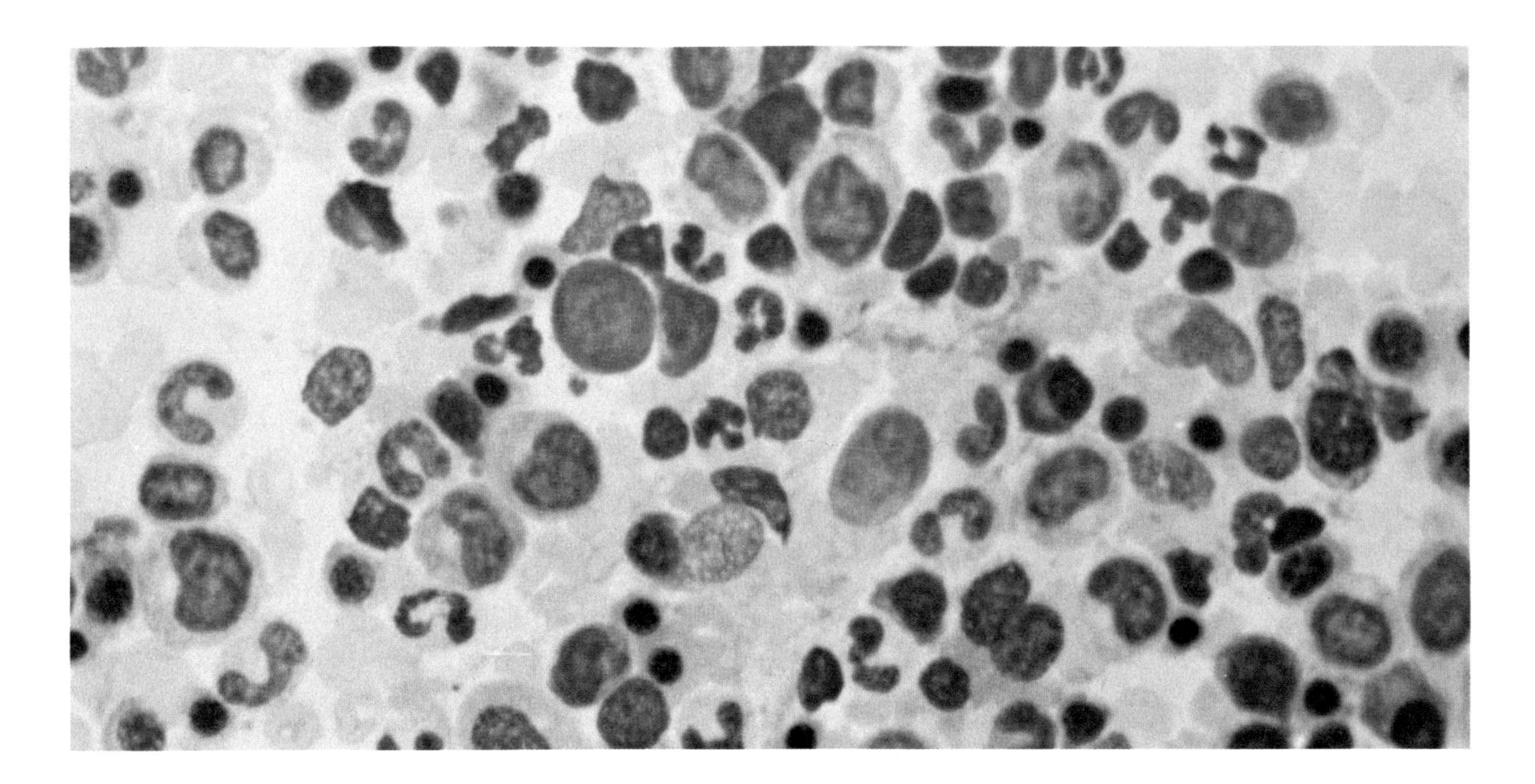

Bone marrow is the "factory" where most blood cells are made. Stained "parent" cells in the marrow are shown above.

without adverse symptoms. Air hunger, associated with extreme shortness of breath, will occur only when the hemoglobin level drops to 5 g per 100 cc, or a third of the normal value. Death will be imminent when the level drops to around 3 g, or a fifth of the normal value.

Iron Deficiency Anemia

Hemoglobin is a protein that is rich in iron and, like the red cells themselves, is formed in the bone marrow. A continuous supply of iron is needed by the bone marrow if it is to form sufficient hemoglobin for the blood. Iron is absorbed from food in amounts that vary in accordance with the body need, but iron is not eliminated or lost from the body in anything but trace amounts. When erythrocytes break down, the hemoglobin released from them is reabsorbed by the body in such a way that the iron content is saved for future use. Thus, a person normally needs very little additional daily iron from his food. If, however, there is a blood loss from the body, the iron lost along with the blood must be replaced by additional dietary intake. Such an additional need for iron occurs in women of childbearing age who experience an average monthly menstrual loss of 35–70 cc of blood containing 17–35 mg (milligrams) of iron. These women would require approximately an additional 1 mg of iron per day to keep up their proper iron balance. Patients suffering from chronic blood loss, such as from stomach ulcers or bleeding hemorrhoids, must also compensate for the loss of iron.

Additional iron is also required for growth. Thus, children and pregnant or nursing mothers require more iron than would otherwise be usual. If the additional dietary iron is not provided, iron deficiency anemia, which is the most common form of anemia in women, can develop. Foods rich in iron that can supply this additional need include cereals, red meat of any kind, and eggs. Milk products are poor in iron; infants fed entirely on them usually develop iron deficiency anemia in a few months.

The amount of iron absorbed from food is variable and may be affected by stomach acidity and other factors. If, with an average diet, the amount of iron absorbed does not replace the iron loss, supplemental iron would be advisable. On the other hand, for a person who is not losing iron, supplements are clearly not necessary. Chronic excessive intake of iron over a period of years may result in excess deposits of iron in body tissues, leading to a condition known as hemosiderosis. In some cases the iron may be harmful to such body organs as the pancreas and the heart. This complication is termed hemochromatosis (*hemo* = blood and *chromo* = color in Greek), which describes a brown discoloration of body tissues that have excessive deposition of iron. (The suffix "osis" usually means "a state or condition of.")

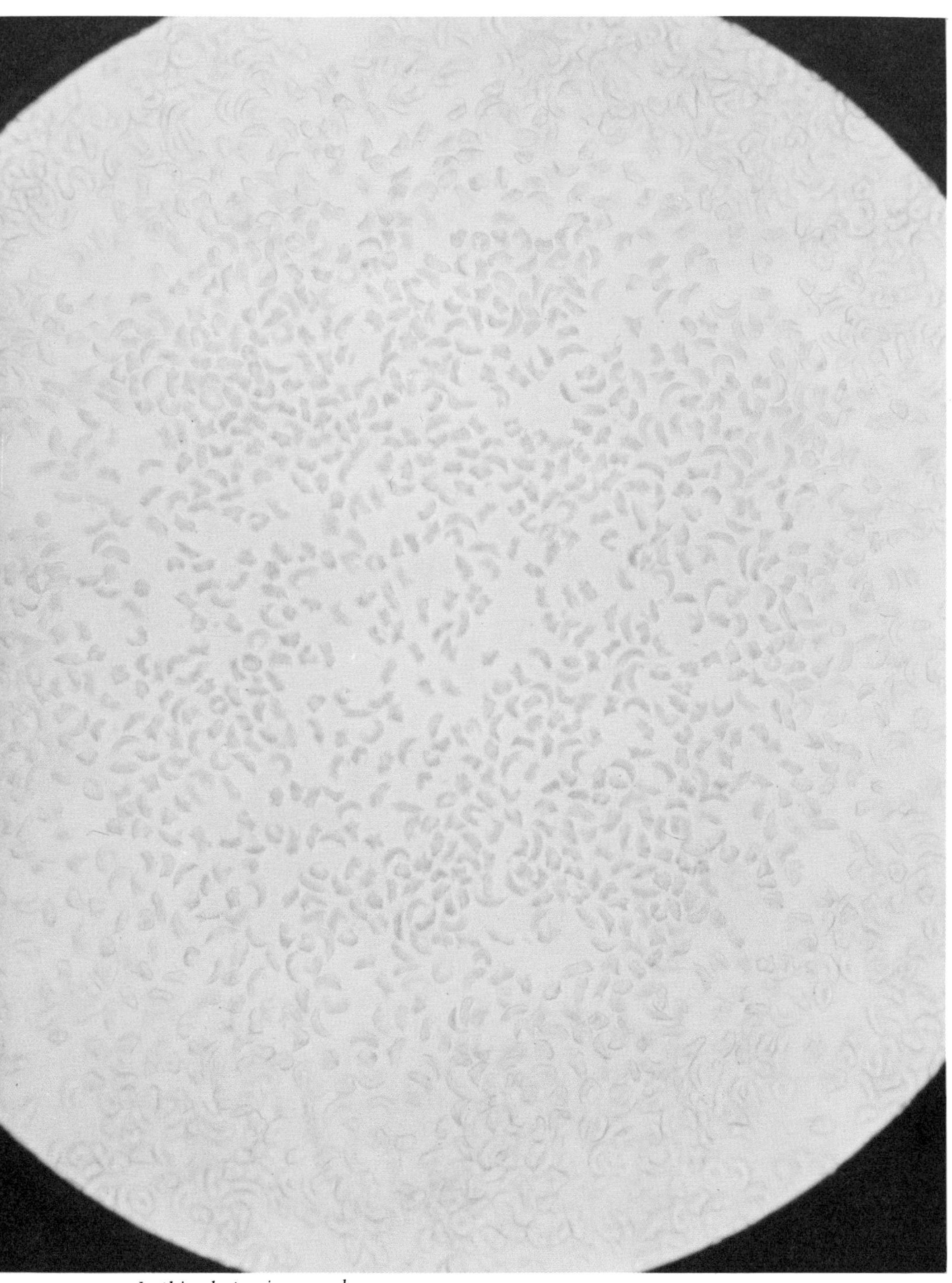

In this photomicrograph of blood from a person with sickle-cell disease, the altered red blood cells exhibit their characteristic shape.

Sickle-Cell Disease

The protein part of the hemoglobin molecule is made up of smaller units, known as amino acids, chemically linked in a definite pattern to form the protein chain. The structure of any protein depends on the number, type, and specific way in which the amino acids join together as they follow the instructions dictated for them by the individual's genetic makeup. If as a result of a genetic mutation some of the amino acids in a protein are misplaced, the biological properties of the resulting molecule will be significantly altered. One example of such an abnormally structured protein is hemoglobin S. At two points on this chain of 574 units, the amino acid glutamine has become substituted for the normal amino acid, valine. These changes cause an abnormal alteration of the physical properties of the hemoglobin. Normal hemoglobin exists in solution within the disc-shaped red blood cell. Hemoglobin S, however, turns into crystals that stretch the cell into a curved, pointed, angular, elongated structure resembling the shape of a sickle. This phenomenon manifests itself only when the red cell is exposed to a low-oxygen environment, producing a condition known as sickle-cell anemia. It may occur in small blood vessels in body organs where blood flow is slow and oxygen may be extracted and used up by tissues. Or it may occur when the patient is exposed to low atmospheric oxygen, such as at high altitudes. The abnormally shaped, sickled, cells cannot pass through the small blood vessels, and the delivery of blood and oxygen to the tissues is obstructed. Severe pain may result; this complication is termed sickle-cell crisis. This process is reversible. If oxygen can be taken up again by the sickled cells, much of the crystallized hemoglobin S may return to the solvent state and the cells may reassume their normal shape and function.

Since the abnormality is a genetic one, patients with sickle-cell disease often have family histories of similar disease. As many as 1 in 400 blacks born in the United States are affected by sickle-cell disease. This means these subjects either carry the sickle-cell trait or have sickle-cell anemia. The latter form of the disease affects 25,000–50,000 patients. Use of a recently developed, low-cost technique for detecting sickle-cell trait in conjunction with genetic counseling might help to decrease the incidence of actual sickle-cell anemia cases in the next few generations. Understanding of the abnormal amino acid structure of hemoglobin S is also enabling physicians to develop chemical methods for the treatment and prevention of crises.

White Blood Cells

Leukocytes, or white blood cells, play an entirely different role in the blood. Essentially, they defend the body against foreign invaders in the bloodstream. Two types of white cells, monocytes and neutrophils, specialize in killing bacteria that gain entry to the blood. Built-in chemical receptors enable these cells to detect and move toward foreign bacteria. The white cells then actually change shape, sending out cytoplasmic armlike tentacles to engulf the intruder in a process known as phagocytosis. The trapped bacteria are subsequently digested within the cell. The white cells can also leave the bloodstream to fight bacterial invasions of body tissues. In these battles between the cells and bacteria, large numbers of white cells may perish. The accumulation of dead bacteria and cells in the tissues will form pus.

In response to an infection the bone marrow can greatly increase its output of white cells. An elevated white cell level in the blood signifies that such a process is taking place. Thus, the white blood cell count is a standard laboratory test that physicians rely on in the diagnosis of infection. The normal white blood cell count is 5,000 to 8,000 per cubic millimeter of blood; the number may increase to 10,000–20,000 in the presence of an infection.

In a number of disease states, the bone marrow may fail to manufacture white cells. Victims of these disorders would, therefore, be highly prone to infection. This failure may be the result of any of a wide variety of causes, including exposure to toxic chemicals, individual sensitivity to drugs and antibiotics, or invasion of the bone marrow by cancer cells. Recently developed techniques make it possible to use white cells donated by normal persons to counteract this deficiency. Unfortunately, this process of "white blood cell transfusion" is available only at a few highly specialized medical centers.

Another form of white blood cell, the lymphocyte, occupies a key position in the body's immunity system. Lymphocytes are well equipped to defend the body against foreign materials of all types. These cells also cause the body to reject grafted or transplanted foreign tissues. A major complication following organ transplant surgery is the fact that the recipient's lymphocytes recognize that the donated heart, liver, or kidney comes from a different, or foreign, source and attempt to reject the transplant. Various chemical means must be used to suppress this lymphocytic function.

Leukemia

The white cell, like many other kinds of cells in the body, may multiply under abnormal conditions in

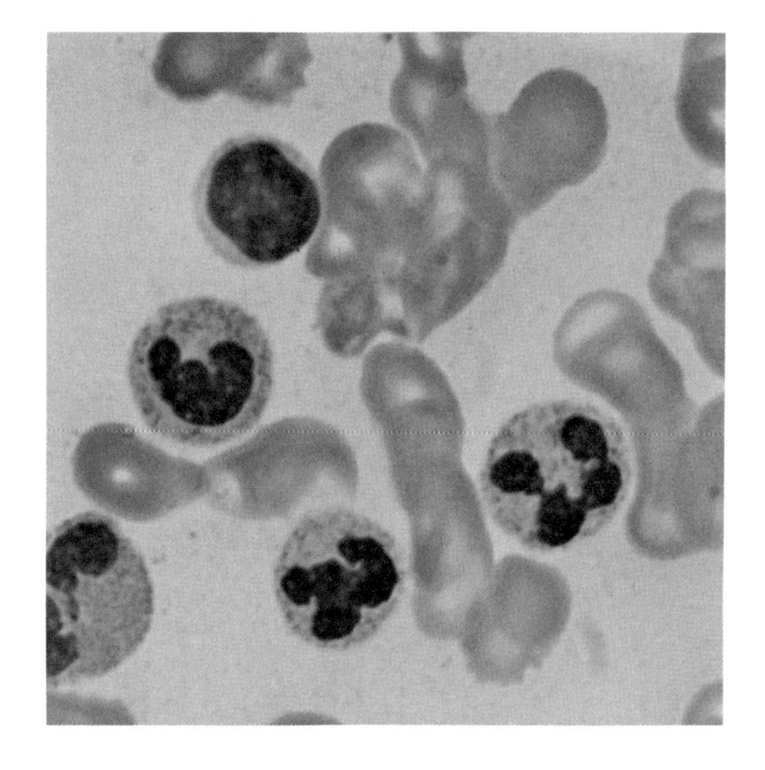

White blood cells are difficult to see under a microscope unless they are stained with special dyes.

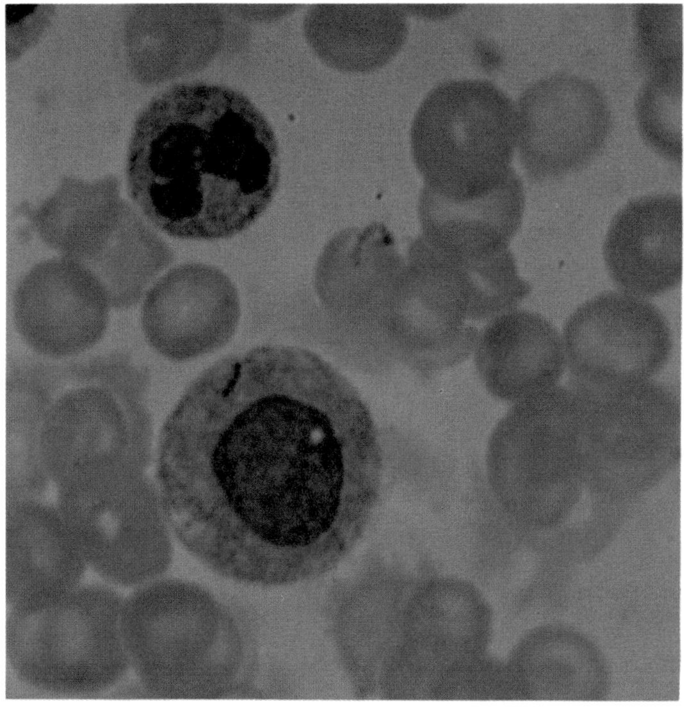

The large white cell in the lower center of the photo can be seen engulfing bacteria.

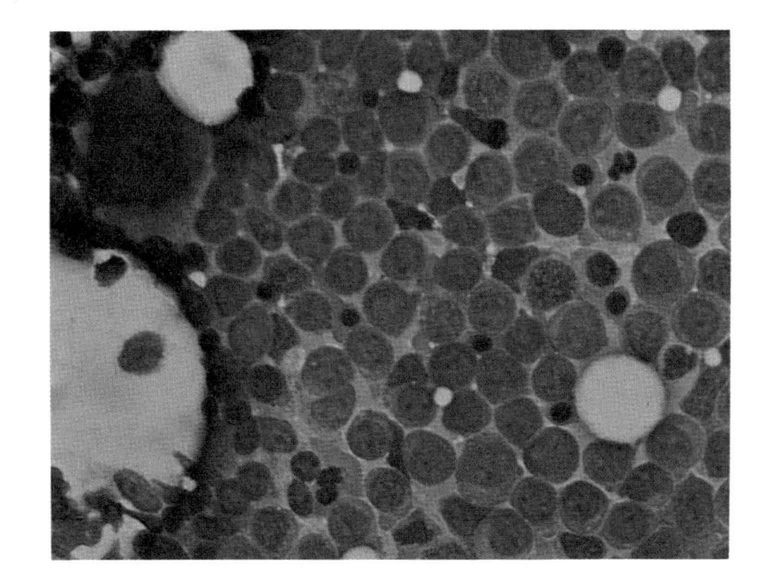

This photomicrograph of bone marrow from a person with acute leukemia shows nearly all cells to be of the leukemic type.

an uncontrollable manner. When this occurs in white cells it is known as leukemia, one of the most dreaded forms of cancer. In the more severe (acute) form of this disease, leukemic cells grow in excessive numbers and dominate the bone marrow and the bloodstream. By their sheer mass, they displace the other forms of blood cells, resulting in the development of severe anemia and bleeding. The latter complication occurs because one of the cell types that is displaced is the platelets, which are important factors in preventing bleeding.

Exactly what causes white cells to turn leukemic, or cancerous, is as yet unknown. It is well recognized that exposure to ionized radiation is associated with an increased incidence of leukemia, as occurred among the survivors of the atomic bomb explosions in Nagasaki and Hiroshima, Japan, and among patients with arthritis who have undergone prolonged deep X-ray treatment. In recent years much attention has been given to the possibility that a virus may be the cause of leukemia. Certain viruses are definitely known to produce cancers, including leukemia, in experimental animals, but no definite virus has been isolated from human cases. Despite the possibility of a viral causative agent, human leukemia cases are definitely not transmitted as an infection. Transfusion of leukemic blood from one individual to another does not result in the disease in the recipient.

Leukemia can affect all ages. The acute form, however, is most common among children. Its first manifestation may be bleeding from any part of the body, with or without fever, and flulike symptoms. As recently as the early 1960s a victim of the acute form ran a rapid downhill course, usually leading to death in six to twelve weeks. With the help of recently developed drugs the prognosis has been dramatically altered. As many as 50 percent of the patients can be expected to respond to early treatment. Many of these patients will ultimately suffer relapses and die. But with continued discovery of newer anti-cancer therapies, more and more cases in relapse are being brought under control again. By the early 1970s a number of children with the acute form of leukemia had remained alive and well for as long as five years from the time of diagnosis of their disease.

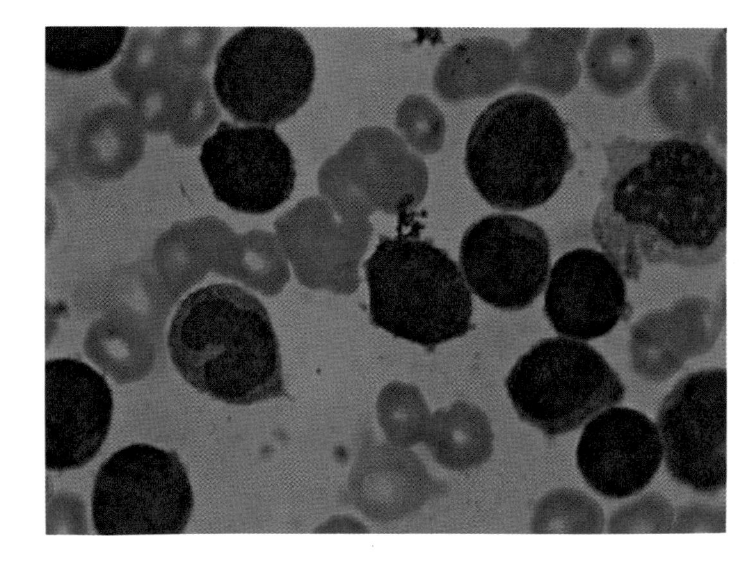

Cells of the leukemic type outnumber other types of white blood cells in this view of an acute leukemia sufferer's blood.

Platelets

The platelet, a thrombocyte, is the smallest of the white cells. Shaped like a round plate with a diameter approximately one-third that of a red blood cell, it is concerned mainly with the prevention of bleeding. Minor injuries occur in the blood vessels all the time. When vessels are injured, blood is prevented from escaping through the injured area by the contraction of the vessels themselves, and by the buildup of the platelets at the site of injury. The inner lining of blood vessels is normally smooth, and platelets can pass through even the smallest capillary without adhering to it. When the lining is disrupted by injury, however, platelets in the circulating blood stick to the roughened, injured area. Within seconds after injury, thousands may gather and form a platelet clump, which serves as an effective plug preventing leakage of blood from the vessels.

Without the help of platelets, minor injuries to the vessels would cause hemorrhages, manifested as bruises and bleeding from body surfaces. The clumped platelets also release chemicals that help to form blood clots. Thus, the first defense against bleeding is the contraction of blood vessels and formation of a platelet plug, while the formation of blood clots can be considered as the blood system's second line of defense.

Platelet Loss

The blood platelets are produced in the bone marrow and subsequently released into the bloodstream. They live in the circulating blood for approximately ten days, after which they are removed, primarily by the spleen. Whenever the bone marrow fails, or when a diseased spleen causes increased destruction of platelets, the number of circulating platelets may be drastically reduced. A normal person has around 200,000 platelets in each cubic millimeter of blood. When this number falls to between 50,000 and 100,000 some bleeding manifestations will occur. At even lower levels (less than 50,000) much more serious bleeding usually happens. This bleeding can be fatal, particularly if it takes place inside a vital organ, especially in the brain. Improved transfusion techniques enable physicians to replace selectively only the platelet component of blood without overloading the other blood components.

Platelets and Thrombosis

While a platelet deficiency may result in failure of the control mechanism for bleeding, excessive functioning of the mechanism, caused by increased numbers of platelets or by an increase in their stickiness, can lead to the formation within the blood vessels of blood clots rich in clumped platelets. These clots may obstruct the normal blood flow to a body organ. Such an obstruction is called a thrombosis. It occurs most frequently in arteries, with the result that the blood supply to the body tissues is cut off. Perhaps the best known example of this type of blocking is coronary arterial thrombosis, commonly called heart attack. This form of arterial obstruction gives rise to damage and death of the heart tissues supplied by the coronary artery. In any thrombosis, if vitally important tissues are affected, serious illness can result, even when the obstructed artery is a small one. The blockage of tiny arteries in the brain, for instance, may result in a permanent damage to nerve tissue, leading to extensive paralysis of the body. Attention has been focused on people with intermittent attacks of temporary paralysis, a condition known as transient ischemic attacks. These attacks are believed to be caused by repeated, chronic blockage of small arteries in the brain. Each attack is brought on by a shower of platelet clumps originating in diseased and hardened arteries in the neck.

Much research work has been done in the hope of understanding the various factors influencing platelet sticking and the tendency of platelets to form thrombi, or clots. To the surprise of many scientists, the commonly used painkilling drug aspirin was found to have an additional chemical effect on the membrane of the platelet cell that reduces stickiness. Platelet-filled thrombi experimentally produced in the arteries of animals were found to be very similar to the arterial thrombosis seen in man. When aspirin was given to the animals before experimentation, no thrombi were produced. Knowledge obtained from such research as this has been applied to the treatment and prevention of a number of conditions in which increased platelet stickiness is believed to play an important role. It remains to be seen whether these treatments can effectively reduce the incidence of coronary artery thrombosis, strokes, and other results of blood clot formation.

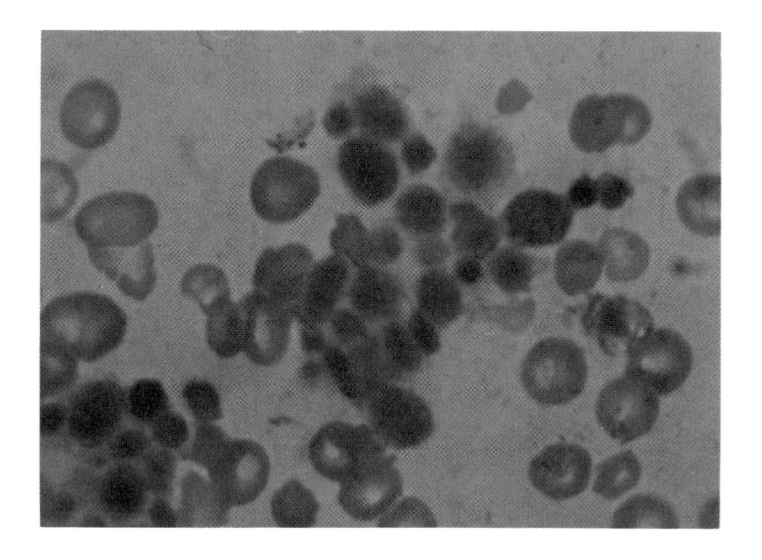

Platelets, important in bleeding prevention, are about a third the size of red blood cells.

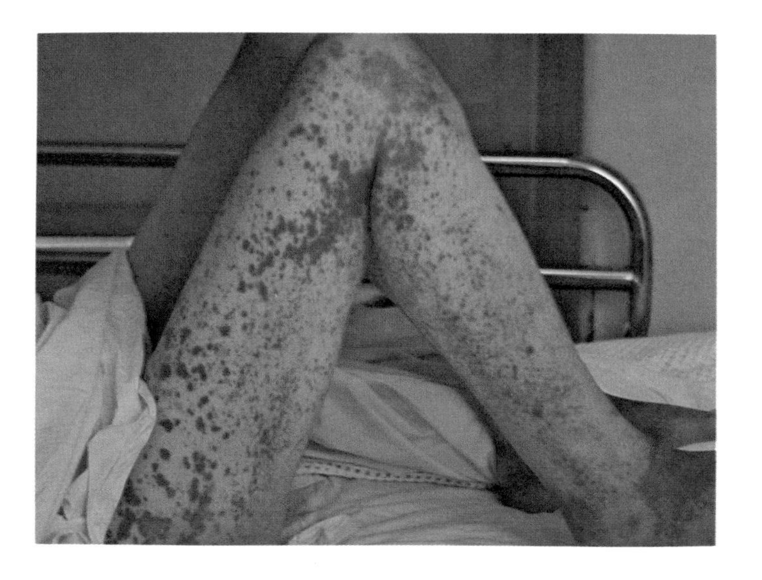

Platelet deficiency can result in hemorrhage in the skin. Sites of hemorrhage appear as pinhead-size spots.

Plasma

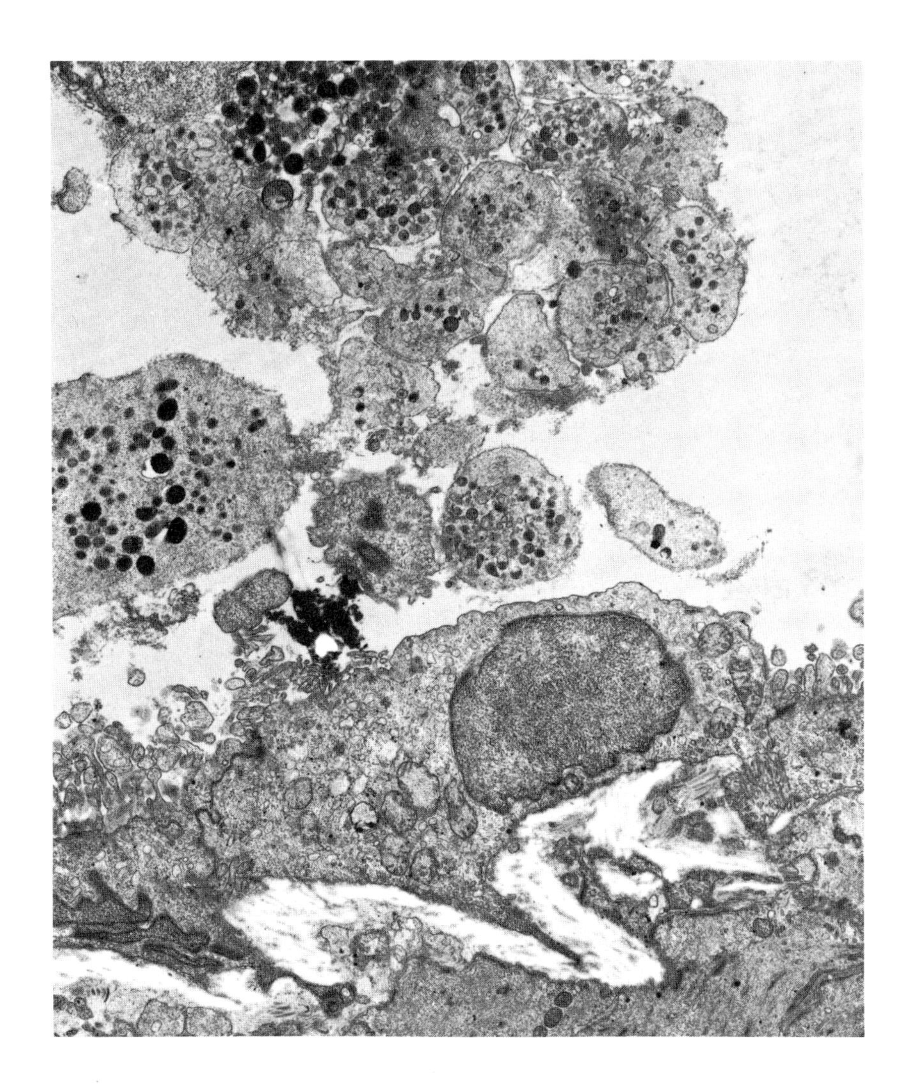

A pile of platelets can be seen sticking to the injured part of a blood vessel wall. Taken under an electron microscope, this view is magnified 1200 times.

The fluid part of the blood, in which important life-sustaining proteins float in solution, is the plasma. Among the proteins is a series of substances that, like platelet clumping, are essential to the formation of a blood clot. Some of the clotting factors are sensitive to a foreign surface or become activated by a roughened surface. When blood is exposed to a foreign surface other than the natural lining of blood vessel wall, the activity of these factors stimulates other reactions that ultimately convert the liquid blood into a jellylike state. The gel hardens and forms a toughened fiberlike clot called fibrin. In an injury, the fibrin clot formation follows platelet clot formation and supplements the latter in plugging the leak in any injured blood vessel. Twelve such clotting factors are known in man. Most of them are produced in the liver.

The ability of the liver cell to form blood-clotting proteins is provided by specific genes. Sometimes abnormalities will occur in these genes and an individual will lack one of these clotting factors, which will manifest itself as a congenital bleeding abnormality, such as hemophilia.

Hemophilia

Hemophilia is a condition in which one of the blood-clotting substances, designated factor VIII, is missing. The first sign of this disease in an infant may be excessive bleeding from minor surgery such as circumcision, commonly the first surgery for the newborn male. The ancient Talmud recognized hemophilia when it forbade the ritual of circumcision to a child if there had been a history of excessive bleeding in his siblings. Later in childhood, the hemophiliac patient will bruise more easily than his peers. Soon, he may be disabled as a result of repeated bleeding into the joints.

Modern extraction procedures permit the factor VIII protein to be extracted from normal blood, prepared in concentrated form, and given to hemophiliacs to replace their deficiency. Many persons have been taught the technique of intravenous injection of this material at home. If they are injured accidentally, they can immediately inject factor VIII and reduce the amount of blood loss that might otherwise occur while awaiting medical aid.

The gene for hemophilia is carried by the chromosome that determines sex in the embryo. The disease itself occurs almost exclusively in males. When the abnormal gene is present in a female, bleeding does not occur, but the female is a carrier and can transmit the disease to her male offspring. Queen Victoria of England, for example, was a hemophilia carrier. Many male members of her family were afflicted with this disease, and, because of intermarriage, hemophilia became prevalent among the royal families of Europe. The manifestations of the bruising caused by the disease among kings and princes is believed to be the origin for the term "blue blood," referring to royal families. Modern techniques for analyzing blood clot factors enable physicians not only to diagnose the disease early in the victims but also to identify the carrier state in the symptom-free female. With proper genetic counseling, it has become possible to prevent transmission of the disease to further generations.

Fibrin Clot and Thrombosis

In addition to the platelet abnormalities that may give rise to abnormal clotting or thrombosis, increased amounts of some clotting factors may greatly increase the tendency to form blood clots rich in fibrin content. Physicians and surgeons have come to realize that increased levels of clotting factors occur during many common diseases, such as heart failure, cancer, and chronic infection, and during the few days immediately following any form of major surgery.

The early recognition of an excessive clotting tendency permits proper preventive treatment before clot formation can take place. All clots begin small and, under favorable conditions, may grow into huge sizes fairly rapidly. Parts of a clot may break loose and be carried by the bloodstream toward the heart and lungs, causing a serious form of illness known as pulmonary embolism. This is a complication dreaded in persons who have heart ailments or who have undergone major surgery. Much publicity was given to one of the earliest heart transplant recipients who perished from such a complication soon after successful transplant surgery. Oral anticoagulant drugs are commonly referred to as blood thinners. This form of medication acts to remove a number of clotting factors from the blood and is often used to prevent clotting. Another form of medication is a clot-dissolving drug. Two compounds found to be effective in this respect are streptokinase and urokinase. Both are capable of dissolving away even large fibrin blood clots.

Plasma Protein as Antibodies

A major part of the protein in plasma is the globulins, which because they are concerned entirely with immunity are usually known as immunoglobulins. Immunoglobulins are also antibodies, structures designed to act against specific materials foreign to the body. An antibody against streptococcus bacteria, for example, may be an immunoglobulin that acts on streptococci only. These antibodies are produced by blood cells residing in the bone marrow and in lymph nodes. When called upon to produce an antibody, they may multiply and put out larger quantities of the specific type of immunoglobulins needed. Not only can the antibodies kill the bacteria directly by clumping them together, but the clumping also helps white blood cells in the process of phagocytosis. The white cells can engulf the bacteria with greater ease.

Immunoglobulins can sometimes cause extensive reactions in the body at the site of their contact with foreign bodies. For example, persons sensitive to certain flower pollens develop eye and nose allergy symptoms when these parts of the body are exposed to the pollen, as a result of the interaction of the antibody and the pollen.

Hyperlipidemia (Excess Fat in Blood)

In recent years the possibility that excessive fat in the blood may cause hardening of the arteries (arteriosclerosis) has aroused tremendous public concern. Fat is normally present in the plasma fraction of blood in small quantities while it is being carried back and forth from one part of the body to another. Fat derived from food is absorbed from the intestine and carried to the liver for further processing. The liver, in turn, not only assimilates the dietary fat but also may make additional fat from other kinds of material such as glucose. The excess and unassimilated fat is stored in fat tissue. It is only when the blood fat level rises, in a condition known as hyperlipidemia, that some of the fat particles may become deposited on the

A huge clot in the main artery of the lung is shown here. The clot caused a severe block to the flow of blood to this vital organ.

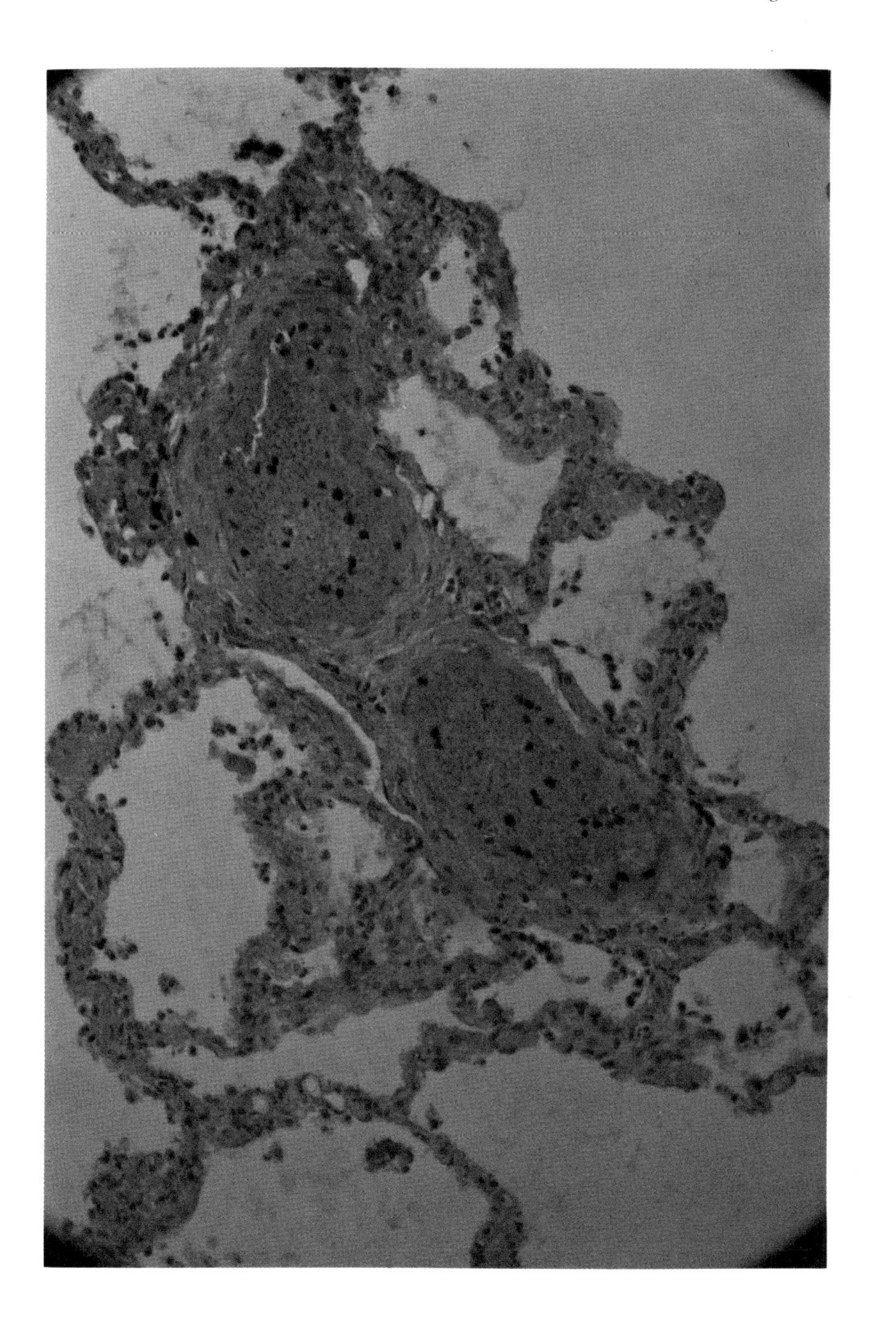

arterial wall. Over a long period of time this can lead to hardening and narrowing of arteries.

The fat in plasma exists in many different forms. The simplest is a fine suspension of small globules known as chylomicrons. An excess of these particles can give the plasma a milky appearance. Another important fatty substance is cholesterol; the quantity of cholesterol in the plasma serves to indicate the degree of fat excess. Its importance was first pointed out by a Russian scientist who observed that rabbits fed large quantities of cholesterol developed serious thickening of the arteries. This finding has been confirmed by many other scientists.

Hyperlipidemia may be the result of many conditions. A common one is the intake of a diet high in fat or cholesterol. Such dairy products as cream and eggs are rich in both fat and cholesterol. In fact, the amount of cholesterol in an average yolk is as much as is present in the animal fat that an average person eats in a week. In many cases, including diabetes, deficiency of thyroid hormone, and certain metabolic disorders, the liver may produce an excess of cholesterol and plasma fat. Hardening of arteries may occur early in life in persons with these diseases.

The arteries that most commonly become hardened are the ones that supply vital organs—the cerebral arteries supplying the brain, the coronary arteries supplying the heart, and the renal arteries supplying the kidneys. Physicians pay great attention to blood cholesterol and blood fat levels, which can be measured routinely in the course of regular health checkups. Most hyperlipidemias respond well to careful dietary regimens. Public health authorities are also concerned about the possible long-term harmful effects of the high intake of cholesterol-rich foods characteristic of the average diet in Western countries. Public education programs that discuss properly balanced diets are important efforts to combat heart diseases and strokes.

Blood Types and Transfusions

Although the first successful cross-transfusion of blood was done in dogs in 1665, and 347 transfusions in man had been reported in the medical literature by 1875, the procedure was associated with such high mortality that it was used only under grave conditions. One important reason for the high death rate was the failure to recognize that there are different types of blood in man and that severe reaction or death may occur in a person who receives blood that belongs to a group different from his own. A historical milestone was the establishment in 1900 by Karl Landsteiner of the ABO blood-group system, which permits persons to be identified as belonging to one of four blood groups. Persons in group A have A-type antigens on their red blood cells; persons in group B, B-type antigens; persons in group AB, both A and B antigens; and persons in group O, no antigens on their red blood cells. A person who does not have a specific antigen on his red blood cells will carry antibodies against the antigen in his blood plasma. Thus, if a person in group A, for example, has group B blood transfused into his circulation, his antibodies will attack the foreign cells, causing severe complications.

Landsteiner was awarded a Nobel Prize in 1930 for his discovery. Since then as many as 24,000 blood types have been established, all of less importance in transfusions than the ABO groups. The presence of blood types in a person, however, follows definite inheritance patterns. It is therefore possible to use different blood grouping identifications in medico-legal cases where such things as proof of parenthood are required.

Rh Disease

Of the other known blood-group systems the Rh, or Rhesus, system, so named because it was first noted in the antibody response of test animals immunized with red blood cells of Rhesus monkeys, is particularly significant because an Rh incompatibility can cause certain serious anemias and jaundice in newborn babies. Approximately 85 percent of the Caucasian population show the presence of the same Rh antigen in their red blood cells and are identified as being Rh positive. Rh disease, or erythroblastosis fetalis, is most commonly seen when an Rh-negative mother, a woman lacking the Rh antigen, is pregnant with an Rh-positive fetus. Some of the Rh antigens from the fetus may cross over into the mother's circulation, where antibodies against the Rh antigen will be formed. If the maternal anti-Rh antibodies pass into the fetus, they cause destruction of the Rh-positive fetal red blood cells. If the destruction is massive enough, the fetus may die within the womb. In less

Blood donation is a quick and painless procedure. Blood is stored for a while in special banks from which it can be drawn when needed.

severe cases the newborn baby may show severe anemia and jaundice. The jaundice is caused by an accumulation of large amounts of biopigments derived from the destroyed red blood cells.

Rh disease does not usually occur during the first Rh-positive pregnancy of an Rh negative when she has not yet built up anti-Rh antibodies. The high mortality rate for any subsequent Rh-positive baby has been reduced in recent years by the use of several techniques. Large amounts of compatible blood can be pumped into the baby as soon as he is born in an attempt to wash out as much as possible of the maternal Rh antibodies as well as of the unwanted biopigments. A serum that can prevent the Rh antibody response in the maternal blood when injected shortly after an Rh-positive birth was developed in the late 1960s. In addition, the routine use of blood-group identification of all expectant mothers and their spouses locates potential cases of maternal-fetal incompatibility, and newer techniques of immunological manipulations can prevent or reduce the severity of the anemia and jaundice when an incompatibility exists.

Transfusion Complications

In recent years large amounts of blood have been used for the treatment of many diseases, creating a worldwide demand for blood that is often difficult to meet. The greatest demand is for the replacement of the blood volume in cases of massive blood loss, such as is seen in serious accidental injuries. Transfusion is also used to sustain life in those patients whose bone marrow has failed to produce blood cells. In addition, modern fractionation procedures can accomplish the separation of blood into its different components for the replacement of specific deficiencies, as in the extraction and use of factor VIII to prevent and to treat the blood-clotting deficiency in hemophiliacs. Such procedures create even further demands for blood.

The high mortality and complication rates encountered in the early blood transfusions were as much the result of transmittance of infections during transfusion as they were of blood incompatibility. The most serious of these infections is serum hepatitis, a virus infection of the liver. Since there are no known sterilization techniques that can kill the hepatitis virus in blood taken from the donor, the only way to prevent this complication in recipients is to screen carefully to ensure only healthy donors. Because serum hepatitis has been found to be much more common among professional blood donors than in the general population, a legal ban on the use of blood from these paid donors is one possible means of reducing the incidence of the disease.

The Lymphatic System

The lymphatic system is a channel of small vessels that serves as an accessory drainage system, in addition to the blood vessels, for the removal of body fluids. The lymph channels drain into rounded little nodules of tissue called lymph nodes, which serve as filters for the lymph fluid. The channels then join each other to form larger ones. The largest trunk is called the thoracic duct, which ends in a vein near the heart. In addition to draining body fluids, the lymphatic system and the lymph ducts also drain the nutrient materials, particularly fat particles, as they are absorbed from the intestines. This drained lymph fluid is a creamy-looking one, especially after a fatty meal.

Role in Fighting Infection

The lymph fluid also contains lymphocytes, most of which are produced in the lymph nodes. These lymphocytes play the very important role in the body's defense against infection that was described previously. Very recently, scientists established that there are two main kinds of lymphocytes. One kind, called T cells, originates from the thymus gland, a small organ located in the chest behind the breastbone. The other kind is derived from the various lymph nodes and spleen and is known as B cells, designating the primitive bursa of the chicken, where they were first found to be produced. B cells are concerned primarily with the cellular form of immunity in the body, which, in contrast to antibody activity, responds to invasions of foreign material by mobilizing lymphocytes to infiltrate the cells around the foreign material, ending in destruction of the invader. This reaction occurs in response to the grafting of a foreign tissue. If the tissue belongs to a family of tissue groups different from that of the recipient, a heavy infiltration of lymphocytes will take place in the grafted material, resulting in the death of the graft. Understanding of this phenomenon has led to great advances in the technology of organ transplantation. The use of drugs or sera that destroy lymphocytes can result in blockage of this form of rejection.

Important in body defense, lymph nodes are located along the body's lymphatic system. Each node has cells capable of destroying bacteria that the lymph flow carries into the node.

A Lymph Node

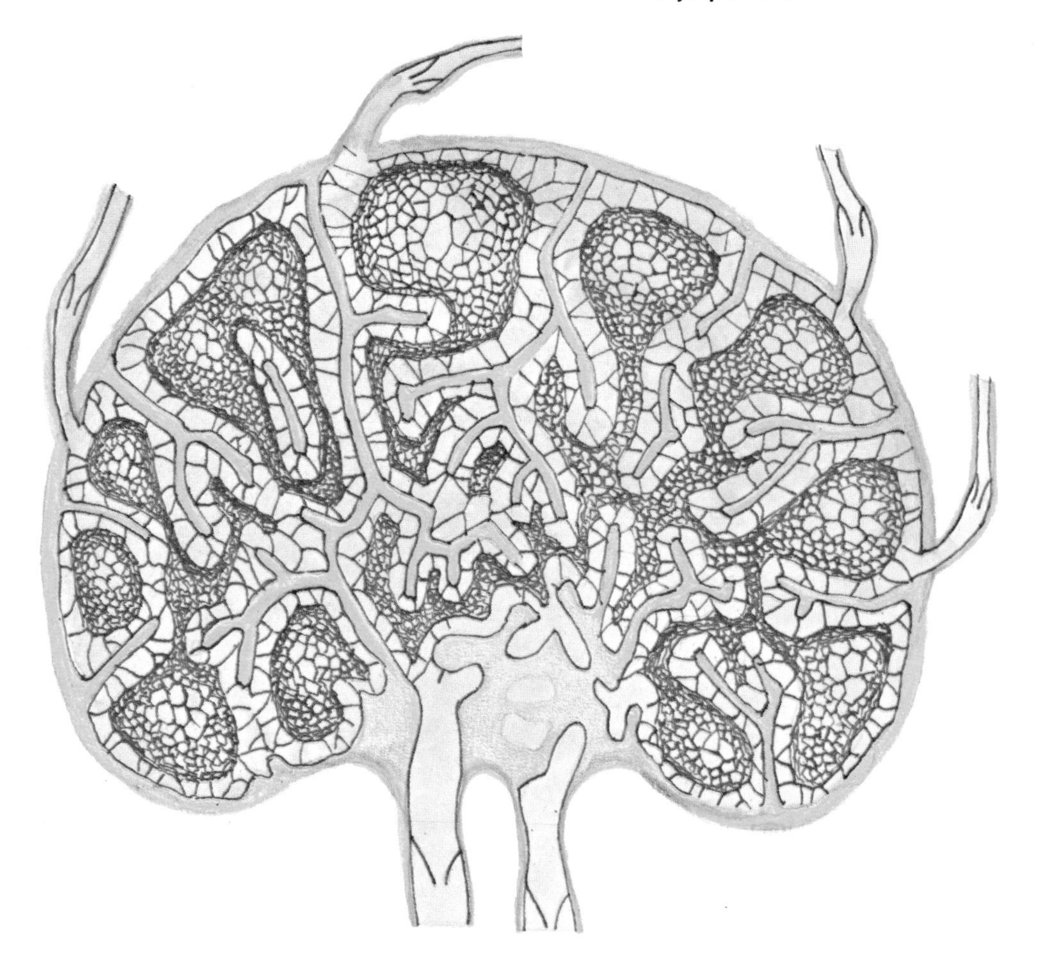

Importance of the Spleen

The spleen can be considered a part of the lymphatic system because the spleen develops prior to birth from the same tissue of origin as the lymph nodes. In an adult the spleen is the size of a closed fist, weighs a quarter of a pound, and is located next to the stomach, just under the left side of the rib cage. It contains mostly lymphocytes, many of which are formed within the spleen. These lymphocytes, like other lymphocytes in the body, have a role in defense against infection. The main function of the spleen, however, is the removal of aged and unwanted blood cells from the circulation. An intricate system of small arteries within the spleen acts as a filter for blood pumped through the organ. The younger and healthier blood cells pass through intact, while the older cells will not survive the passage.

In disease the spleen may enlarge many times its normal size, a condition termed splenomegaly. Instead of doing its normal function of removing only the aged blood cells, the enlarged spleen will also destroy younger and normal blood cells. This activity will lead to anemia and to a decreased defense against bacterial invasion, if excessive numbers of white blood cells are also destroyed. Likewise, circulating platelets may be consumed by the spleen, resulting in impaired protection against bleeding from minor injuries. Under such circumstances, it is essential to arrive at an accurate diagnosis, one that the lack of one or more of these blood cells is the result of an abnormal function of the spleen. Modern diagnostic procedures enable the physician to use radioactive tracers to label the blood cells so that their fate can be followed accurately through the body. If there is premature destruction of these cells, it can be detected. Furthermore, the site of destruction of these cells, for example, the spleen, can also be identified. The obvious approach to the treatment of a malfunctioning spleen that destroys blood cells is its surgical removal. The loss of the spleen does not interfere with normal body function. Other organs, such as the liver, can take up the role of removing old blood cells from the circulation.

In fetal life, the spleen also produces blood cells, a role taken over after birth by the bone marrow. If, however, at any time a great need arises for the body to produce blood cells, for example when the bone marrow is destroyed by disease, the spleen can resume this function. The spleen may also enlarge as a result. It may also start to destroy more cells while at the same time it is producing blood cells. The decision to remove the spleen under such circumstances is not an easy one, because the operation may deprive the body of a new source of production of blood cells. Thus, it is necessary to determine whether the spleen is producing more cells than it is destroying before an operation of this kind is contemplated. Fortunately, through the use of radioactive tracer techniques, it is also possible to estimate the amount of cell production occurring in the spleen.

Blood, Vital and No Longer Mysterious

Complex but vital, blood supplies the body with the oxygen and nutrients needed for life. Blood, supplemented by lymph, also helps defend the body against infection. Pumped by the heart, it is circulated throughout the body by an array of blood vessels, carrying vital substances to and removing wastes from the tissues.

Blood was long held to be a mysterious substance, imbued with the power of transmitting noble attributes. It was not even thought to circulate in the body until William Harvey, a British physician, proved otherwise in 1628. Even today, the Masai warriors of Africa bleed their cattle and consume the drawn blood for food and for its imagined quality of imparting courage and strength.

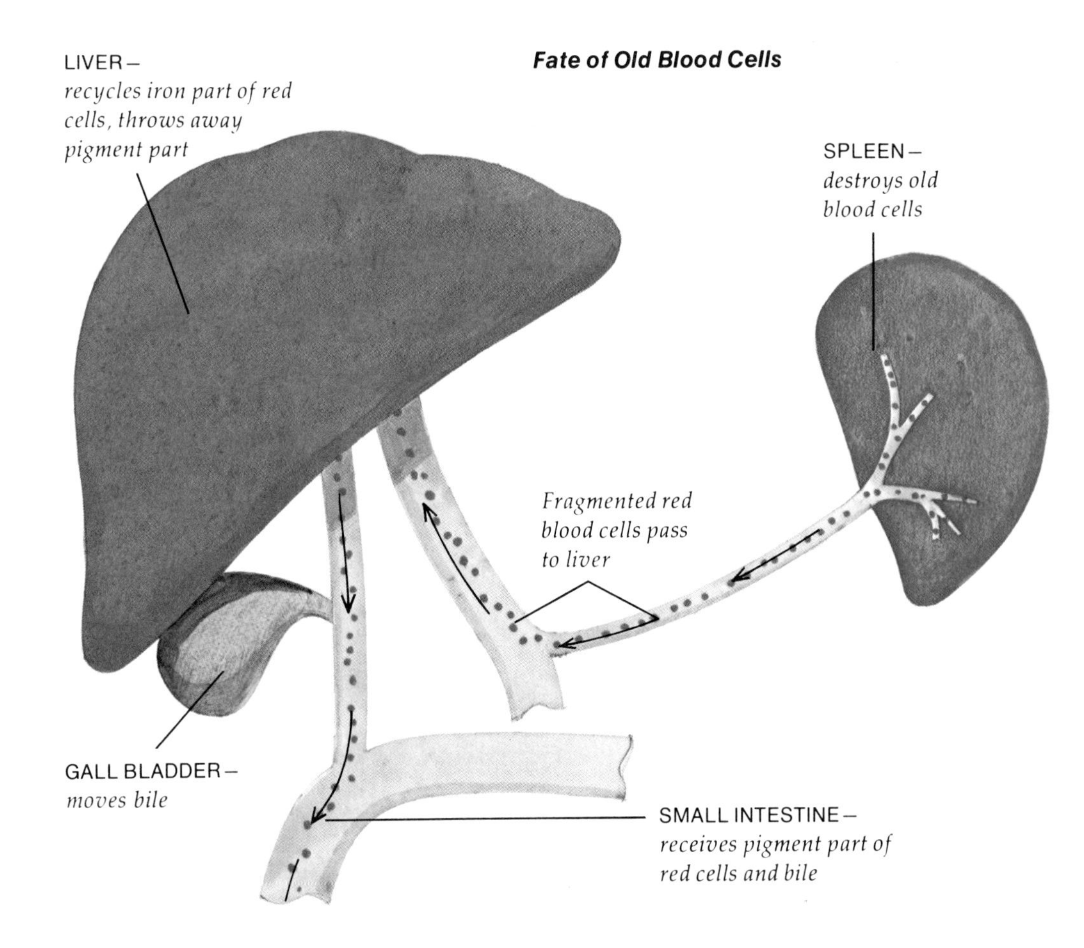

Fate of Old Blood Cells

For further information on the subjects covered in this chapter, consult ***Encyclopædia Britannica:***

Articles in the Macropædia	*Topics in the Outline of Knowledge*
BLEEDING AND BLOOD CLOTTING	421.B.3
BLOOD, HUMAN	422.A.3
BLOOD DISEASES	424.C.4
BLOOD GROUPS	422.A.3.c
LYMPHATIC SYSTEM, HUMAN	422.B.1,2,3, and 4
LYMPHATIC SYSTEM DISEASES	424.D.1,2, and 3
RETICULOENDOTHELIAL SYSTEM, HUMAN	422.C.1,2,3,4,5, and 6

Each kidney is composed of more than a million renal tubules, some of which are shown here in cross section. These tiny tubes are key parts of the body's filtration system. Photo, Melvin Oster.

Van Doren. *Apparently the renal system is the workhorse of the body. I hadn't realized that my kidney is so important to me.*

Kessler. *It sure is. Remember I said, in talking about the skin, that it keeps the outside outside and the inside inside. Well, the kidney maintains the volume and composition of the "internal environment," as the great French physiologist Claude Bernard called it. You know, this internal environment that bathes our cells has roughly the same composition as that of the Devonian sea from which our ancestors emerged half a billion years ago. That fact, to me, is one of the more intriguing in all of medicine.*

9

Renal System

BY MURRAY L. LEVIN, M.D.

The urinary system is designed to rid the body of certain waste products of metabolism. In other words, what is no longer needed after the body has made full use of what it has taken in is eliminated in the form of urine. (The undigested residues of food that the body could not use are removed through the large intestine, which is part of the digestive system.) If allowed to accumulate in the body, the waste products of metabolism would cause illness and, ultimately, death. Additionally, the kidneys, which are essential parts of the urinary system, regulate the amount and chemical composition of the body's fluids. Thus the urinary system plays two roles in preserving health: it permits the excretion of potentially toxic metabolic wastes, and it regulates the body's internal fluids, maintaining what is called the internal environment. In males, the lower portion of the urinary tract also serves as the means of sexual intercourse and procreation.

This chapter, then, considers the way urine is formed by the kidneys, how it is transported through the ureters to the urinary bladder, and how it is excreted. The various signs and symptoms of diseases of the urinary system, as well as the function of the male genital system and some of the diseases affecting it, are also discussed.

The Kidneys

In most persons two bean-shaped kidneys are situated in the upper portion of the abdomen against the rear wall of the abdominal muscles so that the concave portion of the kidney faces the midline of the body. The kidneys vary in size with the size of the individual but in adults are generally about 14 cm (centimeters), or 5½ inches, long, 6 cm, or 2½ inches, wide at their midpoint, and 2.5 cm, or 1 inch, thick. Each kidney weighs about 150 grams, or a third of a pound, and constitutes only 0.4–0.5 percent of the body's weight.

The kidneys receive 20–25 percent of the heart's output of blood every minute via the renal, or kidney, arteries, which are large branches off the abdominal aorta, a portion of the major artery leading from the heart. This extremely high blood flow per gram of tissue highlights the great importance of the kidneys in maintaining bodily health. There are two reasons for such a high blood flow to the kidneys. First, since the main function of the kidneys is to cleanse and regulate the blood, it is quite reasonable that they be exposed to as much blood as possible. Second, since the kidneys perform a great deal of work in carrying out their jobs, they require large amounts of foodstuffs, especially fats, to perform their functions. The renal

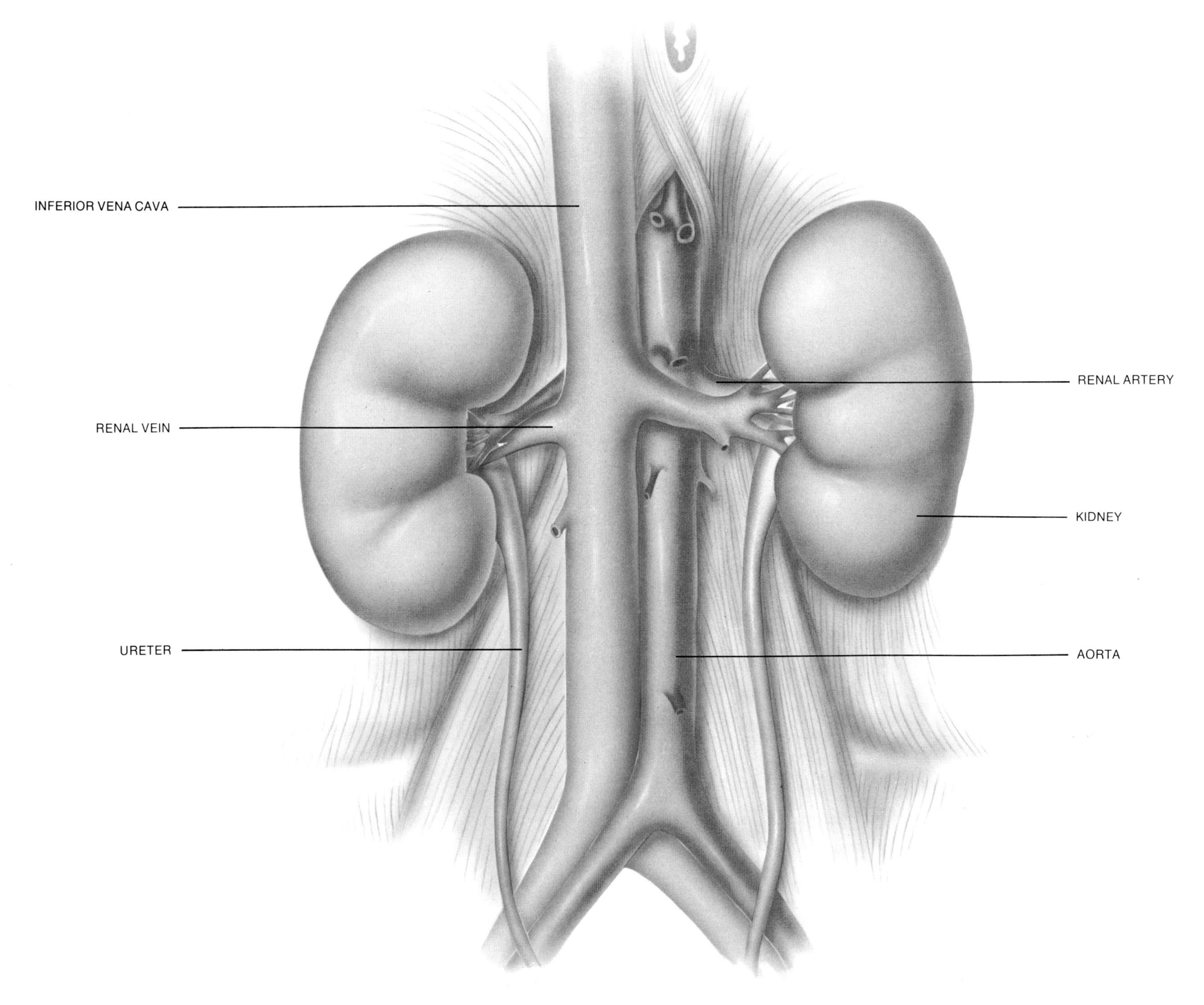

veins, which carry blood away from the kidneys, empty into the great vein of the abdomen, the inferior vena cava.

The kidneys also need nerves, principally to help the renal arteries constrict and relax under special conditions. The renal nerves come from that part of the autonomic nervous system called the sympathetic division. Several of these very thin nerves surround the arteries as they enter each kidney. If the blood pressure falls for any reason, these nerves probably cause the renal arteries to dilate, thereby preserving renal blood flow at a fairly constant level. Similarly, the nerves probably cause the renal arteries to constrict when blood pressure rises above normal levels.

Structure and Function of the Kidneys

The outer shell of each kidney is called the renal cortex. It is here that the filtering portion of the kidneys and much of their regulatory functions are located. The inner portion of the kidneys is the medulla, which regulates the degree of concentration of the urine. If a person has been drinking too much water for his needs, the medulla allows a large amount of water to be passed to the urine. Upon discharge such urine will be very dilute, appearing like water with very little coloration. On the other hand, if a person becomes dehydrated, the medulla, under the influence of certain hormones, removes a large portion of water from the urine and returns it to the body. The discharged urine will be concentrated, appearing quite dark and scanty in amount. By varying the degree to which it removes water from the urine, the renal medulla regulates the amount of water lost from the body to a very fine degree.

The function of the renal cortex is quite complicated. The renal arteries on it branch into ever smaller blood vessels. Each of the smallest of these arteries, called arterioles, enters a structure called a glomerulus, of which there are approximately one million in each kidney. Upon entering its glomerulus, each arteriole divides into several smaller blood vessels, only one cell thick, called capillaries. It is at the capillaries that the first important function of the kidney, filtration of the blood, occurs. As the blood flows through the glomerular capillaries, the blood pressure causes about 20 percent of all the water in the blood and some of its dissolved substances (salt, sugar, waste products, etc.) to be filtered through the walls of the capillaries into the space surrounding them. This process forms about 120 ml (milliliters), or 4 ounces, of filtered solution each minute, or about 180 liters, or more than 180 quarts, each day.

Cross Section of a Kidney

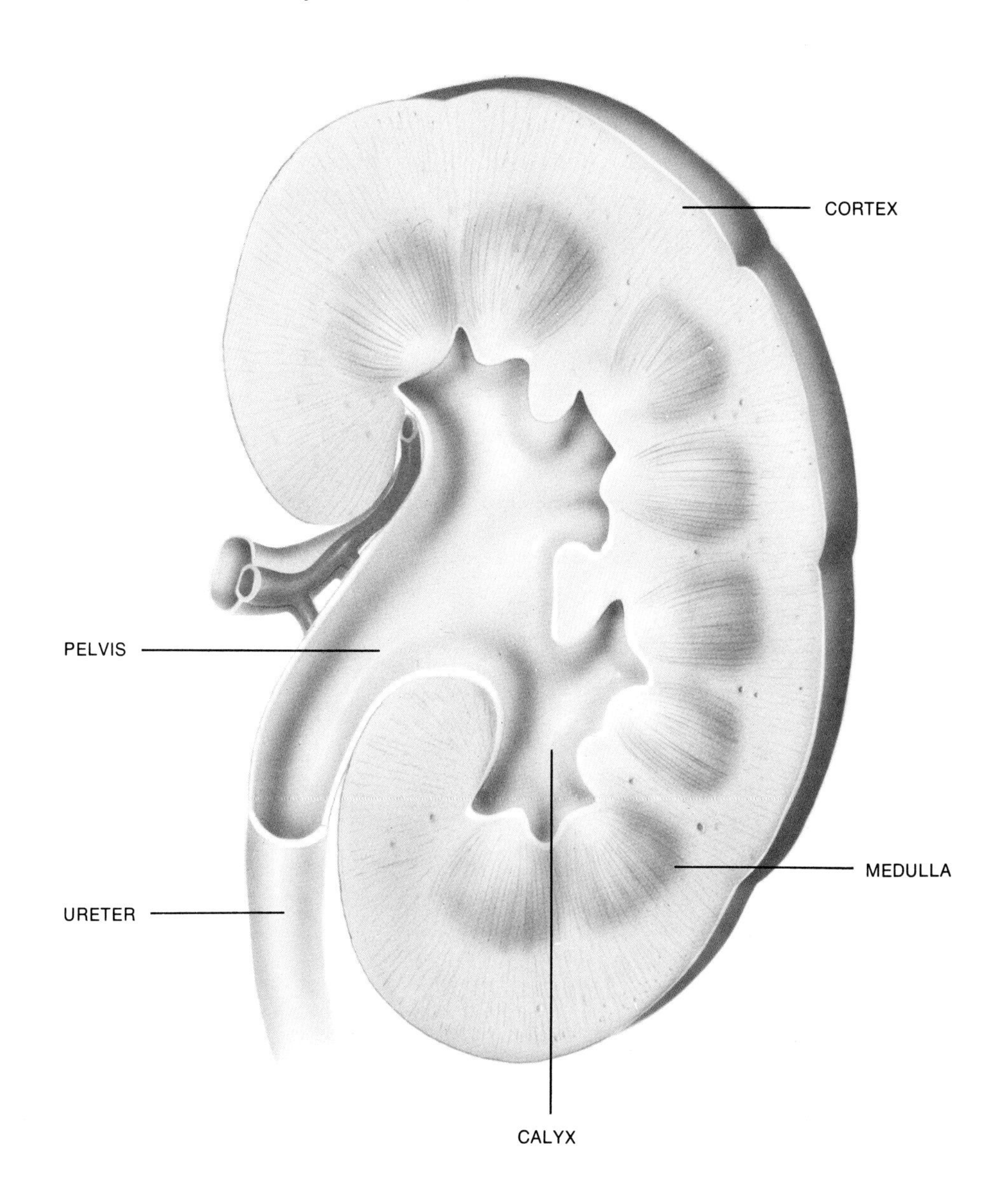

The kidneys filter and return to the body some 180 quarts of water and dissolved substances each day. Parts of the kidney are explained in the text.

This large volume of filtration is a mixed blessing. It allows the body to get rid of enormous amounts of waste products, but it also presents it with a large reclamation project. That is, if all this salt, sugar, water, and amino acids were really lost from the body each day, man would have little time for anything more than drinking more than 180 quarts of poor-tasting liquid daily. Obviously, such is not the case; the average person passes only about two quarts of urine daily. Therefore, the overwhelming portion of the water and dissolved substances that the kidneys filter from the blood must be returned to the body.

The reclamation of the vast majority of this fluid, which is called the glomerular filtrate, is accomplished by the next portion of the kidney, the

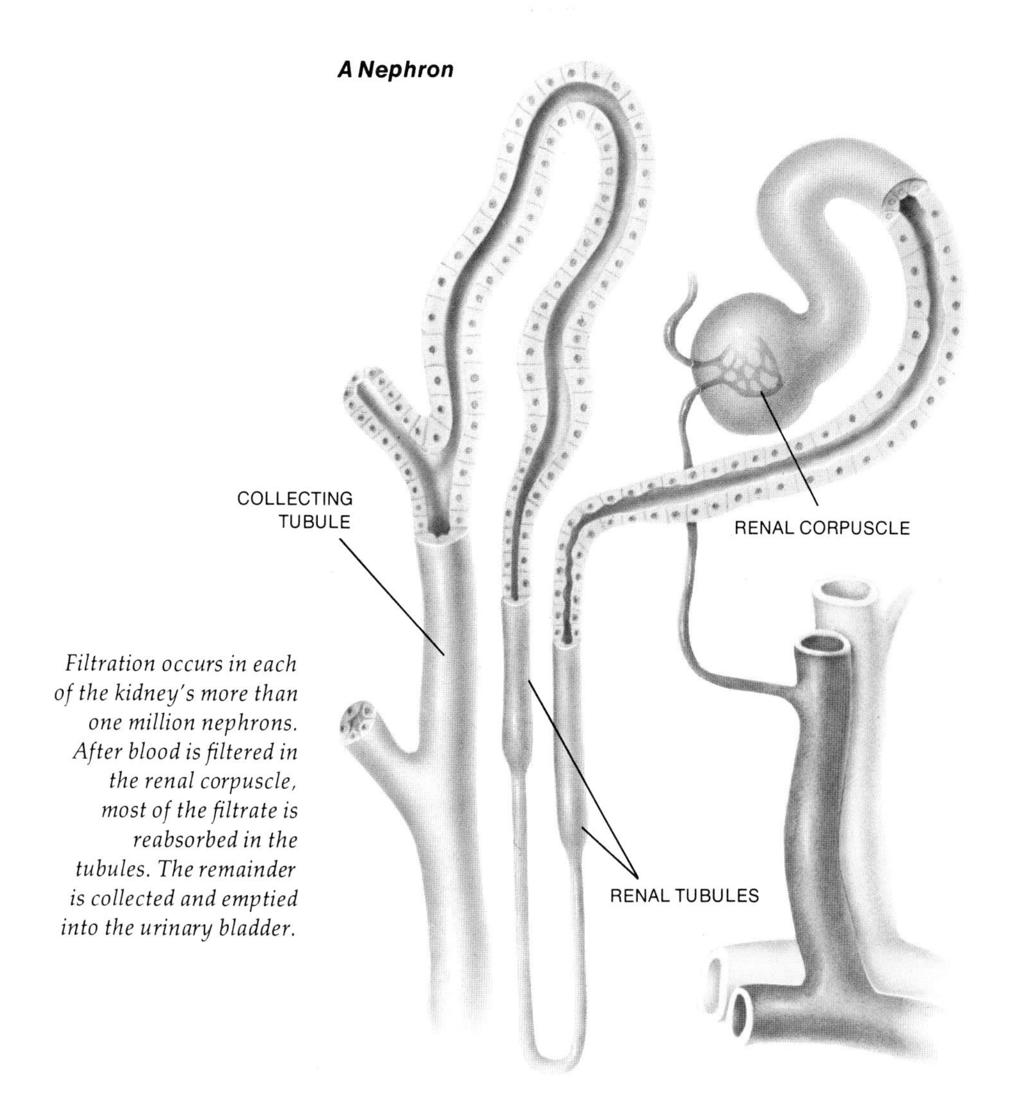

Filtration occurs in each of the kidney's more than one million nephrons. After blood is filtered in the renal corpuscle, most of the filtrate is reabsorbed in the tubules. The remainder is collected and emptied into the urinary bladder.

tubules. Each glomerulus and the tubule leading from it is called a nephron. The filtrate that was formed in the glomerulus enters the first portion of each tubule, where, through the energy derived from fat, glucose, and other sources, the substances the body needs to retain begin to be reabsorbed. By the time the filtrate has passed through all the tubules, 99 percent of the water and salt have been returned to the blood along with all of the sugar, unless the person is diabetic and has too much sugar in his blood and filtrate. Very few of the harmful waste products of metabolism are reabsorbed; these remain in the urine to be passed from the body.

The kidney also has a remarkable ability to regulate the amounts of most of the substances it returns to the blood. For example, if a person has lost a substantial amount of salt and water during severe vomiting, the kidney will be apprised of this fact by the nervous system, by the change of salt content of the blood, and by certain hormones. The tubules will respond by reabsorbing more than the usual amount of salt and water. The converse is also true: if a person eats too much salt, more than the usual amount will be allowed to remain in the urine.

Diseases of the Kidneys

The tasks performed by the kidneys are of major importance in preserving health. It would be expected, therefore, that disease processes impairing renal function would result in significant illness, or morbidity, and, ultimately, in death, or mortality. Such certainly is the case, but neither noticeable morbidity nor mortality occurs until more than 80 percent of normal renal function has been lost. The kidneys have great ability to continue to perform their major functions without any resultant illness even when only 20 percent of renal function remains. In general, there are two groups of disease states associated with kidney failure. Those diseases resulting in rapid loss of kidney function are said to cause acute renal failure, while those that lead to slow, progressive loss of kidney function over periods ranging from months to decades are said to cause chronic renal failure.

The most common cause of acute renal failure is damage to the kidney secondary to a shock or infection in another part of the body that changes the blood flow to the kidneys. The kidneys become damaged as a result of this altered blood flow and cease their filtration of the blood. The patient's urine flow usually falls to less than 450 ml (15 oz.) per day. If fluid intake is not curtailed, the patient retains large amounts of fluid and becomes short of breath from a back-up of fluid in the lungs. Fluid retention may also cause the legs to swell. Other symptoms of acute renal failure include poor appetite, weakness, sleepiness, and rapid breathing, caused by the retention of acid and poisonous waste products that ordinarily would have been excreted by the kidneys. Ultimately, unless treated, the patient with acute renal failure lapses into coma, may have convulsions, and may die from the effects on the heart of the retained poisons and potassium.

Thankfully, the use of modern treatment methods, such as the artificial kidney machine, enables the physician to sustain the patient's life until the kidneys recover from the acute injury. Such recovery usually begins about nine to ten days after the injury but may not begin for several weeks. If the non-renal problem that led to the acute renal failure is controlled, recovery is almost always sufficient for the patient to resume a perfectly normal life.

Other causes of acute renal failure may not have so fortunate a prognosis, or outlook, for the kidneys. In fact, many of them may be associated with destruction of the kidneys and offer no hope of recovery of renal function. These disease states, although acute in nature, are the causes, then, of prolonged, or chronic, renal failure. They can be treated only by diet, long-term use of the artificial kidney machine, or by transplant surgery.

Hypertension and Diabetes

A very common disease, high blood pressure or hypertension, heads the causes of chronic kidney damage. With hypertension the walls of the small branches of the arteries thicken, resulting in decreased blood flow and in insufficient supplies of oxygen and food. When hypertension causes these changes in the renal arteries, the nephrons are ultimately destroyed and kidney function is lost. The fact that hypertension, a disease afflicting 15–20 percent of the population, is the leading cause of renal failure is, in itself, significant. Of greater importance is the demonstrated fact that hypertension can be controlled, and that its complications including renal failure can be prevented without great difficulty in the majority of cases. While the patient with hypertension is taking the prescribed medications, blood pressure may be quite normal, but since most types of hypertension are lifelong, hypertension invariably returns and places the patient's health in jeopardy, if the medication is stopped. There are, however, a few types of hypertension whose causes are known and can be treated surgically. These include various tumors of the adrenal gland, partial obstruction of the renal artery, and congenital constriction of the aorta. But these causes probably make up no more than 5 percent of all hypertension cases. All other high blood pressure patients must be treated medically all their lives.

Diabetes is probably the second most common cause of chronic renal failure. Although the vast majority of diabetics do not develop kidney disease, when diabetes does cause the disease, it does so by obstructing the very smallest branches of the renal arteries (arterioles) and also by replacement of the glomeruli with peculiar deposits of fat and sugar. At present, there is very little evidence that control of the diabetes itself affects the course of diabetic kidney disease. It appears that once the disease process has started, it progresses to renal failure in anywhere from several months to several years.

A Renal Corpuscle

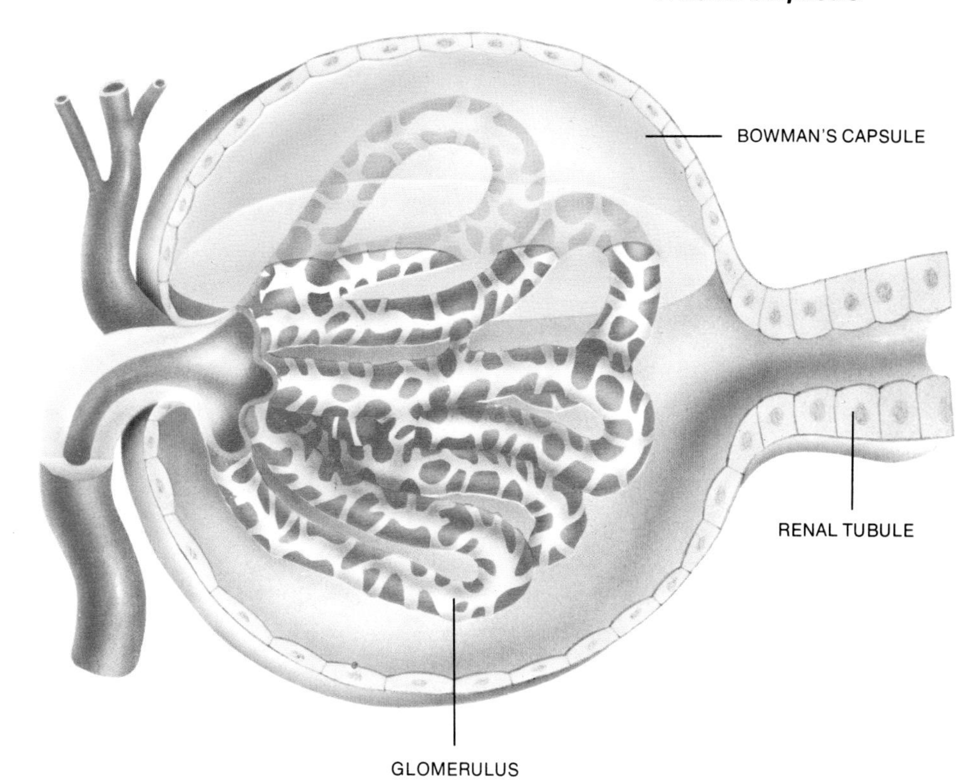

Water and dissolved substances in the blood pass from the capillaries of the glomerulus to Bowman's capsule and then into the renal tubule. Most of the filtered material is returned to the blood.

Glomerulonephritis

Several diseases of unknown causes can result in chronic renal failure. Most of these diseases affect the kidneys and no other organs and are called various types of glomerulonephritis, which means inflammation of the glomerular portion of the nephron. It is presently thought that these diseases result from the production of antibodies against some infectious agent—a virus, bacterium, or fungus. These antibodies combine with part of the infectious agent, or antigen, and the resulting combination, the antibody-antigen complex, is carried via the blood to the kidneys, which are damaged and ultimately destroyed by it. This theory has much experimental support but is by no means proved.

Some causes of glomerular disease can be cured or controlled with medication, especially with cortisone-like drugs. However, many others cannot be controlled at all and go on to chronic renal failure. The patient with glomerulonephritis usually has red blood cells in his urine. At times there are enough red blood cells for the urine to appear bloody or smoky. At other times, the number of red blood cells is so small that it can be detected only under microscopic examination. Patients with glomerular disease also often lose varying amounts of proteins, especially albumin, in their urine. Ordinarily, the glomerulus does not allow blood proteins to be filtered through into the urine. In some cases when it is diseased the amount of protein lost in the urine becomes so great that the body's ability to replace protein is outstripped, and the amount of protein in the circulating blood falls to very low levels. Such patients begin to show generalized body swelling, including the face, hands, and feet. This condition is called nephrosis, or the nephrotic syndrome. It can be caused by any disease that injures the glomeruli, including diabetes and all of the various causes of glomerulonephritis.

Kidney Infection

Occasionally a kidney itself may become infected. Under the age of fifty this occurs much more frequently in women than in men, most likely because women have a much higher incidence of urinary bladder infections. However, over the age of fifty, when men begin to have difficulties with overgrowth of the prostate gland, kidney infections become much more common in men. Acute episodes of kidney infection, called acute pyelonephritis, are usually associated with fever, pain in the involved kidney, and pain or burning while urinating. The urine may appear cloudy and usually contains white blood cells. Bacteria will be cultured from the urine.

Most cases of acute pyelonephritis will heal with proper antibacterial treatment. However, infection is difficult to eradicate and may become chronic, if there is a structural abnormality in the urinary tract. Therefore, when a patient has repeated episodes of urinary tract infection, the physician is likely to order X rays of the entire tract to identify any structural defects. Any significant abnormality is usually corrected surgically.

Chronic renal failure can result from any of several other causes. One, especially, deserves mention. Occasionally, patients develop multiple cysts, or small sacs, in their kidneys, with ultimate destruction of renal tissue. This polycystic renal disease is inherited and is most severe in men, usually causing renal failure by age fifty. Women seldom have the disease with sufficient severity to experience renal failure.

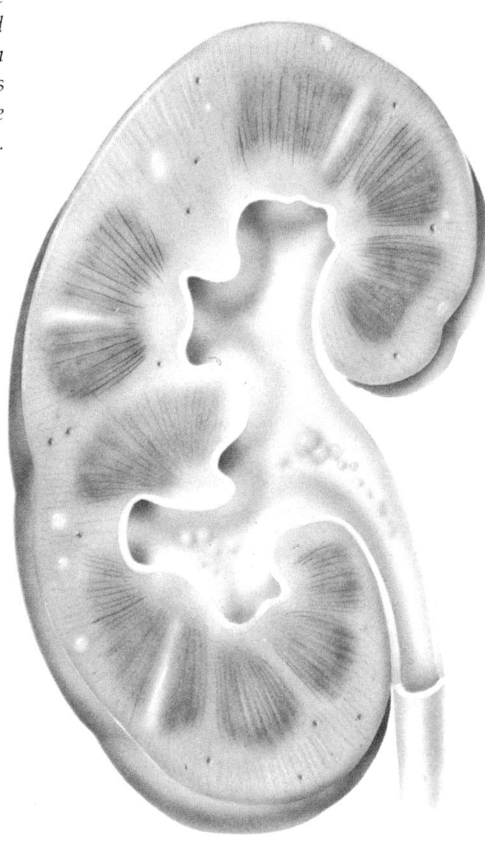

Pyelonephritis is a painful infection of the kidney. The pus formed is sometimes passed in the urine. Antibiotics are used to treat the infection.

Chronic Renal Failure

Except for those patients who develop the nephrotic syndrome and are made aware of their disease process by the retention of fluid and subsequent signs of swelling, or edema, most patients with chronic renal disease feel quite well even when as much as 80 percent of renal function has been lost. Once renal failure has progressed beyond 80 percent, however, most patients begin to experience unpleasant symptoms. Headaches, chest pain, and shortness of breath can be caused by retained salt and water and by hypertension, which is frequently the result of renal failure as well as its cause. Retained metabolic waste products, especially those of protein metabolism, lead to loss of appetite, nausea, and vomiting. Anemia, which most patients experience, produces weakness. Such anemia may be caused by a number of factors; chief among them are the toxic effects of the waste products and the lack of the bone-marrow-stimulating hormone called erythropoietin made by normal kidneys. Since the kidneys are intimately involved in preserving the normal balance of the important minerals calcium and phosphorus, renal failure is frequently associated with bone disease, and ultimately with deposits of calcium in the skin, which may cause severe itching. Finally, if renal function falls below 5 percent, the patient becomes more and more lethargic, lapses into coma, and dies.

Fortunately, this ominous sequence need not always come to pass. Very many of the signs and symptoms can be managed conservatively with proper diet and medications up to the point where only 5 percent of kidney function remains. The mainstays of this management are antihypertensive drugs and a diet that is low in protein and has rather restricted quantities of salt. This combination of therapies relieves the headaches, shortness of breath, chest pain, edema, nausea, vomiting, and lethargy. Adding of certain types of antacids prevents the bone disease and relieves the itching. Maintenance of adequate fluid intake is quite important, as is treatment of any urinary tract infection. Potent drugs called diuretics cause even severely damaged kidneys to lose salt and water and can be used to rid the patient of edema. Finally, alkali can be given to counteract the effects of retained acids.

Although these measures are quite useful in counteracting most of the complications of chronic renal failure, the kidneys usually reach a point where they can no longer perform the task of maintaining a reasonable state of well-being. Without further measures, the patient's health would deteriorate rapidly. Over the past few decades, dialysis therapy has become so sophisticated and so widely available that no patient in the United States need die of renal failure, unless there are other complications.

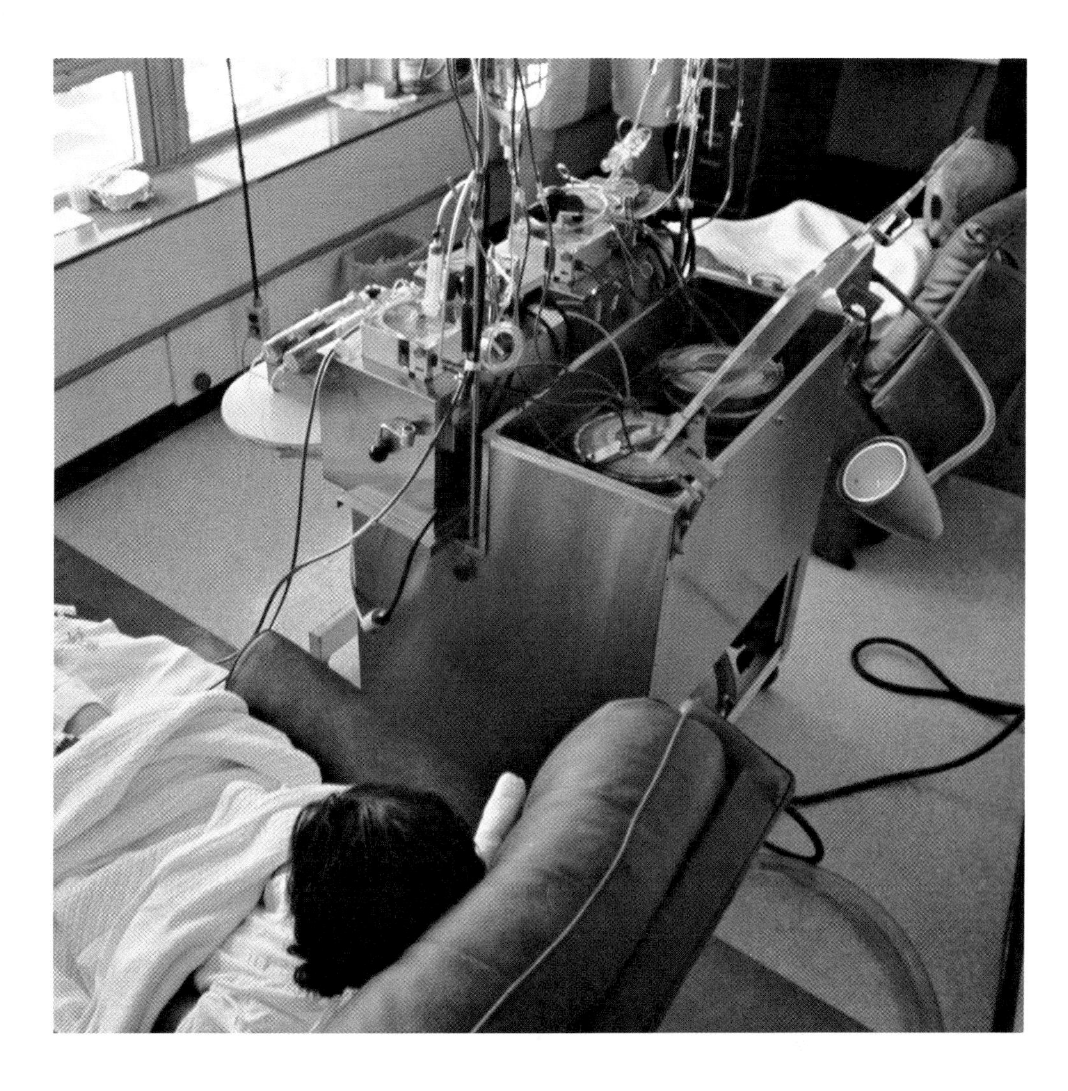

In cases where a person's kidneys do not work, an artificial kidney machine can cleanse blood of toxic materials that build up in the body.

Dialysis

Dialysis allows the toxic materials and excess water and salt to "cross over" from the patient to a machine or a bottle. Currently, two types of dialysis are in common use—peritoneal dialysis and hemodialysis. During peritoneal dialysis sterile solutions are introduced into a patient's abdominal, or peritoneal, cavity through a thin tube called a catheter. The fluid is changed about once an hour, and, with each fluid exchange, more toxic wastes and salt and water can be removed. Peritoneal dialysis has several drawbacks: each dialysis requires forty-eight hours in bed, and the procedure must be repeated weekly; infection of the abdominal cavity is a definite hazard; the procedure is performed with difficulty in patients who have had recent or extensive past abdominal surgery; many plasma proteins and foodstuffs are lost, and many patients have poor nutrition while undergoing peritoneal dialysis; and finally, but by no means of least importance, there is a certain degree of abdominal discomfort associated with the procedure.

Hemodialysis is the term used for cleansing the patient's blood with the artificial kidney. Currently, an artery and a vein in an arm or leg are surgically connected, resulting in dilation of the vein. This connection, called an arteriovenous fistula, can then be used over and over again for months or years. Two needles are placed in the large dilated vein, an arterial needle to withdraw the blood and a venous needle to return it to the patient. The arterial needle is attached to tubing that enters a long, coiled cellophane structure that is immersed in a constantly flowing solution. Toxic substances pass from the blood across the cellophane into the solution and are effectively removed. The blood is then returned to the patient through the venous needle. For best results a patient is placed on the machine about three times a week for about five or six hours each time. The procedure can be performed in a hospital, in a separate facility, or at home by a relative who has been taught how to use the hemodialysis equipment.

The dialyzing procedure, although lifesaving, is not a cure-all. Patients on dialysis remain anemic and do not quite feel healthy. Additionally, they are inconvenienced by having to undergo treatments as often as they must. Therefore, a good many dialysis patients desire another form of treatment—transplantation. Renal transplantation, then, is not a lifesaving procedure but an elective one designed to improve the quality of life for the recipient.

Kidney Transplantation

Kidneys for transplantation can be obtained from living relatives who wish to donate one of their kidneys or from an unrelated person about to die of an incurable disease that does not affect the kidneys. If successful, it can restore the kidney patient to complete health. A patient receiving a transplant, however, must be treated with very potent medications to prevent the body from destroying the kidney as if it were a foreign object. These same drugs also lessen a patient's ability to combat infections, and it is not uncommon for a patient to suffer from repeated infections. As many as 20 percent of the kidney transplant recipients may die from infection alone; as many as 30 percent of the new kidneys may be lost through rejection within a year after surgery. Therefore, the patient has about a fifty-fifty chance of success and runs a significant risk of death. Some patients consider these odds too great and elect to stay on dialysis. Others regard transplantation as their only chance for meaningful life and opt for it. The choice should be the patient's, unless the physician feels that transplantation is too great a risk for the particular individual.

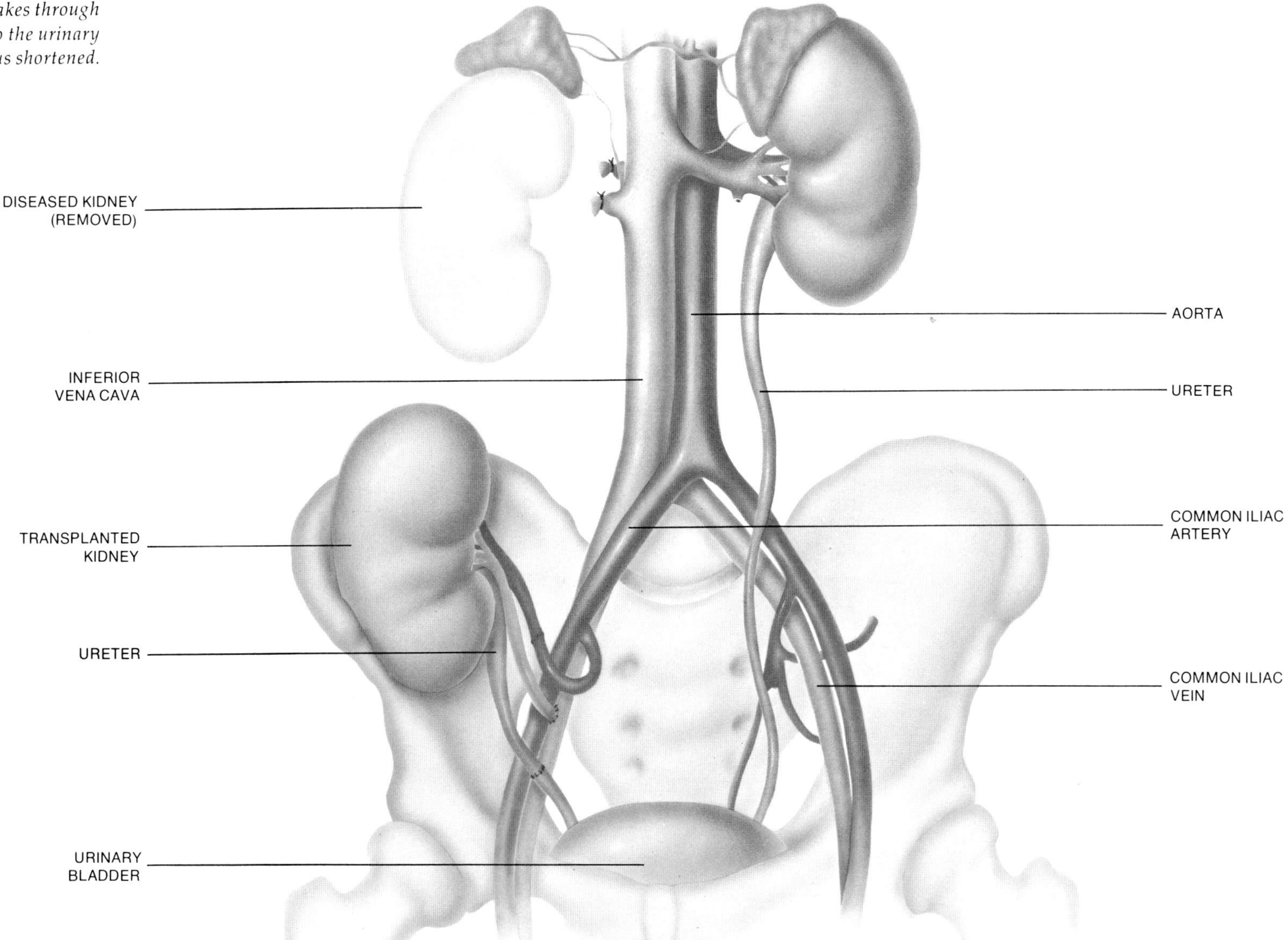

During kidney transplantation, the surgeon attaches the new kidney to the common iliac artery and vein instead of to the aorta and the inferior vena cava. The route that urine takes through the ureter to the urinary bladder is thus shortened.

Once urine is formed within the kidney, it drains into several cuplike structures called calyces. The calyces, in turn, merge into a funnel-like receptacle for the urine called the pelvis of the ureter, which is located near the mid-portion of the kidney, close to the point where the renal artery enters and the renal vein leaves. The pelvis empties into the upper portion of the ureter, which is a long tubular structure leading from the kidney high in the rear of the abdomen down to the urinary bladder low in the front portion of the abdomen. Each ureter is about 25–30 cm (10–12 in.) long and some 7 mm ($^1/_4$ in.) wide. The ureters act as conduits for the urine, which must get from the kidneys to the bladder. They are not just rigid pipes, however. They have muscular walls that contract and relax in waves, propelling the urine toward the bladder but never in the opposite direction, under normal circumstances.

The urinary bladder, in which the ureters end, is in the lowest part of the abdomen just behind the pubic bones. In males the bladder lies directly in front of the rectum, but in females the uterus and vagina separate the bladder from the rectum. In either sex the ureters enter the bladder at the rear of the bladder floor, and the urethra, the tube carrying urine to the outside, leaves the bladder at the front portion of the bladder floor. In males the urethra first passes through the prostate gland, which contributes a portion of the seminal fluid ejaculated during sexual intercourse, and then between a set of sphincter muscles that, when tightened, prevent urine from leaking out of the bladder. After passing through these muscles, the urethra enters the penis, traverses its length, and emerges at the tip, or glans, of the penis. The female urethra is not nearly so long, passing only about $1^1/_2$ inches between the bladder and the urethral opening, which lies just in front of the vaginal opening.

The urinary bladder serves as a reservoir for the urine, which is continuously being made by the kidneys, until it can be conveniently passed. As urine accumulates the bladder wall is stretched, causing a sensation of fullness that increases as the pressure within the bladder increases. When convenient, the person relaxes the urethral sphincter and tenses the abdominal muscles. At the same time the bladder muscles contract, and the urine passes from the bladder and out of the urethra.

Urinary Disorders

The major disorder affecting the ureters is really not a disease but an accident of urinary function. Kidney stones, formed within the kidney and moving toward the bladder, can become lodged in the ureter just as a large rock can lodge in a drainpipe. Most kidney stones form within the portion of the kidney tubules that drains into the calyces of the ureteral pelvis. If a very tiny stone enters the ureter, it usually passes out in the urine. However, if a small stone is caught anywhere in a calyx or in the pelvis, it can attract minerals from the urine and grow. Once a stone that has grown large enough to obstruct the ureter becomes dislodged from the pelvis, there is a strong possibility that the patient will suffer the distinctive, painful symptoms of renal colic. A cramping sensation may be felt in the abdomen on the side of the stone or, occasionally, in the flank. The pain may radiate to the groin and testicle, or the labia in women, or to the penis or urethra, depending on where the stone has lodged in the ureter. The pain usually requires narcotic relief. Injury caused by the stone to the lining of the ureter may make the urine bloody.

The Ureters, the Urinary Bladder, and the Urethra

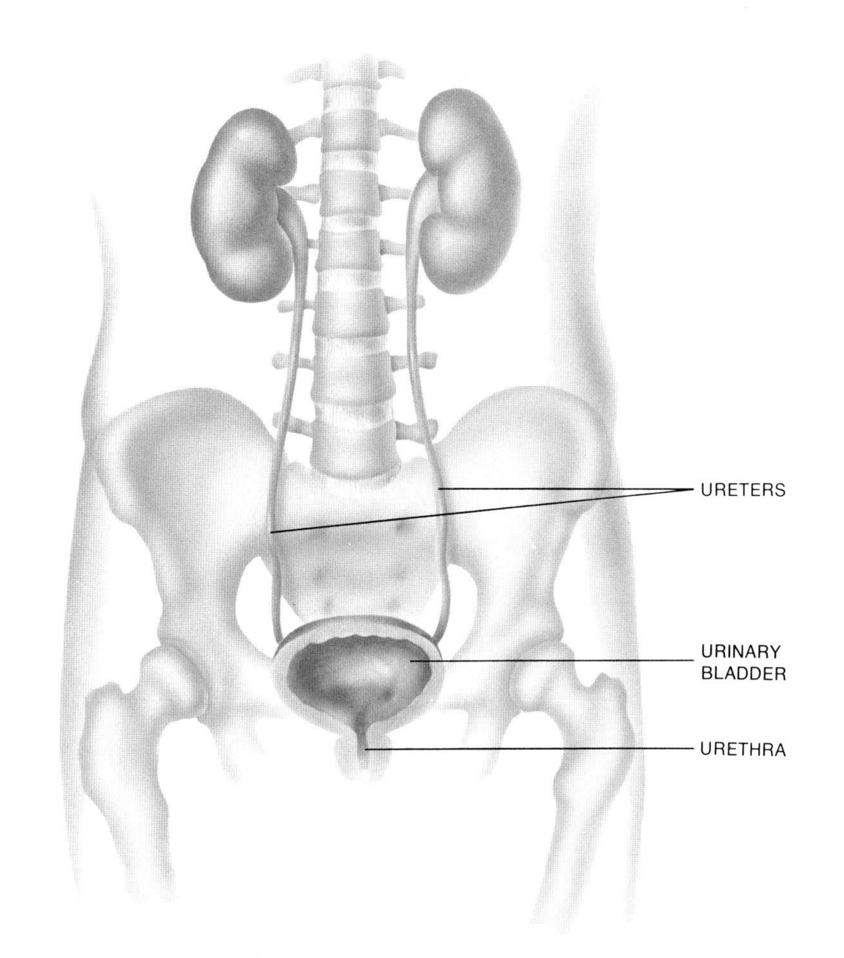

Stones, or calculi, can form in the kidney and lodge in the ureter, where they obstruct urine flow.

Many ureteral stones, fortunately, become dislodged and enter the bladder, to be passed out in the urine. The urine of patients suspected of having ureteral stones is usually strained so that the stone can be found and analyzed. When the chemical composition of the stone is known, appropriate therapy to prevent future stones can be started. If a stone remains lodged in the ureter, the obstruction to urine flow can ultimately lead to loss of kidney function. This, or any other long-term obstruction of the ureter, must ultimately be relieved by the urologic surgeon.

The most common disease of the bladder is bacterial infection, or cystitis, which can cause pain in the bladder or urethral region, severe pain on urination, and cloudy or bloody urine. Cystitis is much more frequent in women because bacteria can enter the female bladder more easily, by swimming up the much shorter urethra. In addition, in young girls the urethra is much more easily contaminated with feces because of its proximity to the anus. Thus, careful hygiene in young girls is mandatory. Women having sexual intercourse are prone to cystitis for many reasons. Frequently, bacteria that are in the urethra can be "milked" back into the bladder by the pressure of the male's penis in the vagina. In addition, manual manipulation of the woman's genitalia by her sexual partner may spread bacteria to the urethra from her vagina or anus. Although quite uncomfortable, cystitis is rarely a very serious threat to the patient's health, unless the infection spreads to the kidneys. It is treated with any one of several agents, and careful attention to personal hygiene and sexual practices may prevent or reduce recurrences.

Like the bladder, a woman's urethra can become infected and cause painful urination for a number of reasons. In men, however, urethritis, or inflammation of the urethra, is usually caused by infection obtained during sexual intercourse. Gonorrhea is the best known of these infections. The symptoms and signs of gonorrhea include painful urination and a whitish discharge, sometimes containing blood. The gonococcus, the organism causing gonorrhea, can be detected in a microscopic examination and culture of the discharge. Urethritis should be treated with an appropriate antibiotic before it scars and partially obstructs the urethra or spreads elsewhere. The other well-known venereal disease, syphilis, does not usually cause difficulty in urinating, but it usually manifests itself first as a painless ulcer on the penis or the labia. It, too, can be diagnosed by microscopic examination, using smears of the fluid from the base of the ulcer, and by any one of several blood tests.

The prostate is also a common site of disease. Infection of the prostate can occur at any age after puberty and is frequently recurrent. It may cause fever, painful and difficult urination, and low back pain. It is rarely serious but can be quite bothersome and difficult to eradicate completely, even with antibiotic therapy.

About 30 percent of all men over fifty are afflicted with a benign overgrowth of the prostate. If this growth is directed toward the rectum, no symptom or disability is likely to be encountered. But if the growth is directed toward the urethra, partial or complete obstruction of urinary flow may occur. Symptoms are decreased force of the urinary stream, difficulty in initiating urination, urinary dribbling, frequent urination in small amounts, and, ultimately, inability to urinate at all. Because the bladder is contracting against an obstruction, its muscles grow and become thickened. Eventually, kidney function can be severely impaired. Treatment is the surgical removal of the obstructing portion of the prostate.

Cancer of the prostate is a very common malignancy that can manifest itself with the same symptoms as benign prostatic overgrowth, or by the pain it causes when spreading to the bones of the spine and pelvis. It is treated by a combination of surgery, radiation therapy, and female hormones.

The testicles—male reproductive structures—are formed during fetal development high up in the abdomen near the kidneys. Prior to birth they begin to descend toward the scrotum through the abdominal wall, pushing part of the peritoneum—the lining of the abdominal cavity—in front of them. At about the time of birth, they descend into the scrotum. The peritoneal outpouching that they have formed usually closes off behind them. If it does not close, any of the abdominal contents, especially loops of small intestine, may enter the scrotum, forming a hernia. Occasionally, a testicle may not descend into the scrotum and must be brought down surgically during childhood.

Each testicle consists of many small tubules lined by cells that form sperm cells. These tubules all join at the top of each testis in a single coiled structure called the epididymis. Each epididymis joins a long tube, the vas deferens, which travels upward from the scrotum. It is this tube that is cut in the vasectomy operation performed for male sterilization. In the lower part of the abdomen, in front of the rectum, the vas deferens links up to a small glandlike structure called the seminal vesicle, which supplies part of the seminal fluid. Each vas enters the urethra at the level of the prostate, and the urethra becomes the vehicle both for elimination of urine and for ejection of seminal fluid.

The testicles and the epididymis may become infected at any age, or may be the sites of benign or malignant tumors. Dilation of any of the ducts may occur, usually with no consequence. Any mass or swelling in a testicle, however, should be shown to a physician promptly for malignancy to be ruled out.

Although the sperm cells are formed in the testicle, they must find their way to the egg cell of a sexual partner for procreation to occur. The erect penis is the only biological vehicle of such a union. Erection is usually brought about by a combination

The Testicles and Male Sexual Function

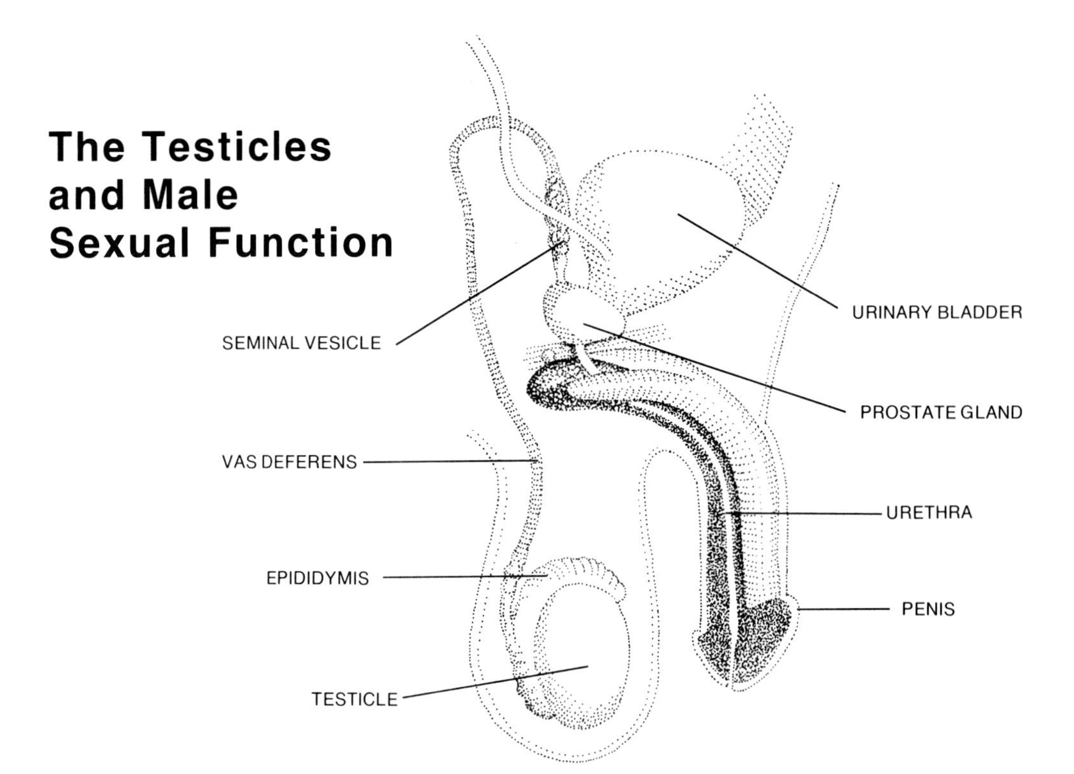

of psychic and nervous impulses that cause the muscles of the penis to become engorged with blood. Such diverse stimuli as a full bladder, accidential penile contact with clothing, or lovemaking can result in erection. Obviously, erection can occur at inappropriate times and be socially unwelcome. Repeated sexual stimulation of the erect penis results in intense nervous system reaction, causing contractions of the pelvic muscles and the sudden release of the sperm-containing semen through the urethra.

Inability to obtain an erection, or impotence, is of obvious social and psychological significance. It may occur as a side effect of certain drugs, as a result of spinal cord injury, as a secondary effect of diseases such as diabetes that affect the pelvic nerves, or following pelvic surgery. Most impotence, however, has a psychological cause. Nonetheless, the physician must explore the possibility of an organic cause and, if none is found, recommend sound psychiatric treatment.

Sperm cells form in the testicles. Together with secretions of the seminal vesicles, prostate gland, and other parts, they make up semen. Sexual stimulation causes semen to pass through the urethra and out of the penis. In males the urethra serves a dual role—a channel for urine and for semen.

For further information on the subjects covered in this chapter, consult ***Encyclopædia Britannica:***

Articles in the Macropædia	*Topics in the Outline of Knowledge*
EXCRETION, HUMAN	422.H.5
EXCRETORY SYSTEM, HUMAN	422.H.1,2,3, and 4
EXCRETORY SYSTEM DISEASES	422.K.1,2, and 3

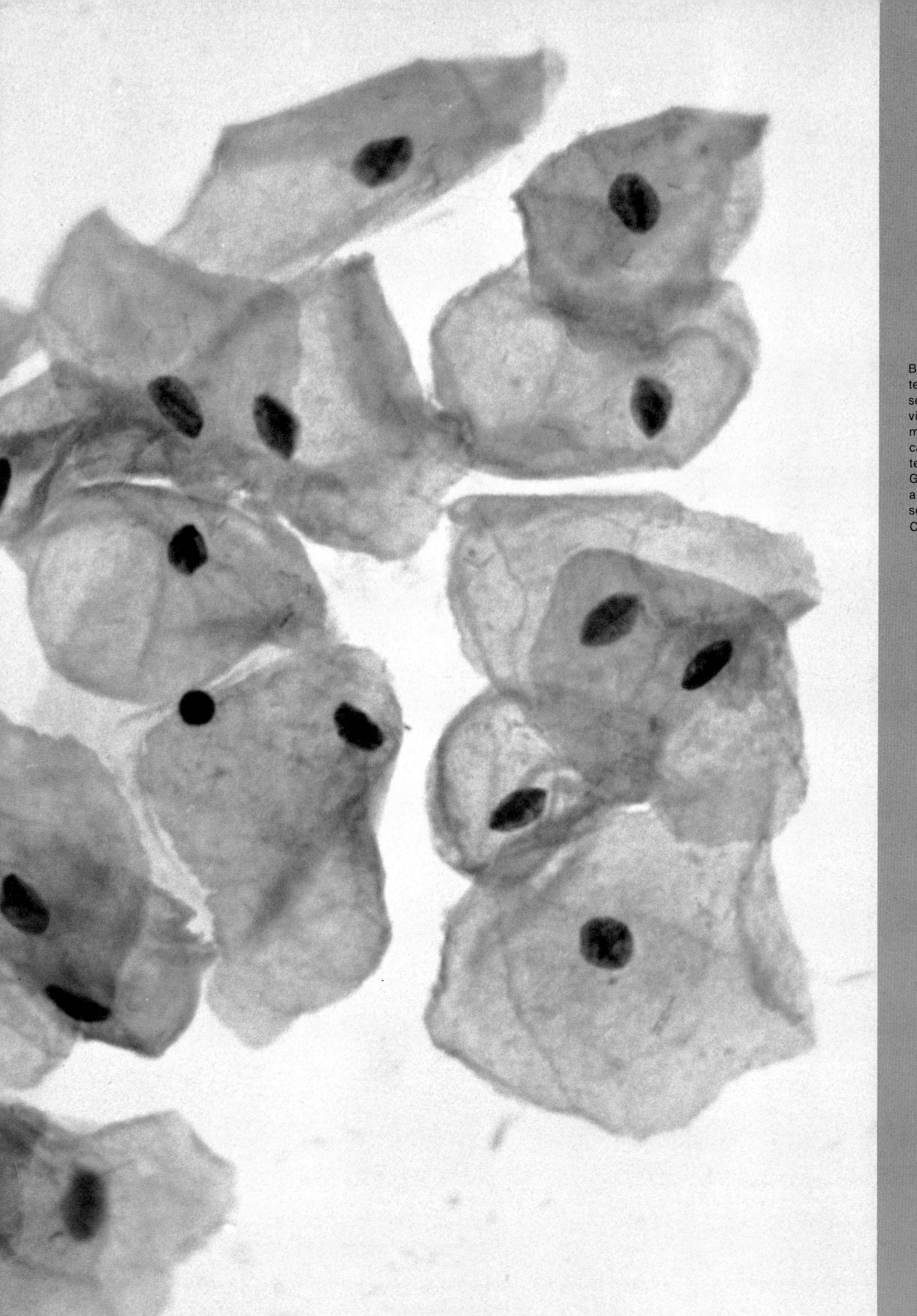

By means of a "pap" test, cells from the female sex tract are stained and viewed under a microscope for signs of cancer. The staining technique was devised by George N. Papanicolaou, a Greek-American scientist. Photo, Catherine M. Keebler.

Van Doren. *I guess this is my favorite chapter.*

Kessler. *Mine, too. Dr. Newton has done a superb job of giving the facts, on the one hand, and keeping alive the mystery, on the other.*

Van Doren. *Exactly. And I want to tell you something very personal. Reading the chapter, I find I envy women the experience of bearing children.*

Kessler. *It's not unusual for men who are honest with themselves to feel that way. The most creative thing that human beings can do, women do, and men can't. We men are forced to take second best, and spend our lives in quasi-creative acts—conducting research, writing poems, painting paintings.*

Van Doren. *That's a profound, and uncommon, thought.*

10

Sex and Pregnancy

BY MICHAEL NEWTON, M.D.

Sex and pregnancy or, more generally speaking, reproduction, are important parts of everyone's life. Understanding the changes that occur in a normal man or woman makes it possible to cope with these changes when they occur and to help oneself or others when something goes wrong with the usual process.

Much of the discussion in this chapter will concern the woman because she nourishes the baby in her uterus and plays a major part in the child's development. But it is important for a man to understand all the events because he will be sharing them and helping with them, whether as son, brother, husband, or father. Sex and pregnancy are family matters.

A new human being is conceived when the father's sperm fuses with the mother's egg. At that moment, the baby's sex is determined. The new being is not totally "new" because it carries the genes, the inherited characteristics, of its parents. The male sperm carries one genes-bearing sex chromosome, either an X or a Y, while the female egg has only an X. If a male X joins with a female X, a female baby is born. If a male Y joins with a female X, a male baby results.

Sex Development Before Puberty

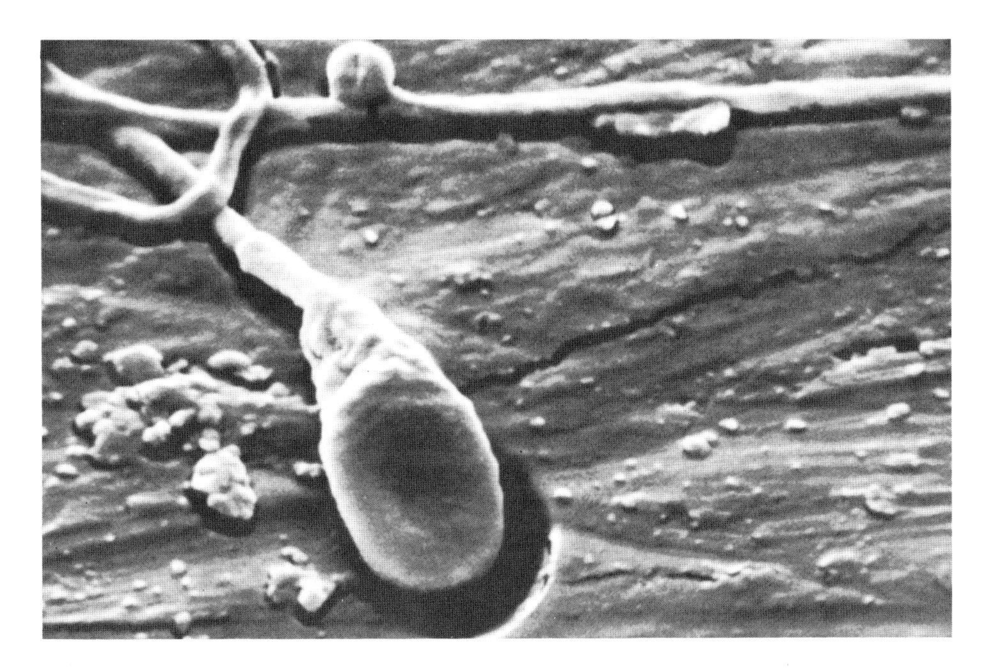

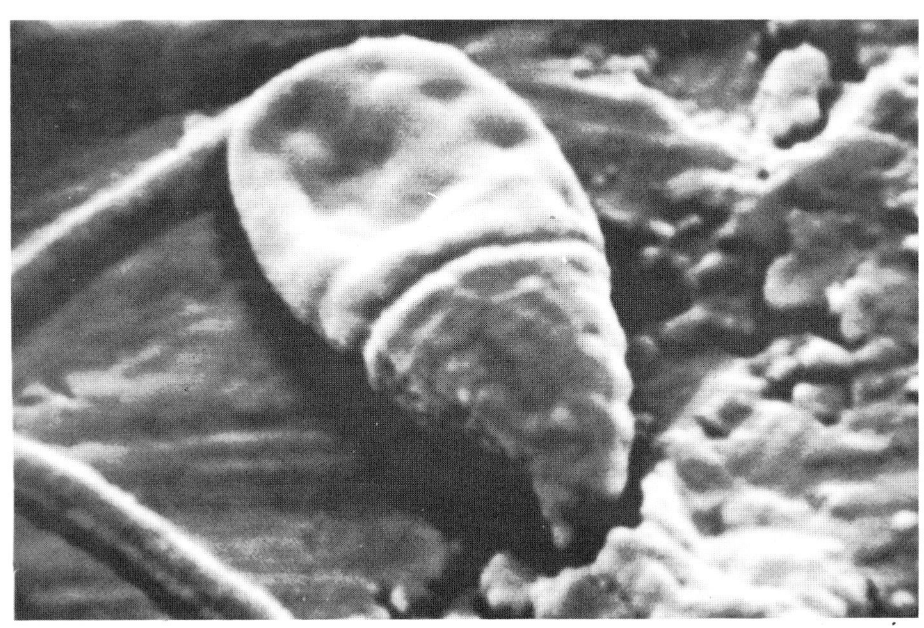

Human sperm cells with their prominent heads are shown as each appears under a scanning electron microscope. The top photomicrograph is enlarged 5,500 times, the bottom one, 11,000 times.

From this first junction, a complex chain of events develops. The new XX individual develops the female primary sex organs, the ovaries, while the XY individual develops the male primary sex organs, the testes. These organs become recognizable under the microscope at a very early time, about the eighth week of pregnancy. The primary sex organs influence the growth of the external genitalia, the penis and scrotum in the male, and the vulva and vagina in the female. The penis and scrotum make it possible to distinguish the sexes by external appearance at about the twelfth week.

Before birth, eggs gradually grow in the ovaries of the developing female fetus so that about forty thousand of them are present in the newborn female child. The eggs are well formed and are virtually in their last stage of development. After the female child is born, her ovaries produce no new eggs, and only a few of her large number eventually undergo final development and take part in the complicated process of menstruation, fertilization, and pregnancy. In the male fetus sperm do not develop in the testes until after the child is born. The greatest growth of sperm occurs during childhood and puberty.

As a result of the changes that occur during fetal life the sex organs, though small, are fully developed in the normal baby at birth in the sense that a child's hand is "fully developed." Maturity does not alter the basic structure of the genitalia. In women the external sex organs consist of the lips, or labia (external, or majora, and internal, or minora), the opening into the vagina—the introitus—partly covered by the hymen, the urethra from the bladder, the perineum, the mons—the soft, fleshy mound covering the front of the pubic bones—and the breasts. The internal sex organs in women include the vagina, the uterus, which is divided into the cervix, or neck, and the body, the fallopian tubes, the ovaries, and the surrounding supporting structures. In the male the external sex organs consist of the penis and scrotum, which contains the testes and the epididymides—elongated cordlike structures along the posterior border of the testes composed of ducts in which sperm are stored. The internal sex organs include the vasa deferentia, the seminal vesicles, and the prostate.

Few other changes related to sexual growth occur that enable one to distinguish a male from a female shortly before or at birth. Anyone looking at a clothed newborn baby has trouble telling its sex. Nor are differences striking as the baby grows into a young child, except that girls and boys are dressed differently and act, or are often expected to act, differently.

The development of the child into an adolescent and adult is the result of a complex set of physiological changes governed by the endocrine glands. The most important of these glands is the pituitary, but the thyroid, adrenal, and others also play important roles. However, sexual changes are only a part of the whole process of maturation.

Female Changes

The basic changes in hormones related to sex in girls involve the increased production of all the pituitary hormones, but particularly those concerned with the ovaries. The pituitary's follicle-stimulating hormone (FSH) causes the follicles—the eggs and their surrounding cells—to complete the last stages of their development. As these follicles grow they produce hormones known as estrogens. There are a number of different estrogens, but as a group they can be described as the "feminine" hormones. Under their influence a girl's external sex organs develop. First, her breasts grow, after which her external genitalia, the labia and mons, become more prominent, and then she develops axillary, or underarm, and pubic hair. Associated with these changes are modifications in height and body contour, including increased prominence of the hips, and deposition of fat beneath the skin.

The event that distinguishes puberty for girls is the onset of menstruation, even though this change is only one of several changes. As the growing follicle in the ovary produces more estrogen, the uterus increases in size, and its lining becomes thicker. The regular breakdown and rebuilding of this lining is termed the menstrual cycle.

Menstruation and the Menstrual Period

Under the influence of FSH one ovarian follicle begins to develop to the exclusion of the others. Its surrounding cells produce more estrogen, causing the lining of the uterus to develop further. At some point, another pituitary hormone, the luteinizing hormone—LH, causes the follicle to break and discharge its egg into the peritoneal cavity. The egg is then picked up by the fallopian tube and carried down to the uterus, where it disintegrates unless it has been fertilized by a sperm. After the discharge of the egg, the cells originally surrounding it form a "yellow body," or corpus luteum. This continues to produce not only estrogen but also progesterone, which can be termed the "nesting" hormone. These hormones cause further growth in the lining of the uterus, making it more receptive to a pregnancy if that should occur. If it does not, the production of estrogen and progesterone drops

Puberty

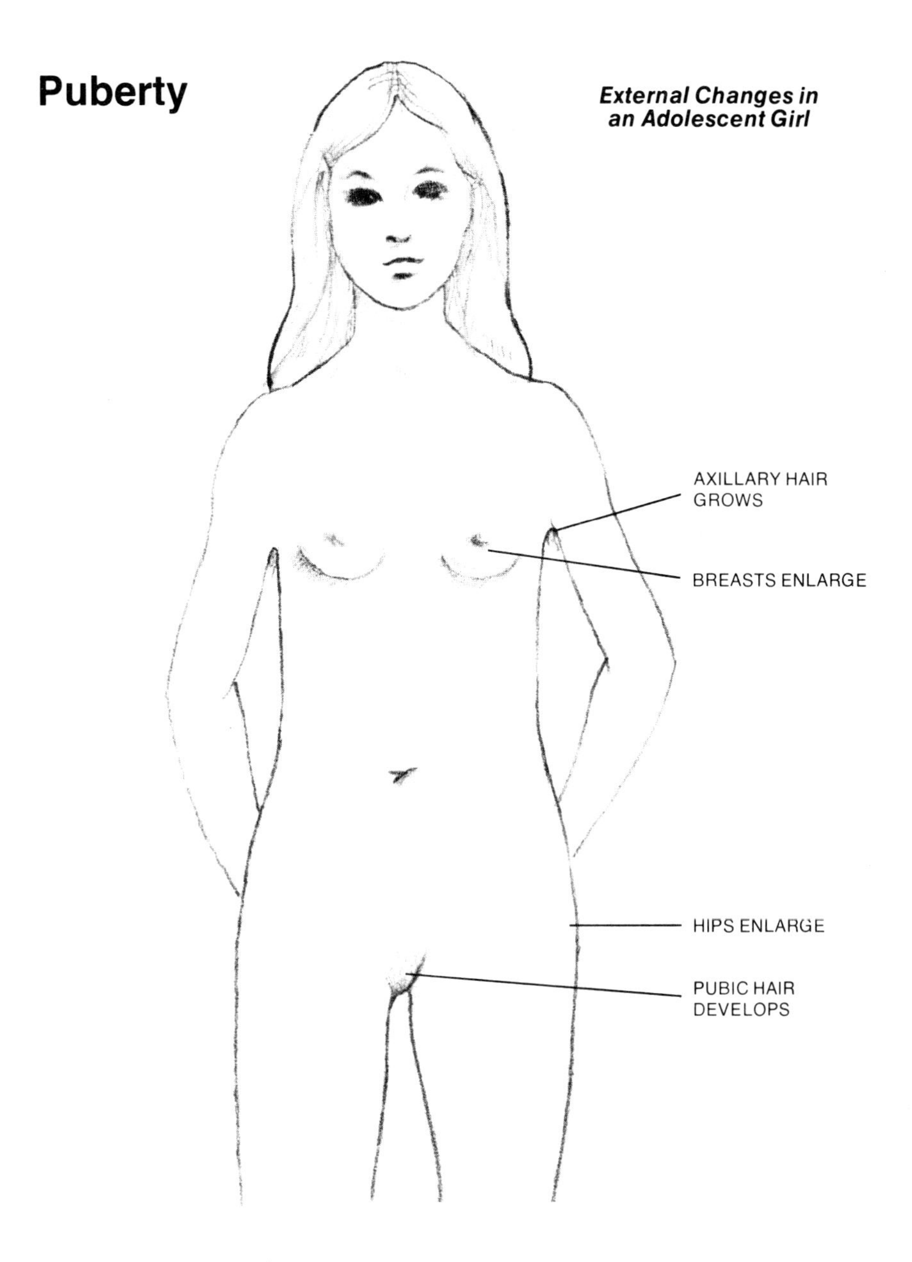

External Changes in an Adolescent Girl

sharply. This decline makes the blood vessels going to the uterine lining constrict. The lining degenerates and is shed as menstrual flow. As the flow proceeds, the FSH increases again, causing the growth of another follicle and the beginning of a new cycle.

Menstruation is a normal process, but it varies somewhat in different women and at different times in their reproductive lives. For example, the age of first menstruation, or menarche, may vary from the eleventh to the sixteenth year. Menstruation now occurs earlier than it did fifty years ago, and this may be related to better nutrition. Also, there may be a hereditary factor: if a mother began to menstruate at an early age, her daughter may also begin early. Menstrual periods—calculated from the first day of one period to the first day of the next—usually occur twenty-six to twenty-eight days apart, but twenty-one to thirty-five day intervals may be entirely normal. When a girl first begins to menstru-

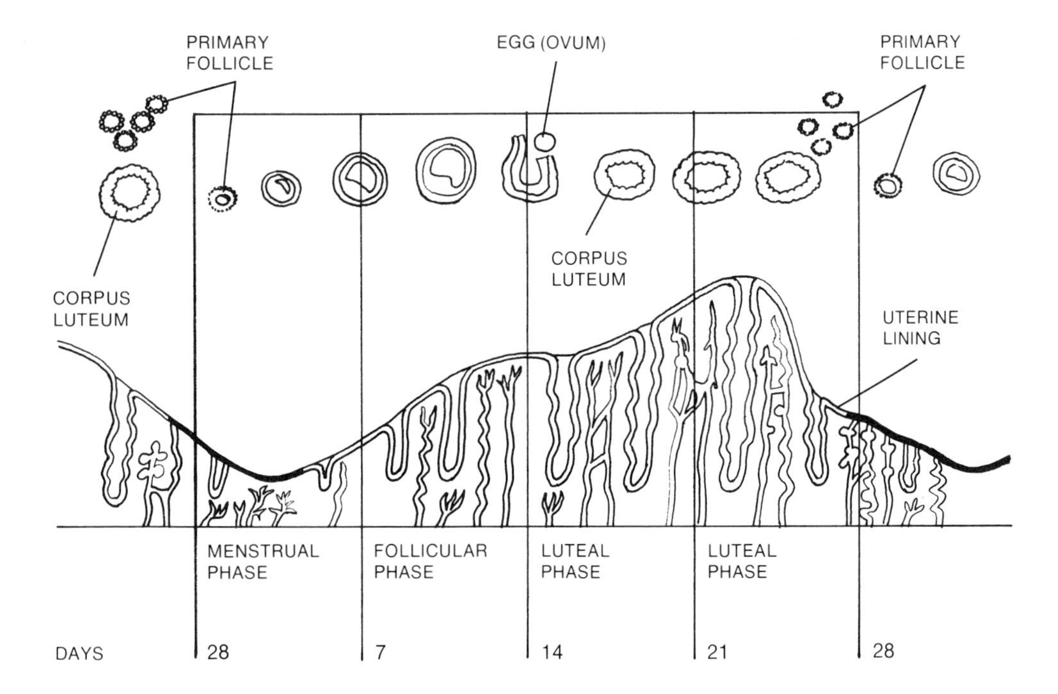

A primary follicle grows in an ovary while the uterine lining bleeds away (on about the 28th day of the menstrual cycle). As the follicle grows it produces estrogen, which influences buildup of the uterine lining (during the follicular phase). On about the 14th day, the egg is released from the follicle, which then becomes the corpus luteum, maker of a hormone during the luteal phase that maintains buildup of the lining. If fertilization does not occur, menstruation begins again by about the 28th day.

ate, her periods are likely to be irregular. Similarly, the duration of the flow may vary from two to seven days and still be perfectly normal. The amount of blood and tissue lost varies from one to two ounces in the average woman but may be less or more than this amount and still be perfectly normal.

Most women notice certain symptoms and/or mood changes in relation to their menstrual cycles. Occasionally in the middle of the cycle—usually about fourteen days before a period—lower abdominal pain may result from ovulation. A little vaginal bleeding may also occur at this time, probably due to the normal fall in estrogen level. Menstruation is preceded in some women by a feeling of bloating and tension known as the premenstrual syndrome and possibly related to the retention of fluids. Menstruation itself is painless in many women, but in some it may be accompanied by minor discomfort, by cramps, or by steady lower abdominal pain, often for the first day or two. Occasionally discomfort is severe and may be accompanied by generalized symptoms such as nausea and vomiting. Immediately after menstruation, some women notice an increase in energy.

Male Changes

Puberty in boys occurs one to three years later than in girls. Again, it is under the direction of the endocrine glands, led by the pituitary. Under these influences, the boy sheds his preadolescent fat and has a sharp growth spurt, both in height and in muscular strength. The sexual glands, the testes, increase in size, and the sex cells begin to go through their last changes leading to the development of spermatozoa, or sperm. At the same time, the penis and scrotum increase in size and other changes occur, stimulated by the male hormones—testosterone and others—that are produced by certain cells of the testes. These changes include growth of hair under the arms and in the pubic area, and later a deepening of the voice. The pattern of hair growth varies; some boys develop a heavy growth of hair on the chest, arms, and abdomen at a relatively early age, while others may develop this hair much later in life or not at all. Boys differ from girls in hair distribution. Boys usually have more hair on the chest and the face, and their pubic hair extends upward in an inverted V toward the umbilicus, whereas in women the top of the pubic hairline is usually flat.

The event that distinguishes puberty for boys is the occurrence of ejaculations. In most boys the first ejaculation occurs between the ages of eleven and fifteen and may result from masturbation, wet dreams (nocturnal emissions), or sexual intercourse.

Adolescence

The emotional changes accompanying puberty have come under much discussion during the last several decades. Basically, adolescence is the time when boys and girls suddenly become more conscious of their capabilities and potential and begin to recognize the existence and attraction of the opposite sex. Such recognition inevitably involves the investigation and discovery of methods of sexual approach.

Many things confuse the adolescent—the restraints of differing family opinions, cultural pressures that limit and question the expression of individuality, and personal and cultural taboos on sexual expression. Also disturbing is the fact that boys and girls mature at different ages. This circumstance gives rise to a situation in which, for a short while, girls are taller, more sexually mature, and more conscious of the other sex than are boys.

Abnormalities at Puberty

Common physical problems related to puberty concern its early or late onset. Some girls begin to develop at the age of nine or ten, and this occurrence is known as precocious puberty; usually, no abnormality is present. When development occurs before the age of nine, medical advice should be sought. At the other end of the scale, delayed

Differences in Growth Between Boys and Girls

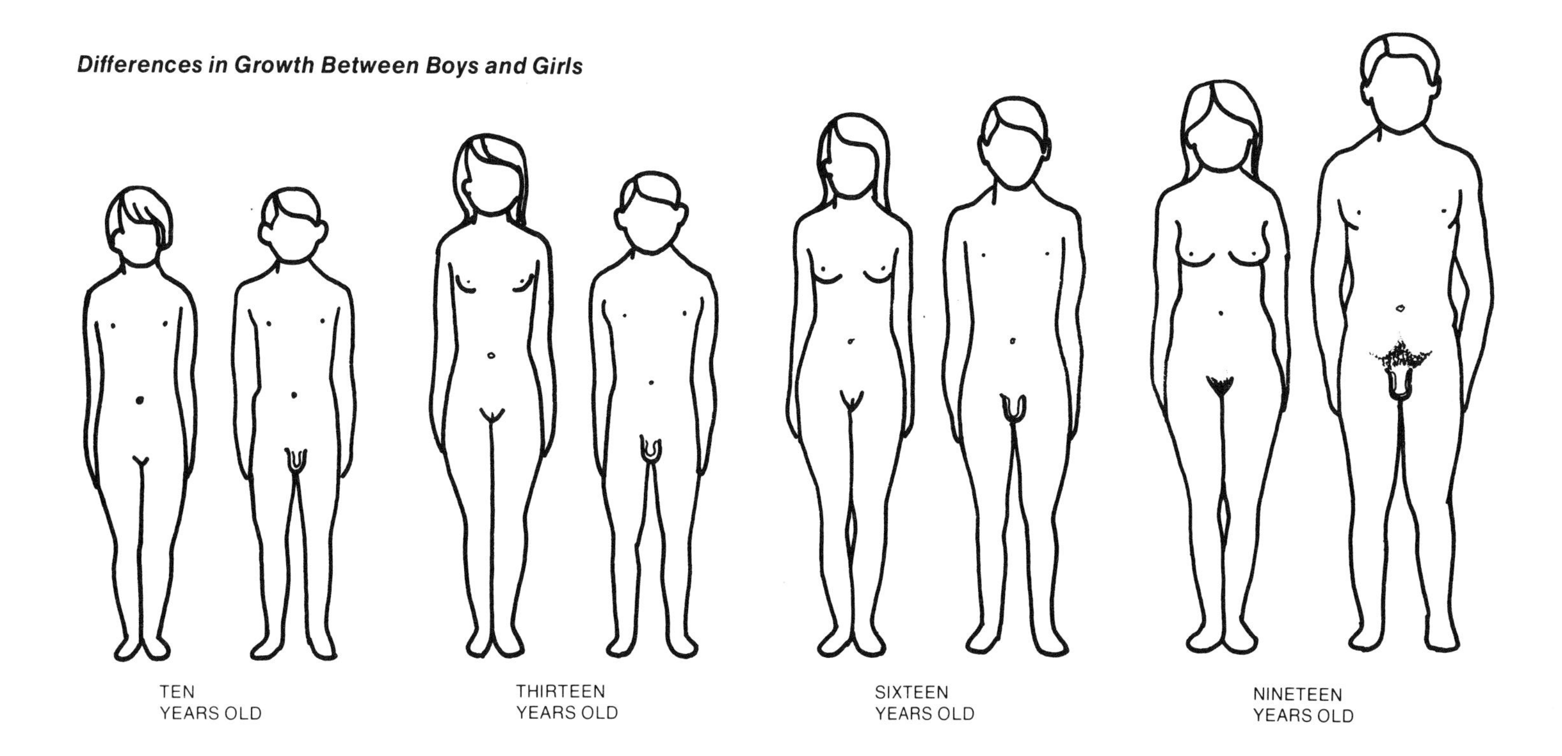

Body Proportion Changes Between Infancy and Adolescence

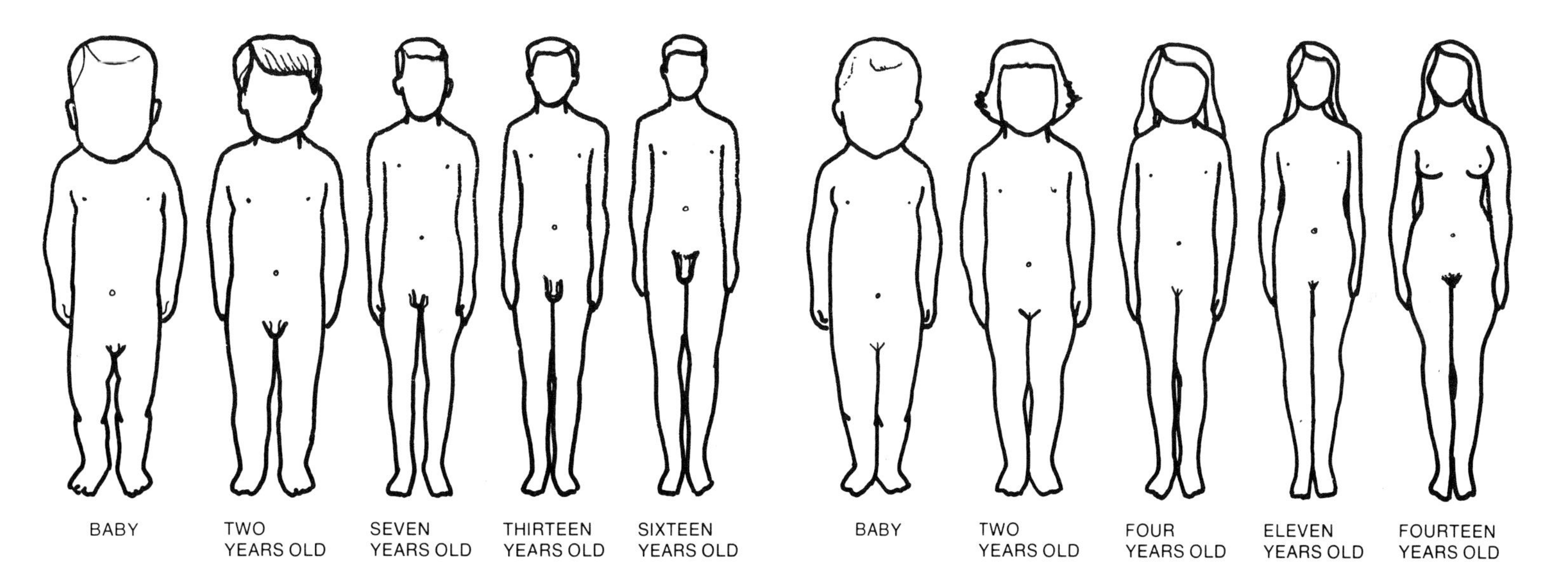

menstruation until the age of sixteen should generally cause no concern. However, failure of any growth spurt or delay of menstruation beyond the age of sixteen is generally an indication that something is wrong, and medical advice should be sought.

With boys, precocious puberty is less common. Delay of the growth spurt or of sexual maturation beyond the age of sixteen or seventeen is reason to seek medical advice.

Adolescents grow very fast, requiring more protein and calories than in later life. Adolescents also need special nutrients, such as calcium, iodine, and vitamins in greater amounts. As girls begin to menstruate, they lose blood regularly and deplete their stores of iron, sometimes becoming anemic. A balanced, nutritious diet is therefore especially important at this time, but persuading teen-agers to recognize the importance of eating properly is often difficult. A complete diet is especially important to girls who are soon likely to become mothers.

The problems of adolescence are often concerned with sexual matters, and these will be dealt with in the following section. But it must be remembered that the behavior of adolescents is a mixture of physical and emotional, sexual and nonsexual components.

Body Response to Sexual Stimulation

During puberty both boys and girls are capable of reproduction. For boys, the reproductive act is simple. Sexual stimuli—sights, sounds, and even thoughts, or a more direct stimulation of the penis—produce an erection. During erection an increase in the amount of blood enlarges the penis, causing it to become hard and erect. The size of the penis when it is not erect varies greatly, depending upon such factors as temperature, exercise, and emotion. On the average, the erect penis measures a little over six inches in length and about an inch and a half in diameter.

Ejaculation

Continued stimulation of the erect penis results in the intensely pleasurable and exciting experience called ejaculation, or orgasm. Ejaculation is a reflex process by which the sperm and the seminal fluid, the latter derived from the prostate gland and seminal vesicles, are ejected in a series of spurts. The time from the beginning of erection to ejaculation varies greatly, from a few seconds to many minutes. Occasionally, a little thin fluid exudes from the penis before ejaculation; this fluid may contain a few sperm. Once ejaculation has occurred, the erection is lost, the male relaxes and requires a variable period of recovery before his penis can again become erect and another ejaculation take place. Then it often takes longer for ejaculation to occur. Men vary greatly in their ability to produce an erection, to maintain it without ejaculation, and to ejaculate two or more times within a short period. In general, younger men have a greater capacity in this regard, but, again, variation is great at different ages, depending largely upon emotional, genetic, and nutritional factors.

Sexual response and ejaculation occur in men and boys during masturbation, wet dreams, and intercourse. Masturbation is a common practice among adolescent boys. Touching the penis is natural, even in baby boys. Later, this exploration extends to the discovery that erection, ejaculation, and a pleasant sensation can be produced in this way. Masturbation is not harmful provided that it

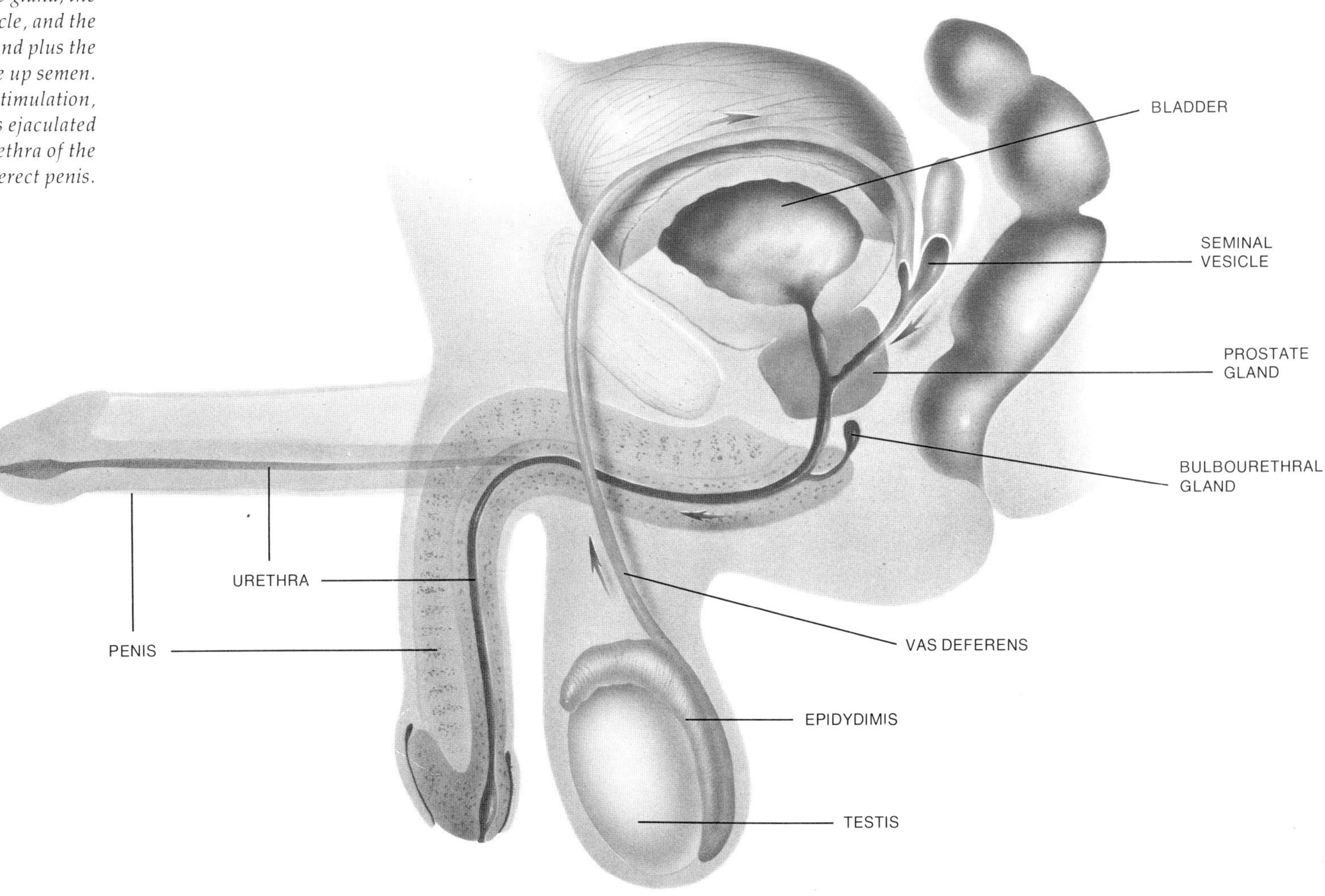

Sperm cells first form in each testis, move through the epidydimis, and travel into the vas deferens. Secretions of the prostate gland, the seminal vesicle, and the bulbourethral gland plus the sperm make up semen. During sexual stimulation, the semen is ejaculated through the urethra of the erect penis.

does not become a major part of the boy's life and divert him from relating normally to girls. Masturbation itself does not have any physical ill effects.

Wet dreams occur irregularly. They may be preceded by erotic fantasy-dreams involving women, things, or events known or unknown to the boy. The first awareness that he has is waking up and ejaculating at almost the same time. A nocturnal emission is a perfectly normal phenomenon and does no harm. Allied with wet dreams are morning erections, often associated with a full bladder; again, these are normal.

Phases of Sexual Response

The sexual response cycle for men as just described consists of four phases: excitement, plateau, orgasm, and resolution. During excitement the penis becomes erect, and muscular tension and other generalized changes such as flushed skin and increased heart rate take place. During the plateau phase these changes are accentuated, and the man or boy becomes ready for ejaculation. Orgasm consists of the discharge of sperm together with an intense response of the whole body and an acute pleasurable effect. A distinct feeling occurs just before orgasm that it is coming and that nothing can be done to stop it. During resolution a gradual relaxation from tension takes place, and normal perceptions return.

In women, sexual response goes through the same four phases but with some differences. Initial excitement results from stimulation by the opposite sex, masturbation, or sexual fantasies. Sexual dreams leading to orgasm also occur in women but less commonly than in men. Women are somewhat slower than men in reaching orgasm, but there are wide variations. Once at the plateau phase a woman, unlike a man, may have several orgasms rather than just one. The resolution phase is slower and may take longer in women.

Sexual dreams and fantasies occurring in girls and women are entirely normal. Masturbation is also common in young girls and usually consists of rubbing the external genitalia. As with boys, it is not serious. The practice is a normal exploratory event of growth, so long as it does not make excessive demands on the girl's life.

The Sex Act

Coitus, or intercourse, begins with the insertion of the penis into the vagina, usually after a varying period of foreplay—touching, kissing, embracing. Foreplay is followed by gradually increasing thrusts of the penis until ejaculation occurs, an event which may or may not be accompanied, preceded, or followed by orgasm on the part of the female partner. The man and woman involved in intercourse seldom have orgasms simultaneously. After an orgasm the penis becomes flaccid, and resolution and relaxation occur. A usual position for coitus is with the man on top, between the woman's spread legs. Other common positions include the woman on top and side to side. An almost endless variety is possible depending upon circumstances and preference of the partners.

Abnormalities of coitus and other forms of achieving orgasm are outside the scope of this chapter. It is worth noting, however, that attitudes toward sex are deeply ingrained in each person and are dependent upon many factors such as religious convictions and outlook on life. Frank discussion between partners is the most important element in making coitus a satisfactory experience for both persons and enabling it to contribute effectively to their overall adjustment and happiness.

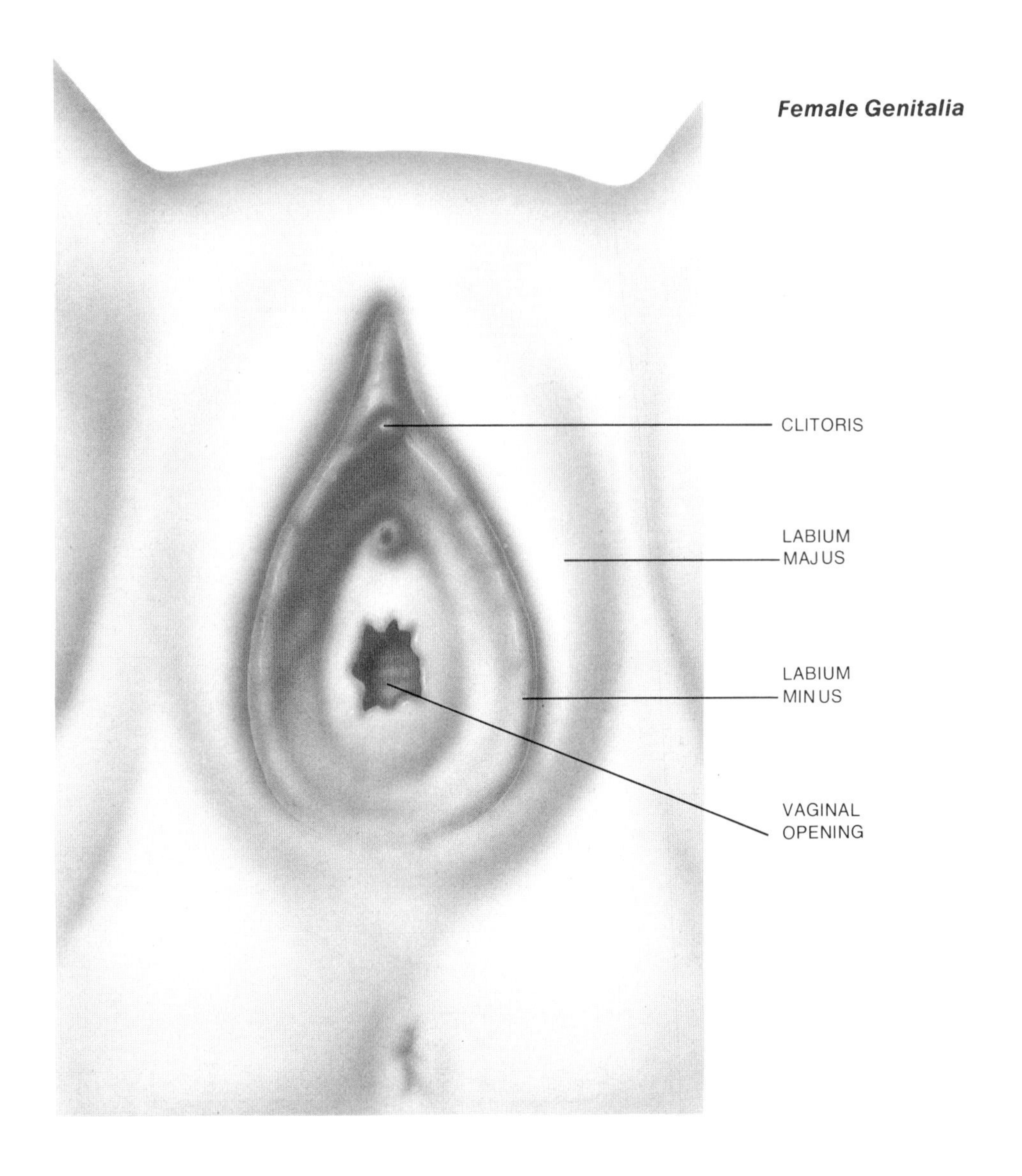

Female Genitalia

Conception

A man's ejaculate—the fluid material produced at ejaculation—usually consists of about one-half to one teaspoonful of spermatozoa and seminal fluid. Each one-fourth teaspoon contains, in the fertile man, sixty to a hundred million sperm; almost every one of these is capable of fertilizing an egg. However, a woman releases only one egg or occasionally two each menstrual cycle, usually about fourteen days before she menstruates.

After ovulation, when the egg is discharged from the ovary, the egg is picked up by the fallopian tube and passed down the tube to the uterus, where, about twenty-four hours after ovulation, the egg disintegrates. On the other hand, sperm may live in the female genital tract (internal sex organs) for up to seven days after they have been ejaculated into the vagina, although it is not known for how long sperm are capable of fertilizing an egg. Thus, although there are only twenty-four hours in each menstrual cycle in which the egg can be fertilized, the fact that sperm survive for up to seven days makes it extremely difficult to predict any specific twenty-four-hour period in which a woman will become pregnant.

The process of fertilization is not fully understood. It is believed to occur usually in the outer half of the fallopian tube. Sperm ejaculated into the vagina reach that part of the tube very fast, probably within a very few minutes, partly by their own motility, since they move at about one-eighth of an inch per minute, but also by the muscular movements of the uterus, which push the sperm upward. When the sperm, usually several of them, approach the egg, subtle chemical changes occur which permit just one sperm to enter the egg. The process of fertilization and cell multiplication for body development then begins.

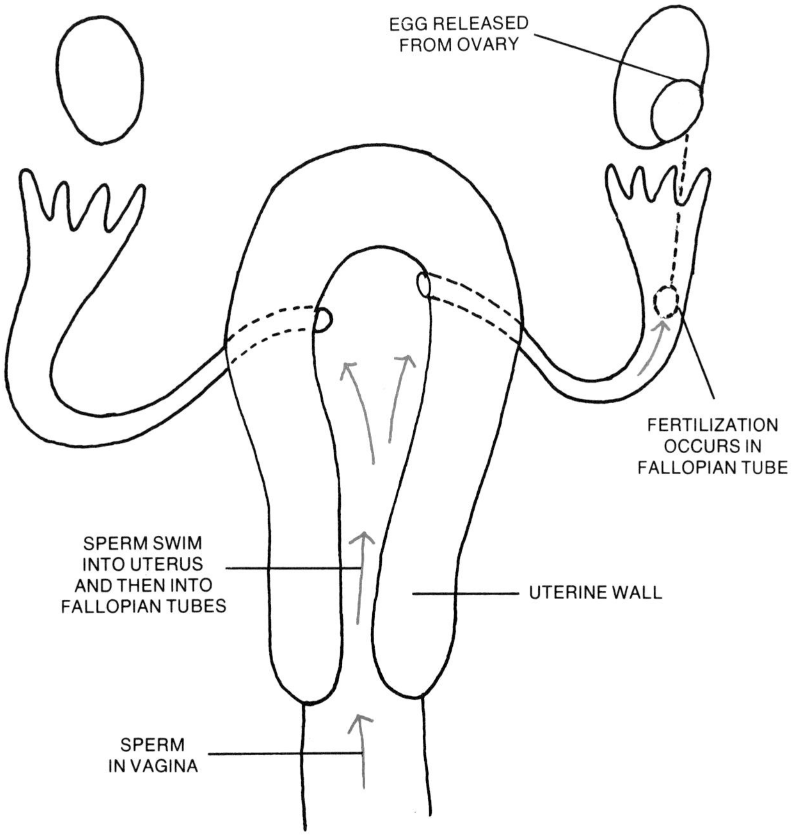

EGG RELEASED FROM OVARY

FERTILIZATION OCCURS IN FALLOPIAN TUBE

SPERM SWIM INTO UTERUS AND THEN INTO FALLOPIAN TUBES

UTERINE WALL

SPERM IN VAGINA

Types of Contraception

For centuries men and women have attempted to prevent or end unwanted pregnancies. When the physiological process was little understood, primitive methods included the uses of various herbs or substances inserted in the vagina or attempts, often violent, at shaking the pregnancy loose. Withdrawal of the penis from the vagina, or coitus interruptus, before ejaculation has also been known and practiced for a long time.

During the last hundred years knowledge of the process of conception has increased enormously. The implications of pregnancy are better understood than they were formerly, and women can now learn much sooner whether or not they are pregnant. Growing concern for the rapid rise in the earth's population has made men and women in most parts of the world aware of the need to plan their families.

Stated simply, family planning consists of stopping, at some point, the process of events leading to fertilization. The most obvious way is not to have coitus, although pregnancy can result from the deposition of sperm at the entrace to the vagina. If a man and a woman have intercourse, the following methods of preventing conception are possible.

One method of contraception for men is the prevention of the passage of sperm from the testes to the penis by blocking or cutting out a portion of the tubes—the vasa deferentia—that convey the sperm. When this operation, called vasectomy, is properly performed, the result is very effective. While complications can follow, they are rare. It is very difficult, sometimes impossible, to join the vasa again so that sperm can again pass. Therefore, a man should not consider vasectomy unless he is willing to run the risk of never again being able to father a child.

Another is the prevention of the ejaculated sperm from entering the vagina by wearing a condom, a "rubber," over the penis. This method is quite reliable provided that it is used carefully.

Several methods of contraception can be employed by women. One is the prevention of the passage of the egg to the uterus by tying or dividing the fallopian tubes. When this operation, called tubal ligation, is properly performed, the result is effective but carries some risk for the woman. Operations to rejoin the tubes are successful less than one-fifth of the time.

Another method prevents the discharge of eggs by the use of oral contraceptives. This method of ovulation prevention is very reliable. It alters normal physiology and carries only slight risks. However, oral contraceptives must be taken regularly.

For contraceptive purposes the lining of the uterus can be altered by using an intrauterine device (IUD) so that the fertilized egg cannot lodge in the uterine wall. This method is quite reliable with some risks to the woman.

Still another birth-control method relies on blocking the entrance of sperm into the uterus by wearing a rubber diaphragm. A contraceptive jelly is often applied to the diaphragm before it is inserted into the upper vagina. Little or no risk is involved, and the method is reliable if used carefully. Proper fitting of the diaphragm is necessary.

Lastly, conception can be avoided by destroying the sperm before they can enter the uterus by the use of contraceptive jelly or foam. This method is also reasonably reliable if the jelly or foam is placed high in the vagina and given time to spread. It entails little or no risk.

Sometimes, conception can be avoided by having intercourse only during the "safe" period, when ovulation is not likely to occur, or by having the male withdraw from the female before ejaculation, coitus interruptus. Both methods are difficult to carry out and unreliable.

Unplanned Pregnancy

Information about family planning is now widely available in the United States through physicians, the Planned Parenthood Association, hospitals, and family planning clinics. Theoretically, at least, it should be possible for any couple to control their fertility and plan their family. However, on some occasions an unplanned pregnancy may occur. Even then, methods are available to end the pregnancy. Either an estrogen hormone may be administered for several days, starting within forty-eight hours after fertilization, or abortion may be performed, preferably within the first twelve weeks of pregnancy. Advice about these two methods should be obtained from a physician.

Neither the "morning after" pill nor abortion is a procedure that ought to be relied upon. Prevention by family planning is much better than having to end a pregnancy after it has begun.

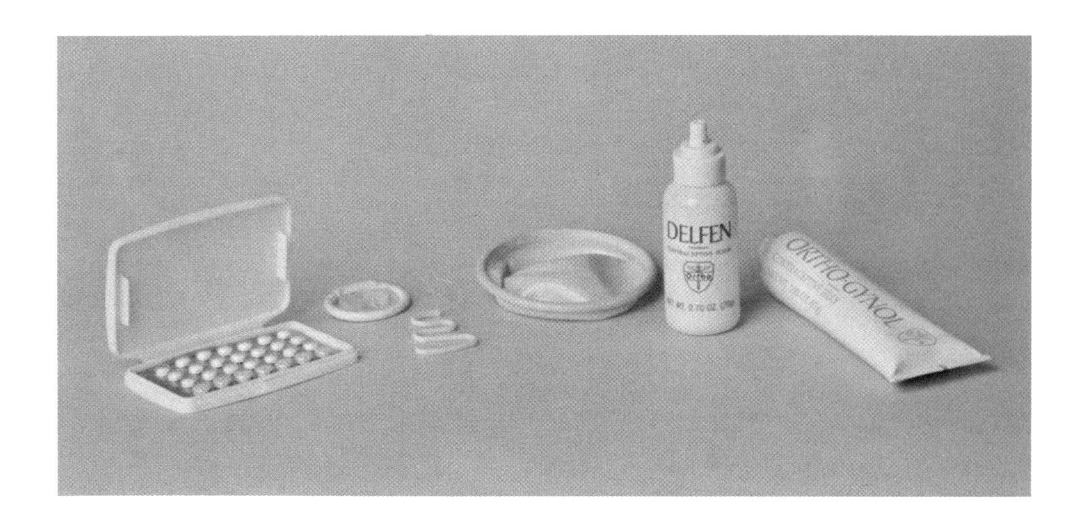

Various contraceptives are shown in the photo. From left to right, they are oral contraceptives, a condom, an intrauterine device (IUD), a diaphragm, contraceptive foam, and contraceptive jelly.

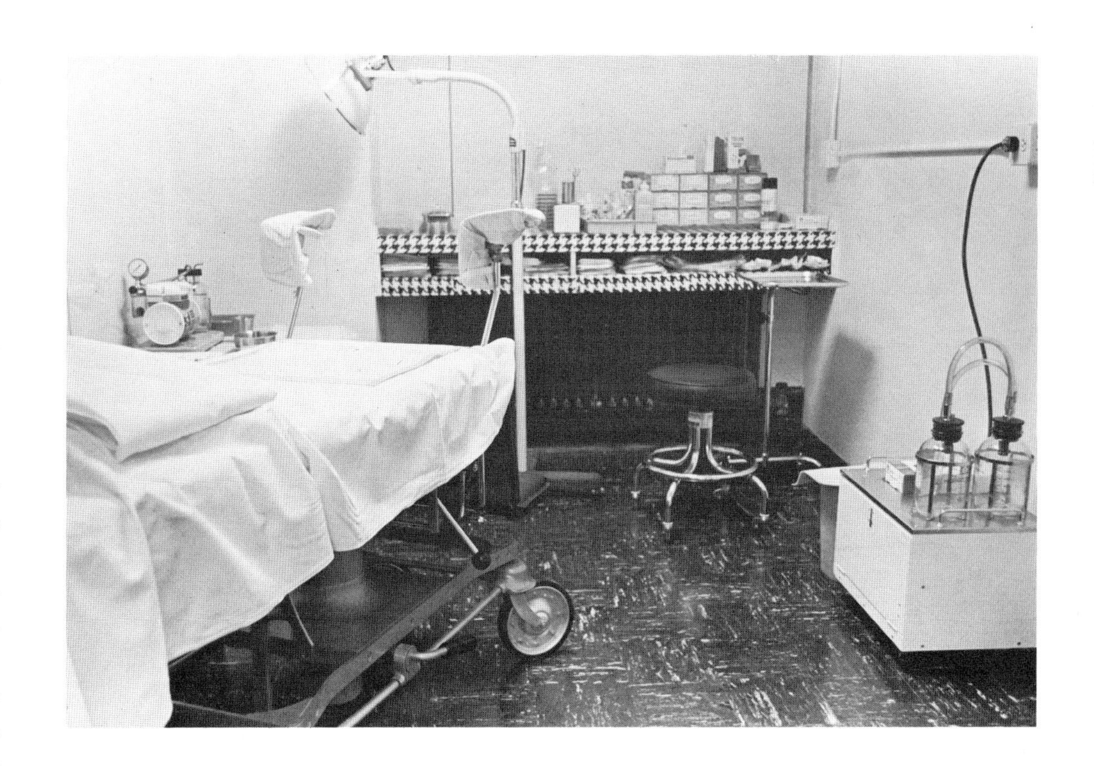

Abortions can be performed in medically-supervised clinics, such as the one above, run by a number of social agencies.

Pregnancy

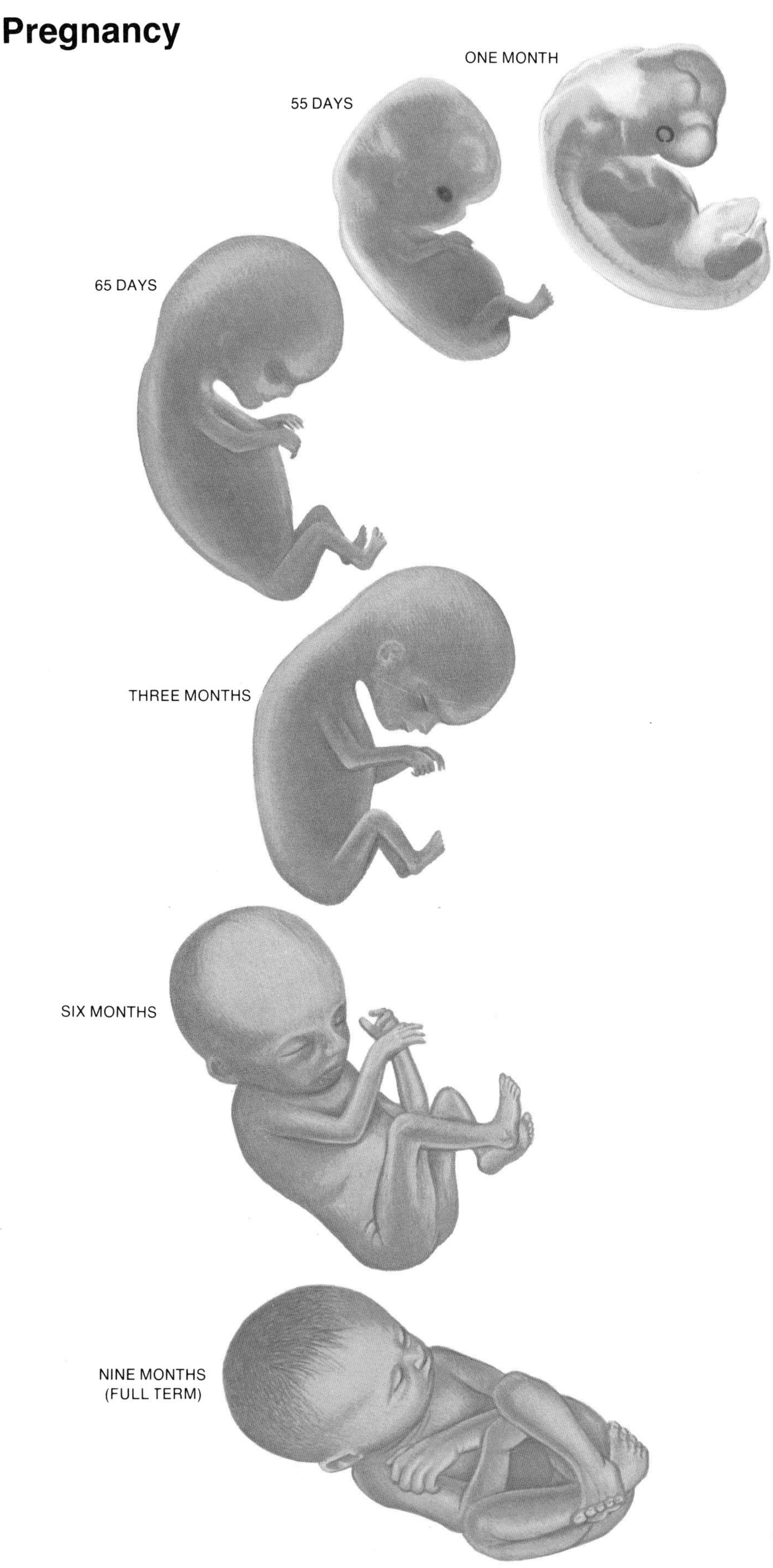

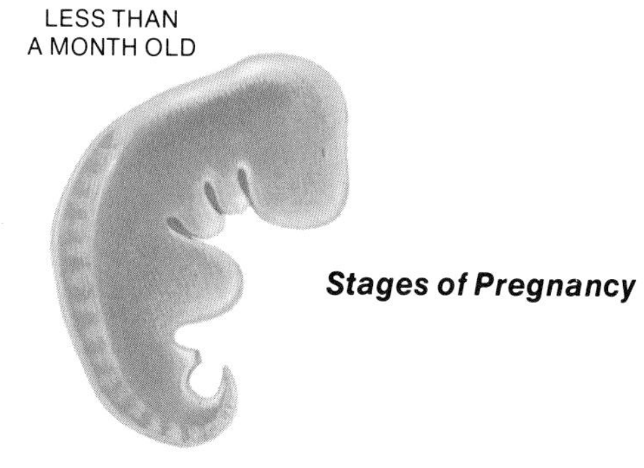

Stages of Pregnancy

Technical advances have made it possible to observe under the microscope the growth and division of fertilized ova. Beginning in the outer half of the tube the fertilized ovum divides into increasing numbers of cells. After about five days it reaches the uterus, and two days later begins to burrow into the wall of the uterus.

Pregnancy is usually calculated from the first day of the last menstrual period and lasts, on the average, forty weeks or 280 days. Actually, however, it does not start until about two weeks later.

The length of pregnancy, therefore, averages 266 to 268 days. The date of the delivery may be roughly calculated by adding seven days to the date of the first day of the last normal menstrual period and subtracting three months.

The growth of the baby is a remarkable progression of events. At first only a small, shapeless blob of tissue, it rapidly acquires human form with a prominent head. The main organs, including the limbs, are formed in the first eight to ten weeks. Sex can be distinguished externally at about twelve weeks. Later development primarily involves maturation of the various organs, especially the nervous system and what are known as the skin appendages—the sweat glands, hair follicles, and so forth.

The cells of the fetus at first extract nutrients directly from the mother's bloodstream. Later a special organ, the placenta, develops and is complete by the sixteenth week of pregnancy. The placenta serves as the means by which food elements pass from the mother to the fetus; in the latter they are used to build up new body tissues. Whatever the mother takes in, including many drugs, passes directly or indirectly into the fetus.

Changes Experienced in Pregnancy

The first symptom of pregnancy for nearly every woman is the missing of a menstrual period. Some women experience breast fullness, nausea and vomiting, tiredness, or frequent urination. As pregnacy progresses, many normal body changes take place. For example, the uterus enlarges and makes the abdomen prominent, usually by the twelfth or fourteenth week. The capacity of the stomach and the bladder decreases. The skin may become darker in some places, and its secretions may increase.

Weight gain normally occurs in pregnancy. It averages about twenty-five pounds but may be more or less. Weight gain results not only from the baby's growth but from breast enlargement, among other reasons.

It is important for a woman to determine as soon as possible whether or not she is pregnant. Although symptoms, such as missing a period, may be helpful and a physician's examination may suggest pregnancy, a pregnancy test is often desirable and necessary. A urine test can be quickly done in a laboratory and, if positive, is usually reliable. Early in pregnancy a woman should have a complete examination to determine if there are any abnormalities that might affect her pregnancy and to help her decide her plans for later care.

The first three months of pregnancy are usually a time of uncertain feelings. Even when the pregnancy was planned and everything is normal, a woman may have a short episodes of not wanting the baby. The baby begins to move as early as its eight week of life, but movements usually cannot be felt until about sixteen weeks. At this time a woman begins to feel more adjusted to her pregnancy. Her general well-being is usually improved. Later in pregnancy the increasing size of the child and anticipation of the end of the pregnancy influence her feelings. She looks forward to having her baby and is eager to have the pregnancy over. Other feelings are also prominent from time to time-anxiety, concern about appearance, worry about finance, and cravings for special foods or activities.

Sexual feelings vary considerably. Sometimes they are increased, especially during the middle third of pregnancy; in some women they may diminish in the first third and again in the last few weeks of pregnancy.

A Healthy Pregnancy

During a normal pregnancy there are several things that a woman can do to help herself. These include proper nutrition, preparation for childbirth, and anticipation of the needs of the new baby. Proper nutrition consists of following a high-protein diet and avoiding high-calorie carbohydrate foods, especially sugars and sweets. Adequate vitamins and minerals are also necessary.

Classes in preparation for parenthood are widely available in doctors' offices, hospitals, and other institutions. They provide an opportunity to find out

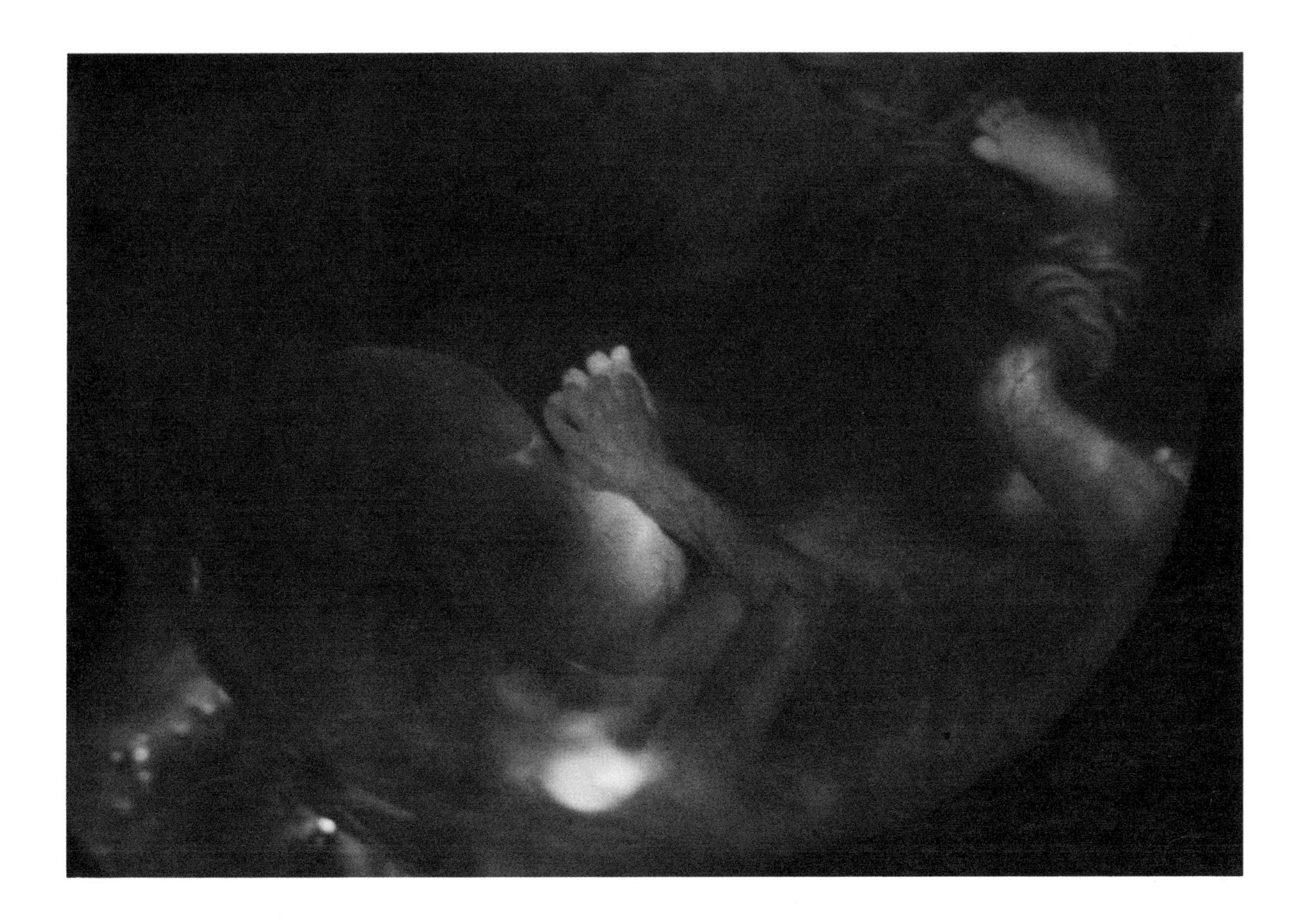

A fetus in about the ninth week of development is shown floating in amniotic fluid in its mother's womb.

in advance what happens during childbirth and to learn techniques of cooperating during the delivery so as to lessen the need for medication and improve the health of the baby. Many of these classes also encourage husbands to participate in the developing pregnancy and actual childbirth experience.

Problems of Pregnancy

Most women feel different during pregnancy because it is an unusual, though normal, female body event. If a pregnant woman is under medical supervision and follows sound principles of nutrition and hygiene, major complications of pregnancy can be avoided.

Morning Sickness. Many women feel queasy at some time during pregnancy and express unusual food likes and dislikes. Some experience occasional vomiting. Morning sickness starts soon after the first missed menstrual period and may last until the twelfth to fourteenth week of pregnancy, rarely longer. Severe, repeated attacks of nausea and vomiting are uncommon.

Morning sickness is probably a result of the hormonal changes that take place in early pregnancy, although a woman's altered metabolism and other factors may be involved. Morning sickness does not necessarily occur in the morning: it may occur in the mid-afternoon or at other times. Simple measures that may relieve it include taking frequent small meals and, especially, eating energy-containing food early in the morning before arising. If the condition persists or is severe, medical advice should be sought.

German Measles and Other Infections. Any infection that involves the mother may also affect the fetus. Of special concern, however, is German measles, or rubella. If the mother contracts German measles in the first trimester of pregnancy—until the twelfth to fourteenth week of pregnancy—there is considerable risk of a miscarriage or birth of a child with a congenital deformity such as heart disease or deafness.

A simple test can determine whether a woman is susceptible to German measles. If the test is positive, she has had or has been exposed to the disease and has acquired immunity. Then she need not worry about getting it again. If the test is negative, however, she is susceptible to German measles. The test is best done before a woman becomes pregnant. If the test is negative, she can be vaccinated, although she should not become pregnant for at least two months after vaccination because there is a slight chance that the fetus may be affected by the virus used for vaccination. If a pregnant woman has a negative test, it is especially important for her to avoid exposure to German measles. If by any chance she is pregnant when she is exposed, she should see a physician promptly.

Bleeding in Pregnancy. The majority of women have no vaginal bleeding during pregnancy, but about a quarter of them bleed at some time. Occasionally, it occurs at the time of the expected menstrual period in the first few months. Rarely, it may be like a normal menstrual period, although shorter and with less blood flow. In late pregnancy slight bleeding, often accompanied by mucus, may precede the onset of labor.

There are many causes of bleeding during pregnancy. Early bleeding poses a threat of miscarriage. Later bleeding may signify that labor is starting prematurely; that the placenta is placed below the baby instead of being attached to the side or top of the inside of the uterus; or that the edge of the placenta has become detached prematurely from the wall of the uterus.

When any bleeding occurs, it should be a matter of concern and should be reported to the physician at the first opportunity. If it is persistent or severe, it should be reported immediately.

Miscarriage. Miscarriage, or abortion, means the delivery of the fetus, placenta, and membranes before the fetus is capable of living outside the uterus—before the twenty-sixth to twenty-eighth week of pregnancy. It most often occurs in the first twelve weeks of pregnancy. Miscarriage is like miniature labor. Bleeding usually occurs first, but this may subside and the pregnancy proceed normally. This event is called a threatened miscarriage. If the miscarriage is really going to occur, bleeding will be followed by uterine cramps and eventually by expulsion of the contents of the uterus.

The majority of miscarriages represent a failure to continue a defective pregnancy. Other causes, such as acute maternal illnesses and even accidents, may occasionally be responsible. However, it is quite unlikely that any physical activity by the mother will cause a miscarriage, so she should not blame herself if it does occur. Nonetheless, bleeding that is followed by cramps should be a reason for consulting a physician because of the possibility of such complications as excessive bleeding or infection.

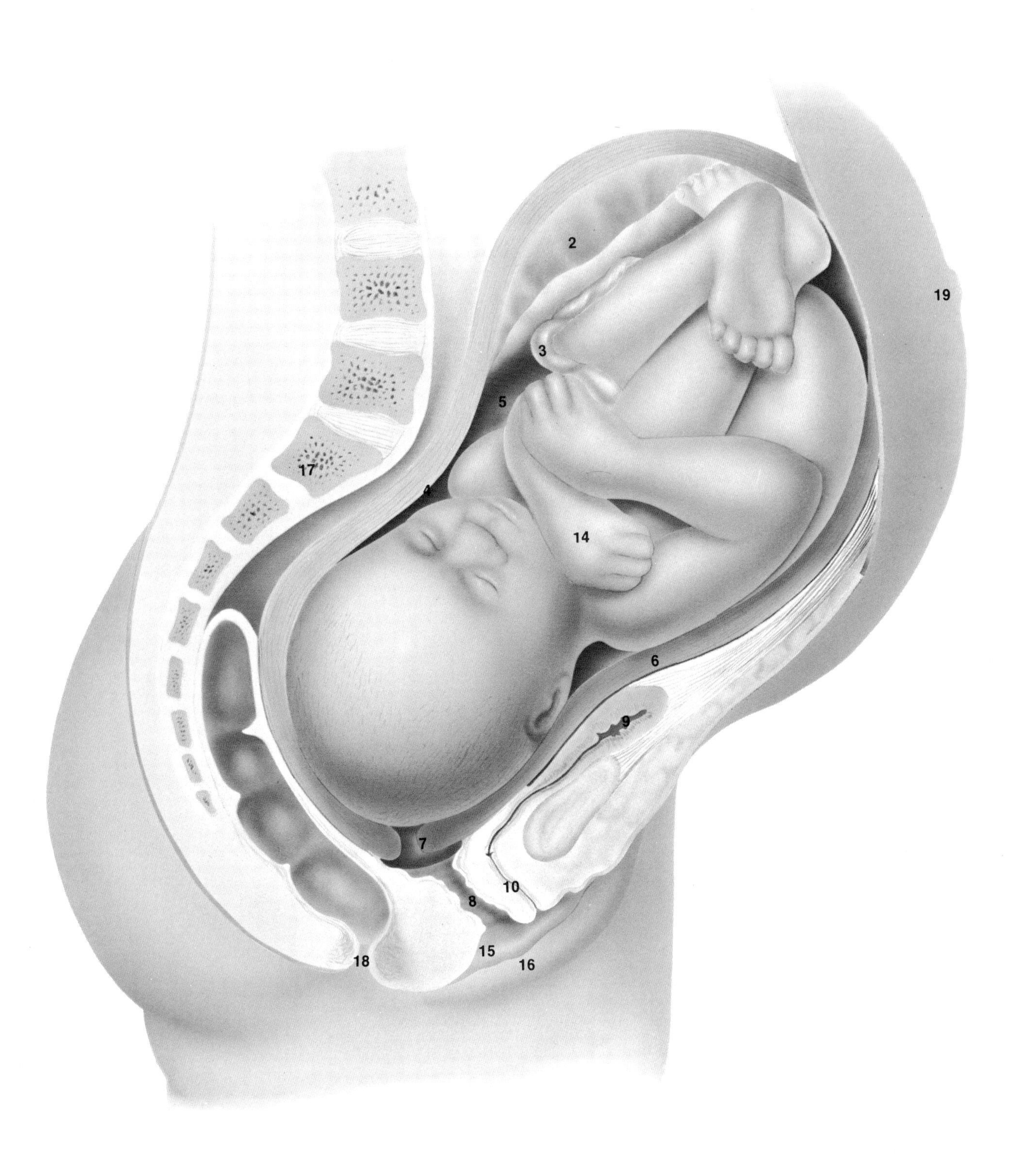
2
3
5
17
4
14
6
9
7
10
8
15
16
18
19

Stages of Pregnancy and Fetal Development

1. Fetus in fourth month of development, 1
2. Placenta, 1, 2, 3
3. Umbilical cord, 1, 2, 3
4. Fetal membranes consisting of an inner amnion and an outer chorion, 1, 2, 3
5. Amniotic fluid, 1, 2, 3
6. Uterine wall, 1, 2, 3
7. Cervix, or neck of the uterus, 1, 2, 3
8. Vagina, 1, 2, 3
9. Urinary bladder, 1, 2, 3
10. Urethra, 1, 2, 3
11. Pubic bone, 1
12. Rectum, 1
13. Fetus in seventh month of development, 2
14. Fetus in ninth month of development, 3
15. Labia minora (only one shown), 3
16. Labia majora (only one shown), 3
17. Vertebra, 3
18. Anus, 3
19. Umbilicus, or navel, 3

Labor and Delivery

No one knows why labor starts when it does. Throughout pregnancy the uterus contracts at irregular intervals, and these contractions are occasionally felt as abdominal cramps. Toward the end of pregnancy, the contractions become more common. Occasionally, they recur regularly for a long time and are then called "false labor." It is sometimes difficult for women to decide whether the contractions signal false or true labor. When real labor begins, it is often accompanied by discharge of bloody mucus or a fluid (rupture of the bag of waters).

First Stage of Labor

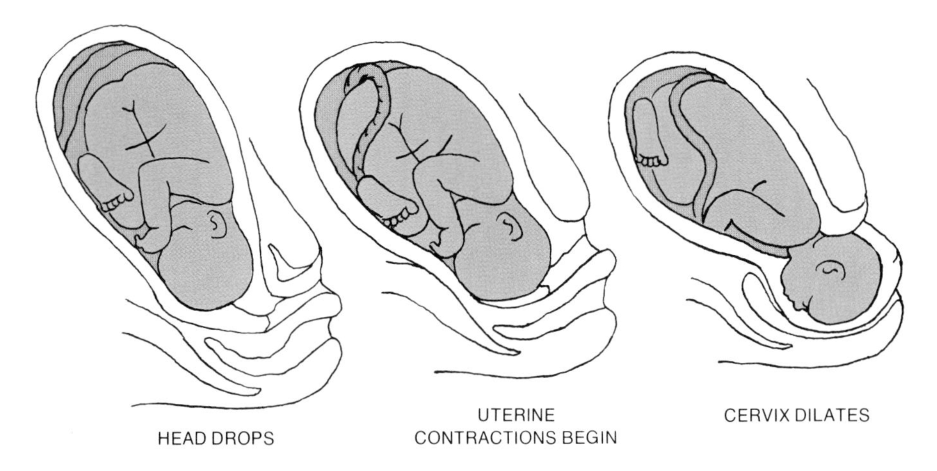

HEAD DROPS — UTERINE CONTRACTIONS BEGIN — CERVIX DILATES

Second Stage of Labor

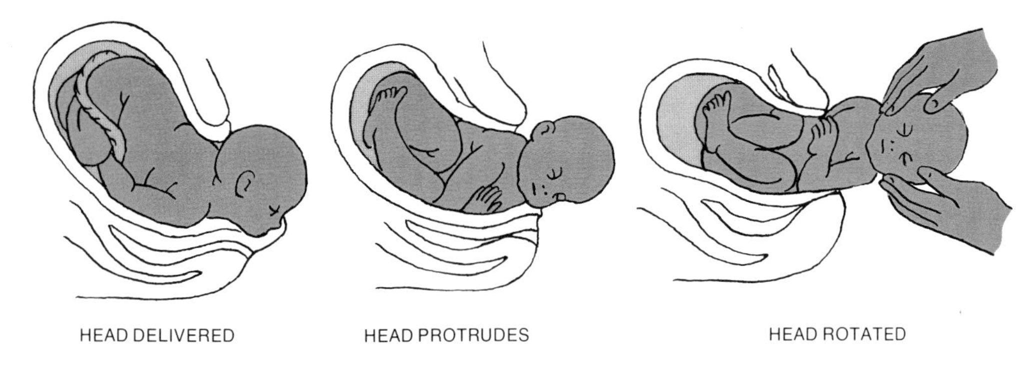

HEAD DELIVERED — HEAD PROTRUDES — HEAD ROTATED

Stage After Birth

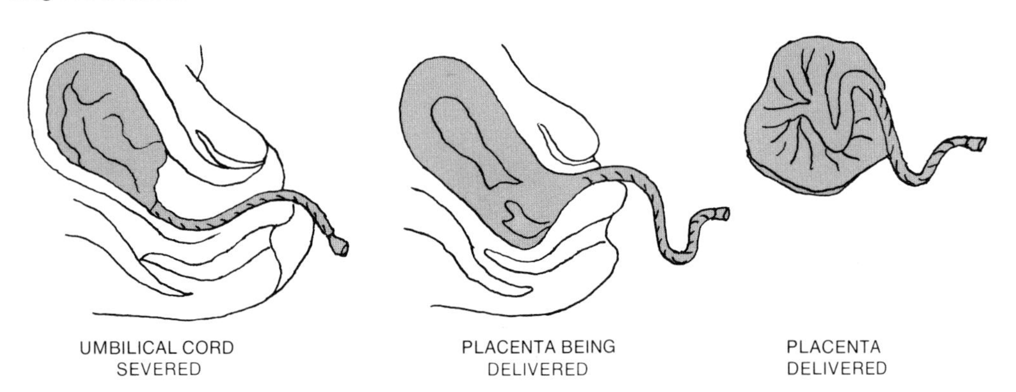

UMBILICAL CORD SEVERED — PLACENTA BEING DELIVERED — PLACENTA DELIVERED

What Happens During Labor

Once established, labor usually progresses to the delivery of the baby. Physically what happens is that contractions of the uterus push the baby against the cervix, causing it to shorten and eventually to open. The first stage of labor lasts until the cervix is fully open, about four and one-half inches. The average length of time of the first stage for a woman having her first baby is about eight hours, and for a woman having later babies, about four hours. During the second stage the uterus, aided by the muscular expulsive efforts of the mother, pushes the baby down through the vagina. The average length of this phase for a woman having her first baby is about fifty minutes, and for a woman having her second or later baby, about twenty minutes. The third stage of labor lasts from delivery of the baby until the delivery of the placenta or afterbirth. It averages about eight minutes in duration.

The emotional aspects of labor are very important. Often, just before labor a woman has a spurt of energy during which, for example, she cleans her house. How she reacts to the beginning of true labor depends upon her individual emotional background and the preparation she has received during pregnancy. As the first stage of labor progresses, most women experience some discomfort with the contractions. During the latter part of the first stage, the transition stage, this discomfort is more noticeable and may be accompanied by hiccups, nausea, and shivering. The onset of the second stage is heralded by the desire to hold the breath and push down to expel the baby. During the second stage the mother experiences a series of physical and emotional changes that have been likened to those of orgasm. After delivery relaxation occurs, only to be interrupted by a further expulsive effort to deliver the placenta.

Help During Labor

In most cultures, women are not left alone during labor. They are usually attended by other women. In some places men serve as consultants if difficulties are encountered. As knowledge developed, midwives—women with special skills—began to give help to laboring women. In Western Europe and in the United States, during the past hundred years, physicians have taken greater responsibility, though midwives retain important functions in normal cases in most countries. In the United States the nurse-midwife is gaining increased acceptance and importance.

The function of attendants during labor is to help the laboring woman help herself, to relieve her discomfort, to watch for and treat any abnormalities that may develop, and to assist at the delivery of the baby.

In the United States, most babies are now born in hospitals. This custom has developed because of the possibility that unexpected complications may occur for either mother or baby in a place where expert help is not promptly available. Hospital care may, however, add to the emotional turmoil that often accompanies labor. The most important function of the attending physician and staff is, therefore, to stay with the mother, to enable her to have her husband or other family member with her if she wishes them, and to comfort and reassure her. Special breathing techniques may be very helpful at this time. Pain-relieving drugs may be necessary, but they immediately pass to the baby and may make him sleepy after birth.

For many mothers who have received preparation during pregnancy, delivery may require little or no anesthesia. For others, various kinds of local nerve blocks or even general anesthesia may be needed. Prepared mothers may be able to deliver their babies spontaneously and without difficulty, but others may need to have instruments used and an incision, called an episiotomy, made at the opening of the vagina to assist the baby's exit.

Delivery of the Baby

A normal baby begins to breathe very soon after his head emerges from the vagina. He moves his arms and legs indiscriminately and responds to stimuli by movements and by crying. He is anxious for warmth and comfort. His breathing gradually becomes regular, and he begins a pattern of sleeping most of the time except for eating and gradually increasing periods of wakefulness and sociability. Excretion of urine and bowel movements usually occur within a few hours.

The normal baby is ready to nurse immediately after birth. If put to the breast, his natural instinct is to suck. This enables him to stimulate the production of milk as soon as possible and provides him with warm comfort. Early and close contact between mother and baby seems likely to be extremely important to the child's later development.

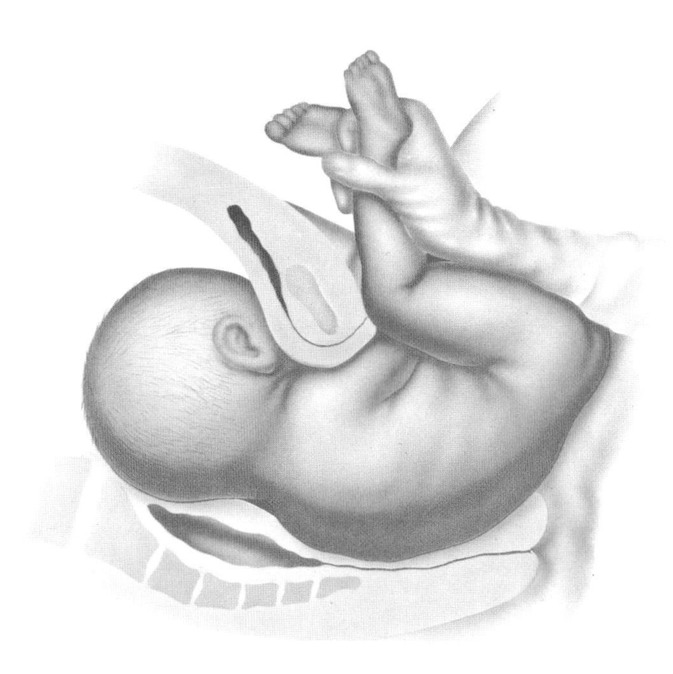

This illustration shows a breech presentation of a baby. A "breech baby" is delivered buttocks first or feet first.

Occasionally, delivery does not occur in the usual manner—the baby coming head first through the vagina. In about 3½ percent of all births the baby comes out buttocks first—a breech birth—and there is slightly greater risk for the baby in this case. When the mother's pelvis is too small or some other abnormality arises that may prevent a normal delivery, delivery of the baby through an incision in the abdominal wall and the uterus, a cesarean section, may be performed. About 5 percent of deliveries occur in this manner. Since cesarean section is a surgical operation, the mother needs anesthesia and all the equipment available in a modern hospital operating room. However, with proper care a cesarean birth is not very risky.

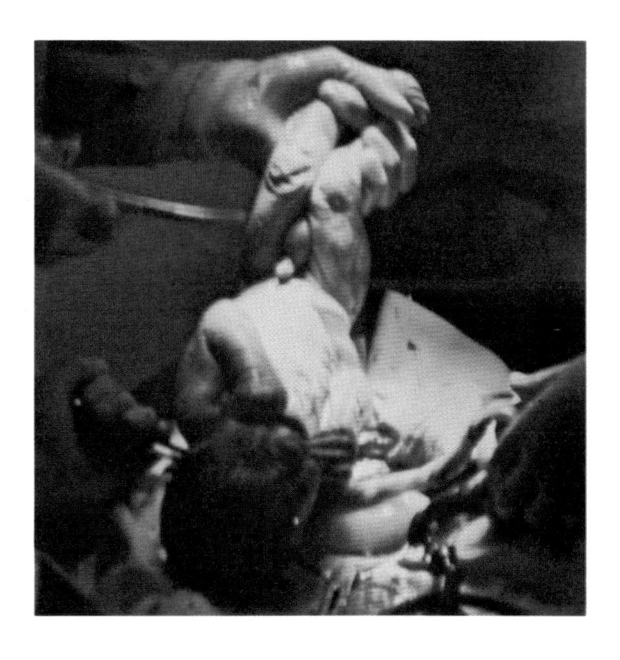

The delivery of a healthy baby is a happy moment for the mother as well as for those who assist the birth.

Postpartum Changes in the Mother

The mother's body undergoes many changes as it returns to normal after the birth of a baby. The whole process takes about six weeks, although most of it is completed in two to three weeks.

The uterus continues to contract regularly after the placenta has been delivered. Contractions are often felt by the mother during the first few days after delivery, particularly when she is nursing the baby. The uterus is about twice normal size at two weeks after delivery, and it takes about six weeks for it to return fully to normal. As the uterus contracts, it expels blood and some tissue from its lining. This is known as lochia, and for the first five to seven days it is bloody. Then it becomes a thin, blood-tinged discharge which may continue for as long as three to four weeks. At the same time, the opening of the uterus at the cervix gradually closes, and the stretched vagina and tissues of the perineum return to their normal state.

After delivery the mother's hormones change abruptly. The placenta no longer produces large amounts of estrogens and progesterone, and the balance of all the hormones is therefore altered. This alteration has one important consequence—the immediate release of the hormone prolactin, which is responsible for the production of milk. Later, the pituitary hormones resume the function that they have in nonpregnant women. The usual changes in the ovaries follow, and the menstrual cycle begins again. The onset of the first menstruation after childbirth varies greatly. In a woman who is breast feeding her child and is not giving him bottles or solid food, resumption of menstruation may be delayed for several months, but in a woman who is not breast feeding it starts, on the average, fifty-five days after delivery. Ovulation may precede menstruation in some women. Therefore, it is possible for a woman to get pregnant even before menstruation has occurred.

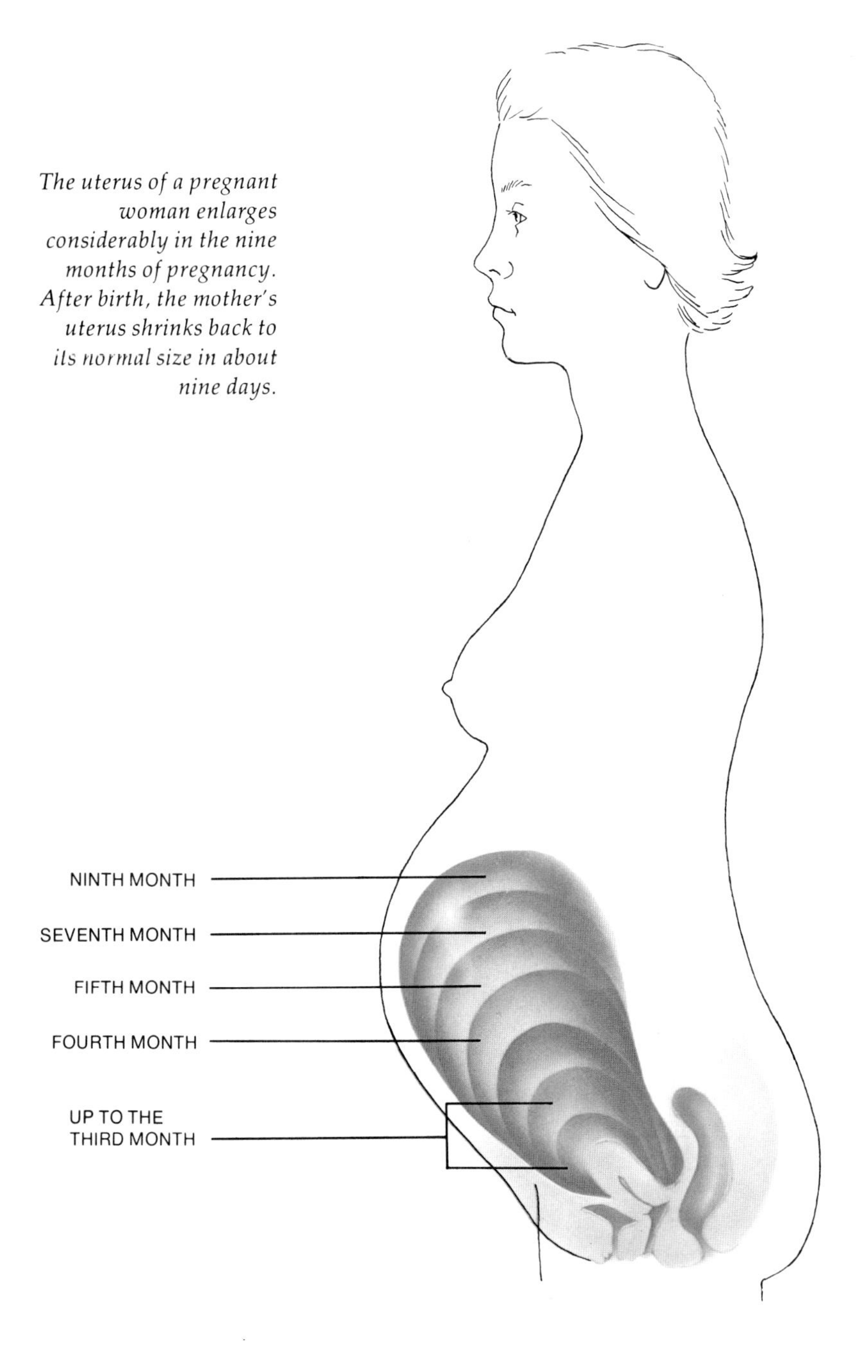

The uterus of a pregnant woman enlarges considerably in the nine months of pregnancy. After birth, the mother's uterus shrinks back to its normal size in about nine days.

Breast Feeding

The secretion or production of milk occurs after delivery in almost every woman, and many notice milky discharge before delivery. This event happens even if a woman has had a miscarriage as early as the sixteenth week of her pregnancy. The onset of lactation or milk production is the result of complex hormonal changes, but especially the release of prolactin.

Once the first milk, often called "colostrum," has appeared, its continuation depends on two factors. First, the breasts must receive adequate stimulation from the nursing baby, and, second, an important reflex known as the let-down, or ejection, reflex must function properly.

The great importance of the baby sucking is shown by the fact that it is possible to produce a plentiful supply of milk by nursing every two hours or ten times or more daily. The let-down reflex occurs when the baby takes hold of the nipple. This results in the discharge from the pituitary gland of a hormone called oxytocin, which makes muscle cells in the deeper parts of the breast contract so that the milk is pushed out and can be more easily obtained by the baby. The natural results of this mechanism can be seen by the fact that when the baby sucks one breast, milk may drip from the other. An important aspect of this reflex is that it can be inhibited or partly stopped by pain, fear, embarrassment, or similar emotions.

A very few women cannot physically nurse their babies. The fact that many do not nurse them is psychological; these mothers do not wish to nurse or are afraid to do so. In the United States, the latest data showed that only 27 percent of women even started to nurse their babies. This trend represents a significant change in child rearing, the result of which will not be fully understood for many years. Although artificial milk has been developed so that babies can thrive on it, it still does not have many of the characteristics of human milk. It is hard for the bottle-fed baby to obtain as close contact with his mother as the breast-fed baby, and the pleasurable aspects of the mother-child relationship that are involved in breast feeding are lessened.

The length of breast feeding depends upon the mother's and the baby's inclinations. Breast milk is a complete food for the baby until four to six months of age; additional food is not required. Then the baby needs more proteins and certain minerals and vitamins, especially iron and vitamin C. Fortunately, by this time he begins to be able to put food into his mouth. In many cultures, breast feeding is continued into the second or even third year. There is no physical reason why this practice should not occur, provided that the baby is obtaining a varied diet from other sources: excellent psychological reasons may exist to continue the close nursing relationship for a long period.

Emotional Changes and Complications After Childbirth

Postpartum physical changes are, in themselves, a stress. In addition, a woman suddenly realizes that she is now responsible for a new being, and she may have little knowledge or understanding of how to handle this responsibility. Depression, or "postpartum blues," as it is called, affects many women in the first few days and weeks after delivery. Closeness to the baby, a loving family, and adequate rest and help at home may make this period easier. As her physical strength increases, helped by a good diet, the mother becomes more able to cope with her new situation, and her mood improves.

For most women, childbirth is followed by an uneventful recovery. Occasionally, complications occur. Soon after delivery, sudden bleeding may take place because the uterus does not contract properly, or because a small piece of the placenta remains behind in the uterus, although this fact cannot always be determined from examination of the placenta at the time of delivery, or because of an injury somewhere in the birth canal. Infections used to be the dreaded curse of childbirth. Although these occur occasionally, they can usually be treated with one of the powerful antibiotics.

Some mothers prefer to nurse their babies. Nursing permits close contact between baby and mother at the same time that the baby gets excellent nourishment.

Infections may occur in the uterus, the kidneys, the bladder, or in the breasts. Minor problems that beset some women at this time include difficulty in getting back to their pre-pregnant weight, (although reduction should not be tried until breast feeding ceases), relaxation of the abdominal muscles, and, occasionally, an unusual amount of hair loss.

Effect of the Baby on the Family

Although it is the woman whose body undergoes many changes after childbirth, the arrival of a new baby has a great effect on her husband and other family members. This is illustrated among some primitive cultures by the custom of couvade (*couver* = to hatch in French), in which the husband mimics the travail of his wife, is segregated, is specially treated, and does no work for a period of time after his wife delivers, because of the great stress that this event imposes upon him. The same feelings beset the American husband, although he does not go into couvade. He may feel useless and displaced in his wife's affections by the new baby. The situation requires more giving and more work around the house than perhaps is usual for him. The same kind of feelings may occur in older children. It is important not to ignore these feelings but to recognize them in advance and make plans to deal with them. In particular, special arrangements should be made to provide the love and affection that all family members need. In this way a closer family unit can be built.

Sexual Changes in Later Years

In most women, the menstrual cycle continues with only minor variations, unless interrupted by pregnancy and lactation, until their mid or late forties. Then a decline occurs, imperceptible at first in most instances, in ovarian function. First, a decrease in the incidence of ovulation may take place and so in the secretion of progesterone; these events may lead to variations in the time and amount of menstrual flow. Eventually, so little estrogen is produced that the uterus does not respond and menstruation ceases. Pituitary FSH continues to be produced and, in fact, blood levels of this hormone are higher than before, since it is no longer "opposed" by estrogen. After menstruation ceases, the ovary and also the adrenal glands continue to produce some estrogens for some time.

The Menopause and Afterward

The terms "change of life" and "menopause" are used interchangeably. The age at which menstruation usually stops is now about fifty years, although considerable variations are found. Any time from forty-five to fifty-five years of age is normal. Most women report that before stopping, menstruation becomes scantier, and the periods are farther apart. For about 10 percent it stops suddenly, while in a few the flow actually increases before it ceases.

Many symptoms are reported to be associated with the change of life. The only typical ones are the so-called hot flashes. These are sudden sensations of warmth and blushing followed by perspiration, especially over the face and upper part of the body. These signs and symptoms may occur frequently or rarely and are not dangerous. Other symptoms, such as nervousness, headaches, or insomnia are not really characteristic of the change of life, although these complaints may often be associated with emotional disturbances that may occur at this time when a woman's family is suddenly grown up and when she needs other kinds of work to feel useful.

After the menopause, as estrogen secretion decreases, a woman's breasts may lose their firmness to some degree. Her pubic and axillary hair become more scanty, and the external genitalia are less prominent. Her vagina may lose some of its lubrication.

Men do not go through the change of life in the same way as women. No sudden change occurs because males do not have the hormonally governed cycle experienced by women.

A good sex life can continue after menopause. The absence of menstruation makes the use of contraceptive pills or devices unnecessary, although their use should probably be continued for six months after the last normal menstrual period. Changes in sexual desire may be due to emotional rather than physical factors. The response patterns of both men and women remain the same, although they may be slower.

The use of hormones in women to replace those that decrease or disappear after the menopause has been widely discussed. Hormones are of real value when the symptoms of hot flashes are very disturbing or if there is great loss of vaginal moisture. Otherwise their use may be unnecessary. It is difficult to continue taking them in the correct dosage, and they may possibly lead to irregular vaginal bleeding. The effects of long-term administration of hormones over many years are not fully known.

The uterus and ovary of a baby girl (A) are shown. Those sex organs of a women who never had a baby (B) are compared with those of a woman who had been pregnant (C) and those of a woman after menopause (D).

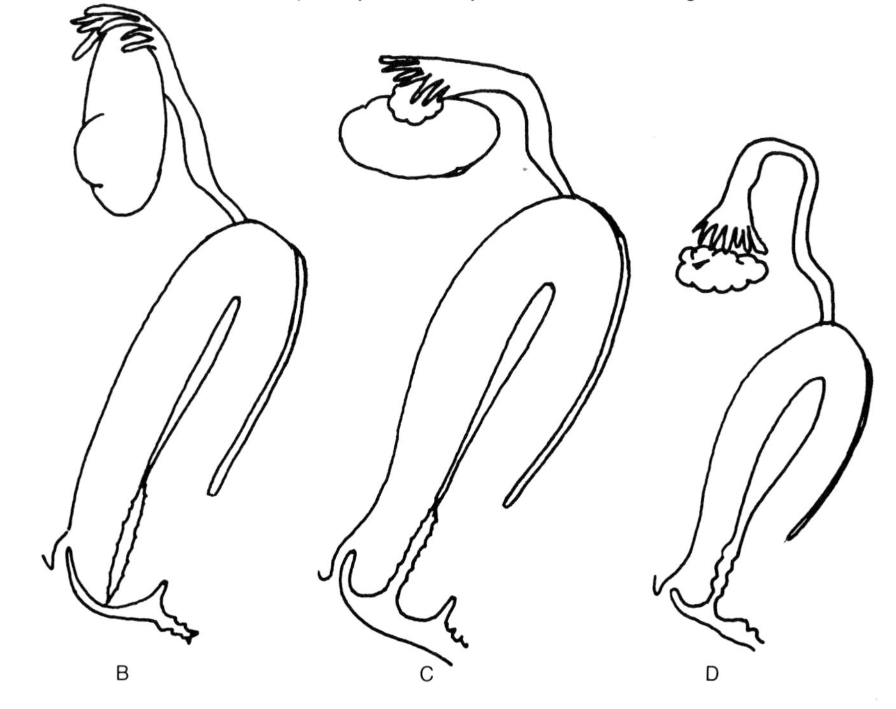

Medical Problems of the Female Internal Sex Organs

During the reproductive years and after the change of life, some women experience certain medical problems related to their genital organs. In the earlier years, infections, both of the lower genital tract and of the uterus and tubes, may occur. Also fairly common is a disease known as endometriosis, in which pieces of the lining of the uterus, or the endometrium, pass out through the tubes, perhaps at the time of menstruation, and lodge and grow on the ovaries and other pelvic organs. This tissue enlarges and may bleed and cause discomfort at the time of menstruation.

As a woman enters her middle years, she may be subject to the growth of fibroid tumors of the uterus. In fact, about 30 percent of women have such tumors by the time of menopause. They are not malignant, are usually small, and cause no difficulty. They frequently decrease in size after menopause.

In the middle years, also, cancer becomes a possibility. The commonest site for cancer is in the cervix. Early stages may appear in a woman's twenties or perhaps thirties, and the fully developed disease is usually seen in the forties. Fortunately, cancer of the cervix can be detected in the preliminary phase by regular examinations and "pap" smears.

In the fifty- and sixty- year age group, cancer of the body of the uterus may occur. Its appearance is usually heralded by irregular bleeding. The possibility of uterine cancer emphasizes the necessity of continuing regular checkups and reporting any irregular bleeding after menstruation has ceased. At this age, or even before, some women may notice protrusion of the uterus or the walls of the vagina out of the external opening. This protrusion is due to relaxation of the ligaments and tissues, which may be due in part to childbirth and in part to the aging process. When discomfort is marked or symptoms are disabling, surgical treatment can provide a cure.

Another concern for women in their middle and later years is cancer of the breast. It is the most common type of cancer among women.

The causes of breast cancer are not known. It is found more frequently in women who have begun their families later in life. Breast feeding does not seem to be related to breast cancer, although, interestingly enough, this cancer is less frequent in countries in which breast feeding rates are high. Recently, a theory has been suggested that virus particles in breast milk may be related to the later development of breast cancer. However, this theory has not yet been proved.

Early detection of breast cancer can reduce its severity. Regular examination by a woman of her own breasts, especially immediately after menstruation, can greatly aid early detection. Examination by a physician at the time of regular annual checkups is also very important.

The exact diagnosis of breast cancer requires a biopsy, or removal of a piece of tissue for examination under the microscope. Surgical removal of a part of or all of the breast is usually required to treat the disorder. Often, X-ray therapy is given as additional treatment.

Sex in Perspective

This chapter has concerned itself in some detail with the male and female sexual organs, their growth during intrauterine life, childhood, and adolescence, and their function during maturity. The comments on the normality of masturbation and nocturnal emission and on the wide limits of normal development at puberty should help set at rest some questions in the minds of readers to whom these subjects have posed concerns.

Also discussed were the physiological basis of intercourse; the phases of sexual excitement; the processes of menstruation, fertilization, conception, pregnancy, labor, and delivery; and the many changes that follow the birth of a baby.

All of this should place the relatively simple act of copulation in its proper perspective. Sex is not merely a simple genital union between a man and a woman but should be a rewarding and pleasurable part of life.

For further information on the subjects covered in this chapter, consult ***Encyclopædia Britannica:***

Articles in the Macropædia	*Topics in the Outline of Knowledge*
BIRTH CONTROL	425.D.3
EMBRYOLOGY, HUMAN	422.G.1
LACTATION, HUMAN	422.F.2.c.iv
MAMMARY GLANDS, HUMAN	422.G.2.c.vi
MENOPAUSE	422.F.2.c.v
MENSTRUATION	422.F.2.c.i
PARTURITION, HUMAN	422.F.2.c.iii
PREGNANCY	422.F.2.c.ii
REPRODUCTIVE SYSTEM, HUMAN	422.G.2
REPRODUCTIVE SYSTEM DISEASES	424.I.1,2,3, and 4

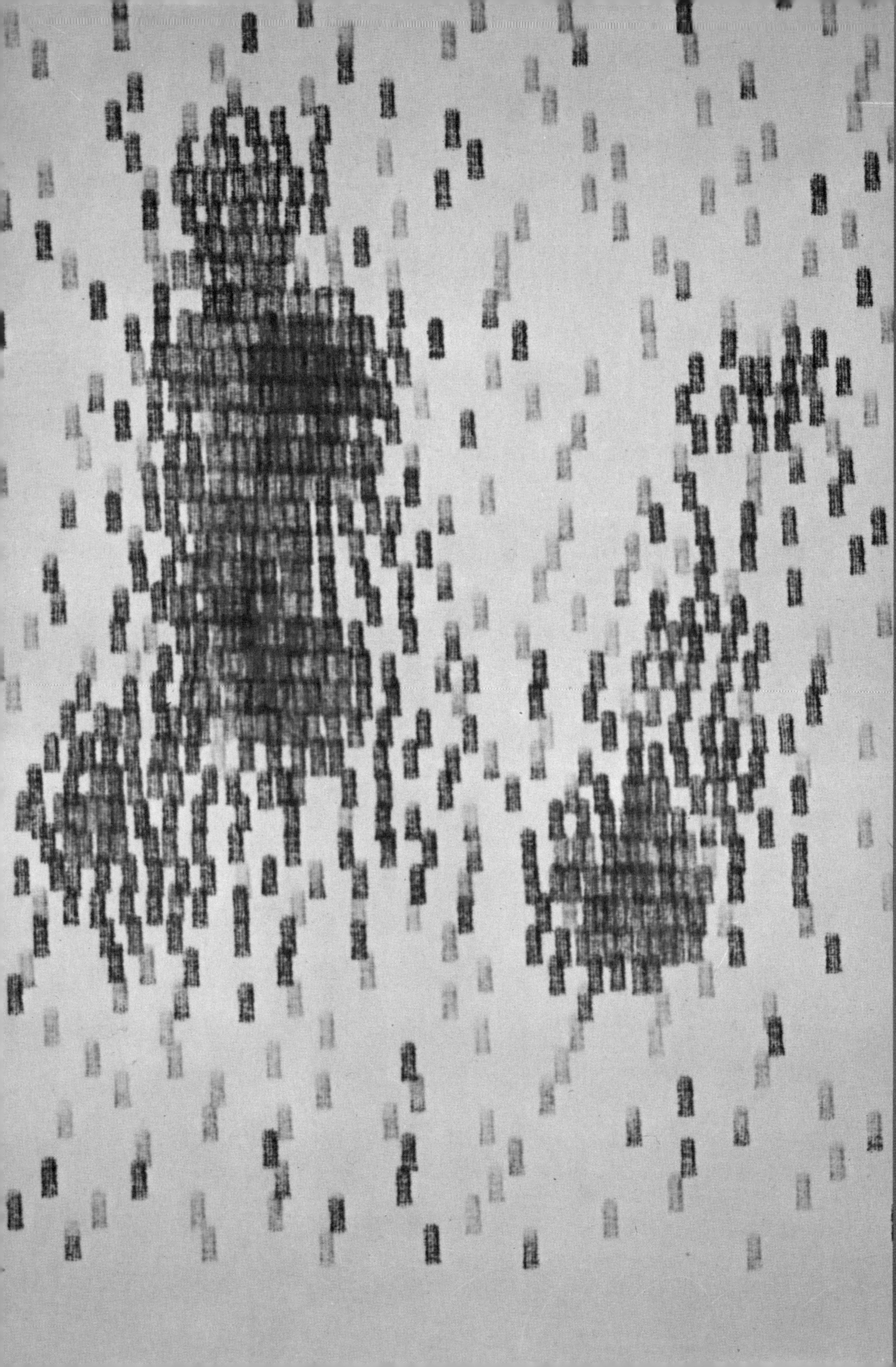

Such modern methods as the use of radioactive tracers in the body help physicians in the accurate diagnosis of disease.
Photo, Camera MD

Kessler. *Charles, this and the next chapter make a pair.*

Van Doren. *About the communication and control systems in our bodies—the systems that see to it that our parts work properly and together?*

Kessler. *Yes. The endocrine system provides a set of chemical controls, enormously subtle, and finely tuned so as to respond to both external and internal stimuli. The nervous system is more like a set of electrical controls, if possible even more subtle and finely tuned.*

Van Doren. *I'm glad I have them—and also glad that I'm not conscious of using them. These automatic systems free me—give me more time for voluntary activities.*

11

Endocrine System

BY WILL G. RYAN, M.D.

Hormones are chemical substances whose primary function is to preserve the well-ordered status of the whole body. They regulate metabolism—the combined chemical processes of the body. Hormones are not known to be utilized in any of these processes, but the chemical substances influence the rates at which metabolism occurs. Hormones are, therefore, catalytic agents causing either the inhibition or acceleration of the activities carried on in the tissues.

Various glands produce substances and release them into the body. Those that discharge their substances through ducts, such as oil glands in the skin or salivary glands in the mouth, are exocrine glands. If, however, an organ produces hormones and secretes them directly into the circulatory system without the use of any ducts, the organ is an endocrine gland. Some organs of the body carry out both functions: the ovaries and testes produce hormones distributed into the circulation and other substances secreted through the ducts. Some organs are occasionally misnamed glands even though they do not secrete: the lymph "glands" are glandlike but are nodes that manufacture certain of the white blood cells and play a role in ridding the body of harmful substances.

Endocrine Glands

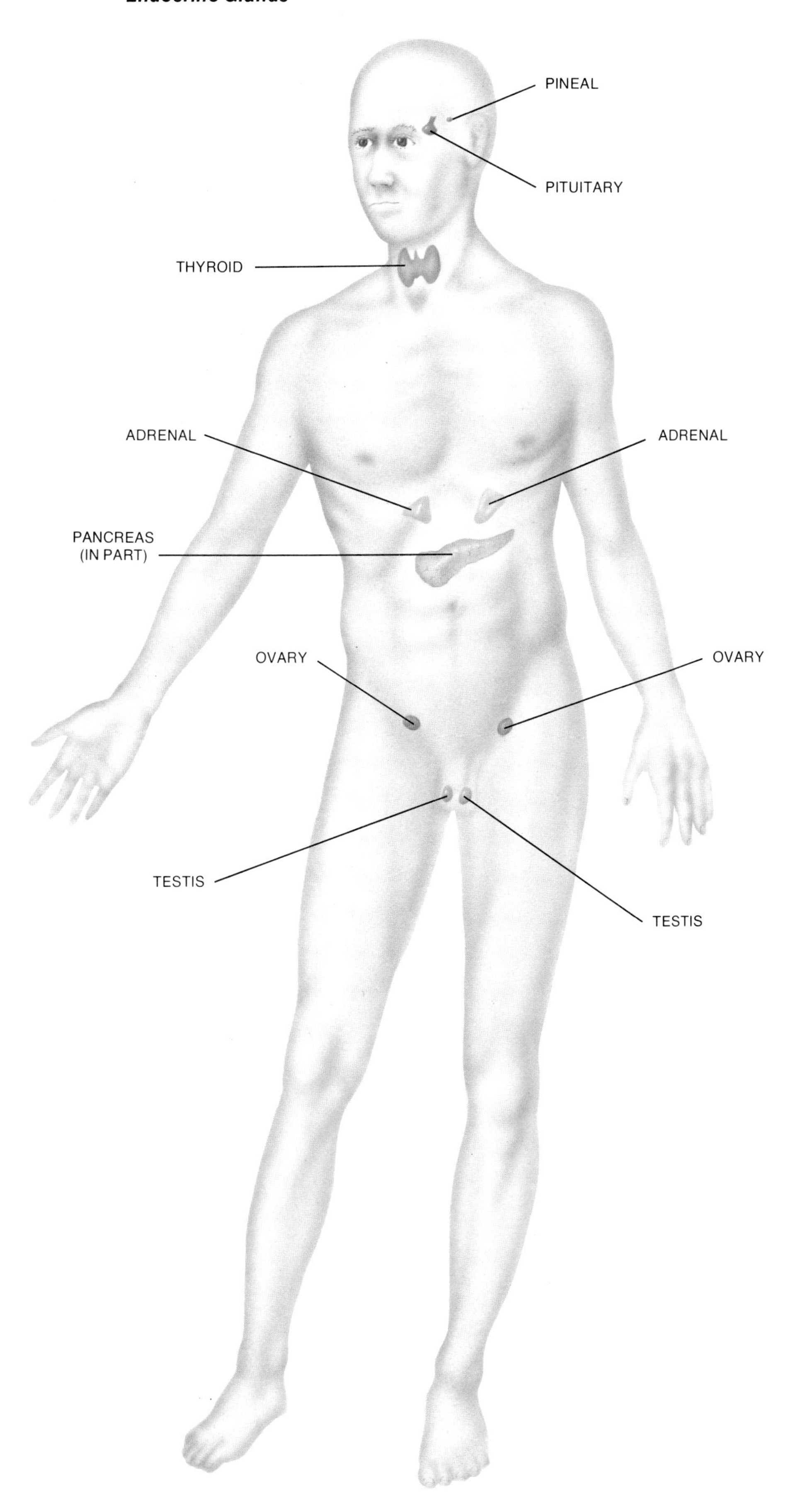

The organs, therefore, that regulate body processes through the action of hormones are the endocrine glands. These organs are all small ones and are located strategically throughout the body. Some are under the control of the nervous system, while at least one, the pituitary, regulates some of the others. All of them, however, are interrelated in their function: the activity of one will in many instances affect the activity of another as they all work together to keep the bodily processes in balance.

The chemicals secreted by the endocrine glands are of several types. Some are small proteins called peptides, while others are steroids, chemicals with ring structures of a special type. These hormones are found in the bloodstream in extremely low concentrations, often less than a millionth of a gram per milliliter of blood, but they are nonetheless very potent. They are transported through the bloodstream to the site of their action, where they combine with something on the surface of the cell called a hormone receptor. This linking sets up a chain of events that often includes the production of cyclic AMP, another body chemical common to many tissues that triggers any of a variety of events. Certain parts of the cell may become active and carry out protein synthesis, or other protein chemicals called enzymes may be activated so that they can regulate certain actions of the cell.

Some endocrine secretions are more necessary for life than others. Lack of insulin or of hydrocortisone, a secretion of the adrenal gland, is incompatible with life for more than a few hours. On the other hand, lack of secretion of the sex glands is not fatal but prevents propagation.

The Pituitary Gland

The pituitary gland is about the size of a cherry and is connected by a stalk to the hypothalamus, the lower part of the brain. The pituitary is divided into two portions, the anterior and the posterior.

The anterior pituitary is under control of the hypothalamus and is connected to it by a special set of veins called a portal vein system. It makes several hormones, most of them of the trophic type, so called because they effect the regulation of certain glands. The trophic hormones of the anterior pituitary are thyroid stimulating hormone (TSH), which controls the activity of the thyroid gland; adrenocorticotrophic hormone (ACTH), which controls the adrenal cortex glands; and follicle stimulating hormone (FSH) and luteinizing hormone (LH), which control the activity of the ovaries and the testes. Each of the hormones of the anterior pituitary is, in turn, regulated by the hormones (a specific one or more) made in the hypothalamus by specialized nerve cells and transported down to the anterior pituitary by the portal venous system. The hormones of the hypothalamus are called releasing hormones because they cause the release of the anterior pituitary hormones. The hypothalamus is, to some extent, under control of the higher centers of the brain, and nervous influences such as anxiety sometimes affect the function of the glands that are under its control.

Growth hormone (GH, or STH for somatotrophic hormone) is also secreted by the anterior pituitary. It indirectly controls growth, and probably other bodily processes. It probably causes the production in the liver of another hormonal substance, called somatomedin, and, through it, acts on cartilage tissue to stimulate bone growth.

Another hormone made by the anterior pituitary is prolactin, which is somewhat similar in structure to growth hormone and controls production of milk by the breast. This hormone also possibly has certain influences on the activity of sex glands.

The posterior pituitary gland is connected to the hypothalamus by a stalk of nerve tissue. The two hormones of the posterior pituitary are vasopressin, which regulates water excretion by the kidney, and oxytocin, which regulates the process of labor in delivering a baby. Both are small peptide hormones composed of about ten amino acids. They are made within the hypothalamus. After their production in the hypothalamus, they travel down the nerve stalk bound to a protein substance called neurophysin and are stored in the posterior pituitary until called upon for use.

The cells of the pituitary gland may have tumors that are either active and produce a hormone or inactive and destroy the surrounding pituitary tissue, which lowers the activity of the gland. One of the common types of pituitary tumors secretes excessive growth hormone. If such a tumor occurs in childhood, it causes excessive growth, and the person becomes a "pituitary giant," sometimes reaching eight or nine feet in height. If such a tumor occurs in adulthood, only the hands, feet, joints, and soft tissues become enlarged because the long bones of the body have stopped growing. This condition is acromegaly.

Other active tumors of the pituitary are quite rare. Inactive ones are more common. Because of the destruction of pituitary tissue that these tumors cause, a loss of the trophic and other hormones results in a lack of proper stimulation of the thyroid, adrenal, or sex glands, which, too, become inactive. These tumors may also create pressure on surrounding structures resulting in severe head-

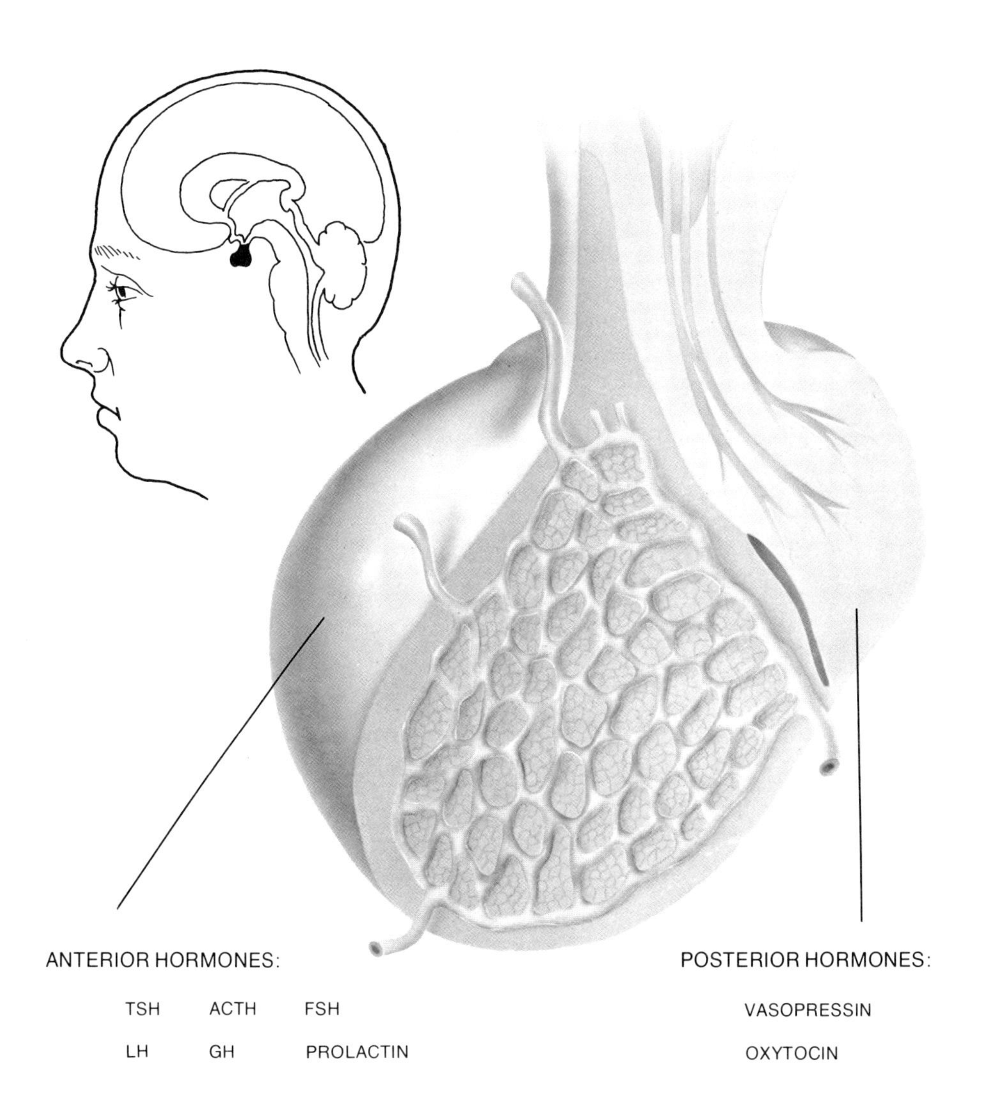

The pituitary gland has two parts: the anterior lobe, made of epithelial (lining-type) cells; and the posterior lobe, made of nerve-like cells. The hormones of the pituitary are described in the text.

aches and, sometimes, partial blindness by placing pressure on the optic chiasma, a place in front of the pituitary where nerves cross as they enter the eyes. Such inactive tumors respond to X-ray or surgical treatment, but if they have caused the destruction of sufficient pituitary tissue, a person will require supplemental thyroid, adrenal, and sex hormones. If the person is a child, growth hormone will also be needed for proper growth. Also, there are other as yet unknown reasons that cause the pituitary gland not to function properly. If they take place during childhood, they can prevent proper growth.

Glands under the control of the pituitary hormones also control these hormones. As the thyroid, adrenal, or sex glands secretions increase in the bloodstream, they cause the suppression of the release of their respective trophic hormones from the pituitary. This influence is called the negative feedback control mechanism. In addition there are other controls on the pituitary hormones. The secretion of growth hormone is very much influenced by the level of sugar in the bloodstream, and deprivation of water causes increased salt concentration in the blood, which, in turn, causes a release of vasopressin, which acts on the kidney to conserve water. Levels of prolactin secretion are stimulated by a baby's suckling of the mother's breast. (The pituitary in a male also secretes prolactin and oxytocin but with no known function.)

The Thyroid and the Parathyroid Glands

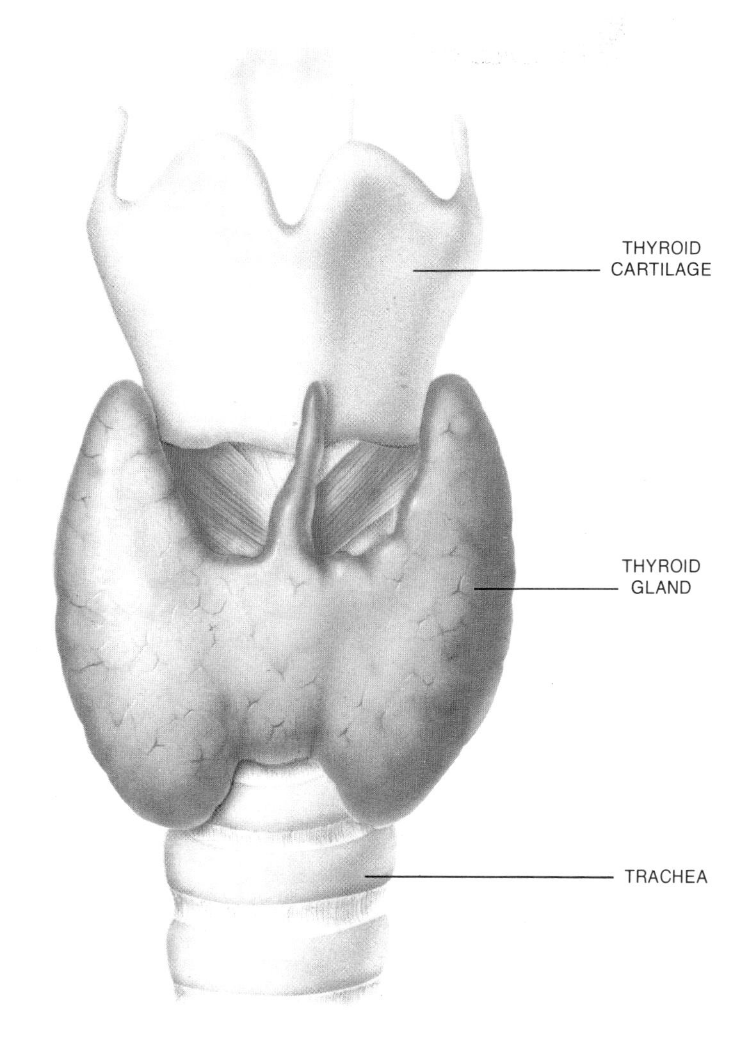

The thyroid gland (*thyreos* = shield in Greek) resembles a shield in appearance. Located in the lower front of the neck, the thyroid has two lobes, each about the size of a large pecan, on each side of the trachea, or windpipe. These are connected by a small piece of thyroid tissue called the isthmus. The thyroid gland makes three hormones, two of which are very similar, differing only slightly in molecular structure. These are made by a group of secretory cells in the thyroid and are called thyroxine and triiodothyronine. Together they compose what is commonly thought of as thyroid hormone. Its main function is to regulate the rate of the metabolic processes of the body, but the thyroid hormones are also essential for development of the brain in early life and for normal growth. The third hormone of the thyroid gland is made by cells that lie between the follicles. It is called calcitonin, because of its role in calcium metabolism.

Goiters

Whenever the thyroid gland is enlarged, the condition is called a goiter, which may be seen as a swelling in the lower, anterior neck. The thyroid gland is dependent on iodine for the production of its hormones. In world areas where diets lack iodine, goiters are frequent. Because iodine is now added to salt in many places, including the United States, iodine deficiency has become an uncommon cause of goiter. Other causes of goiter include pituitary deficiency and exposure to excessive radiation.

Some goiters may be underactive; others may be overactive; and still others may be normally active.

When overactive, goiters may be of two types: one, which is smooth, is associated with a condition called Graves' disease; the other type is nodular and tends to occur in later life. Both conditions cause excessive production of thyroid hormone, or hyperthyroidism, which speeds up the body processes and results in excessive heat production, nervousness, and fast heartbeat. When these conditions are severe, death may result.

When Graves' disease is present, a bulging of the eyes may also exist, associated with an impairment of the eye muscles called exophthalmos, or ophthalmopathy. When severe, this condition may cause pressure on the eye nerves and produce blindness. It can be treated by cortisone-like drugs or by surgery to remove part of the bony orbit around the eye to help to relieve the pressure. Graves' disease may be associated with thickening of the skin over the shins and, rarely, with clubbing of the fingers. In addition, a person with hyperthyroidism is usually hungry and frequently loses weight because the body is burning food faster than it is being consumed.

Middle-aged or elderly persons may get nodular goiters that produce similar signs and symptoms of hyperthyroidism. This kind of hyperthyroidism is not associated with protrusion of the eyes, as in Graves' disease, but is more likely to produce symptoms of heart failure because of the extra load placed on the heart by the excessive rate of metabolism.

Other Thyroid Disorders

Conversely, underactive thyroid function, or hypothyroidism, causes a slowing of the metabolic rate of the bodily processes in general. Persons with such deficiencies are sluggish and appear puffy, with dry skin and brittle hair. A mucuslike substance collects in the skin and brings about a swelling condition called myxedema. Hypothyroidism is caused by a number of conditions, one of which is an inflammation called thyroiditis. Underdevelopment of the thyroid or defects within the gland that result in defective thyroidal synthesis also may occur as a result of genetic defects. When hypothyroidism occurs in infancy, cretinism results. Because thyroid hormone is necessary for development of the brain, these persons are almost always mentally retarded. Characteristically, they also have a wide spacing of the eyes, a flattened bridge of the nose, a large tongue, a puffy appearance, and incomplete growth.

One of the most common causes of hypothyroidism is the treatment of hyperthyroidism with radioactive iodine. Iodine is concentrated in the thyroid

Goiter Formation

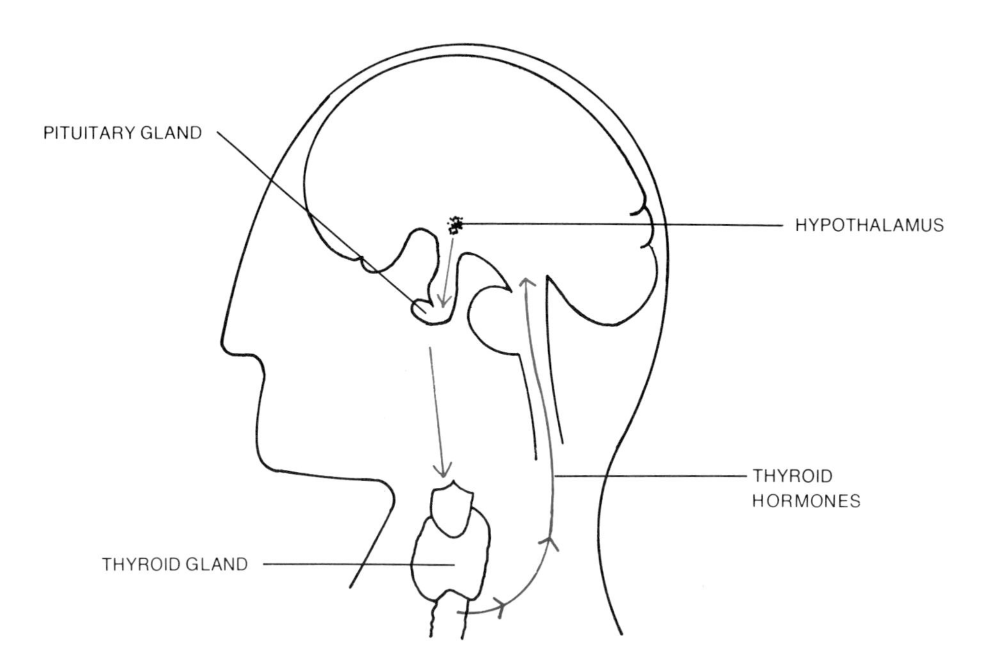

A substance from the hypothalamus causes the pituitary to release TSH, which signals thyroid cells to use iodine for thyroid hormones. At certain levels, they stop TSH production. But when iodine is insufficient, thyroid hormones cannot be made. In their absence TSH continues to stimulate thyroid cells, and they grow larger in the effort.

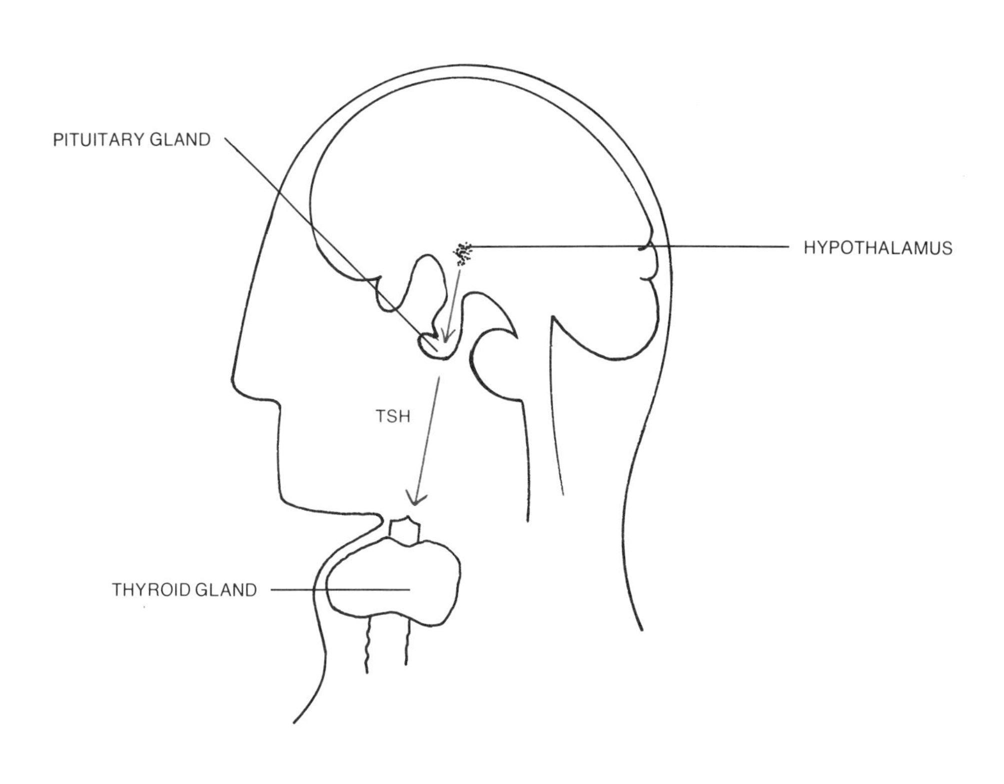

An underactive thyroid gland in infancy can result in cretinism (A). A goiter (B) can stem from several causes, but because of commercial iodized salt, goiter in the United States is rarely the result of iodine deficiency. Hand tremors (C) are common in an adult who produces too much thyroid hormone.

gland for use in the production of thyroid hormone. Radioactive iodine may be given to the hyperthyroid patient in sufficient dose to destroy most of the functioning thyroid tissue without harming other tissues of the body. In about one-third of the patients treated in this manner the radioactive iodine produces hypothyroidism. Surgery of the thyroid gland to treat hyperthyroidism or for removal of thyroid cancers may also result in hypothyroidism. The condition, however, is much easier to treat than hyperthyroidism: dried thyroid extract or thyroid hormone therapy taken orally can satisfy thyroid requirements.

Radioactive iodine is also useful in diagnosing thyroid diseases and in outlining or determining the size of the gland by a procedure called a thyroid scan. In addition, radioactive iodine helps in evaluating whether masses in the thyroid are the result of inflammation or of cancer.

Lumps or masses in the thyroid may be cancerous and are sometimes removed when the physician is not certain of their character. Some thyroid cancers are functional and produce excessive amounts of thyroid hormone, particularly when they have spread to other parts of the body. Some thyroid cancers are very slow growing; patients may live for more than twenty years after their conditions have been diagnosed, even though the cancer may not have been removed by surgery. A history of X-ray treatment in infancy or childhood increases the likelihood that a lump in the thyroid may be cancerous.

Many persons who are tired, sluggish, or somewhat overweight have these symptoms because of causes other than thyroid deficiency. Thyroid hormone has been used in the treatment of overweight, but such large doses are required that they cause undesirable side effects and may be dangerous. Usually, the dose of thyroid in weight control preparations is so small as to be ineffective. Abnormalities of thyroid hormone secretion may cause menstrual cycle irregularities. Thyroid is often prescribed by gynecologists to help women regulate their menstrual cycles, sometimes without a good diagnostic basis.

The thyroid may be underactive because it is not being properly stimulated by TSH. Tests can distinguish this condition from a defect within the thyroid that may cause hypothyroidism.

While circulating in the blood, thyroid hormone is bound to a specific protein called thyroid-binding globulin. The female sex hormones, or estrogens, found in birth control pills cause an increase of this protein in the blood, resulting in a corresponding increase in the amount of circulating thyroid hormone in the blood. This does not mean the patient is hyperthyroid. In making any diagnosis, therefore, the physician does not rely on laboratory findings alone but takes into account the patient's history of medication intake.

Parathyroid Glands

The parathyroid glands, usually four in number, are located in pairs just behind the thyroid and near the upper and lower ends of the lobes of the thyroid gland. Usually the parathyroids are about the size of a small pea. They, along with vitamin D, are the prime regulators of calcium and phosphorus metabolism. These two minerals, found in the bones in the form of crystals called hydroxyapatite, form the substance that makes the bones hard. The proper concentration of calcium in the blood is also necessary for the proper function of the body cells, and the parathyroid glands maintain this concentration of calcium in a very narrow range (9–11 milligrams per 100 milliliters of blood).

Parathyroid hormone is a very small peptide that is secreted in response to a lowering of calcium in the blood. Together with vitamin D, parathyroid hormone enhances the absorption of calcium by the gut and the resorption of calcium from bones. Parathyroid hormone stimulates the resorption of both calcium and phosphorus from bone by acting on certain cells called osteoclasts, which dissolve little parts of the bone and release their components into the bloodstream. Other cells called osteoblasts act at the same time to build up bone by taking calcium and phosphorus from the bloodstream. Thus, a dynamic equilibrium exists between blood and bone that is finely regulated by the parathyroids. Parathyroid hormone also acts on the kidney to retain calcium but steps up the kidney's excretion of phosphorus. Part of this effect results from an enhancement of the metabolism of vitamin D to a form that is active in the body. Food is a source of vitamin D, but it can also be made by the action of sunlight on a cholesterol-like substance in the skin.

Overactivity of the parathyroid glands, caused by tumor formation or enlargement, increases resorption of calcium from the bones, resulting in too much calcium in the blood. This excess of calcium is excreted by the kidneys and frequently becomes associated with the formation of calcium phosphate stones which may block the urinary channels. Therefore, the concentration of calcium in the blood of a patient with a kidney stone is usually checked. If hyperparathyroidism exists long enough, the dissolving of the bones may cause them to be weakened, deformed, and subject to fracture. Advanced kidney disease, which may involve phosphorus retention and lack of calcium absorption, may be associated with a condition called secondary hyperparathyroidism, which may include weakening of the bones, with fractures and deformities. Sometimes in marked overactivity of the parathyroid glands, the level of calcium in the blood rises so high that the calcium causes the heart to stop beating. Heart muscle cells are dependent on the proper concentration of calcium in the blood for their normal function.

Occasionally the cause of deficient parathyroid function, or hypoparathyroidism, will be unknown. More commonly, it is the result of damage during thyroid surgery. This damage is difficult to avoid because of the close proximity of the parathyroids to the thyroid and because of their small size. Hypoparathyroidism leads to a lowering of the calcium in the blood and sometimes causes marked, cramplike contractions of the muscles, a condition called tetany. Skeletal muscles, like the heart muscle, need the correct concentration of calcium in the blood to function properly.

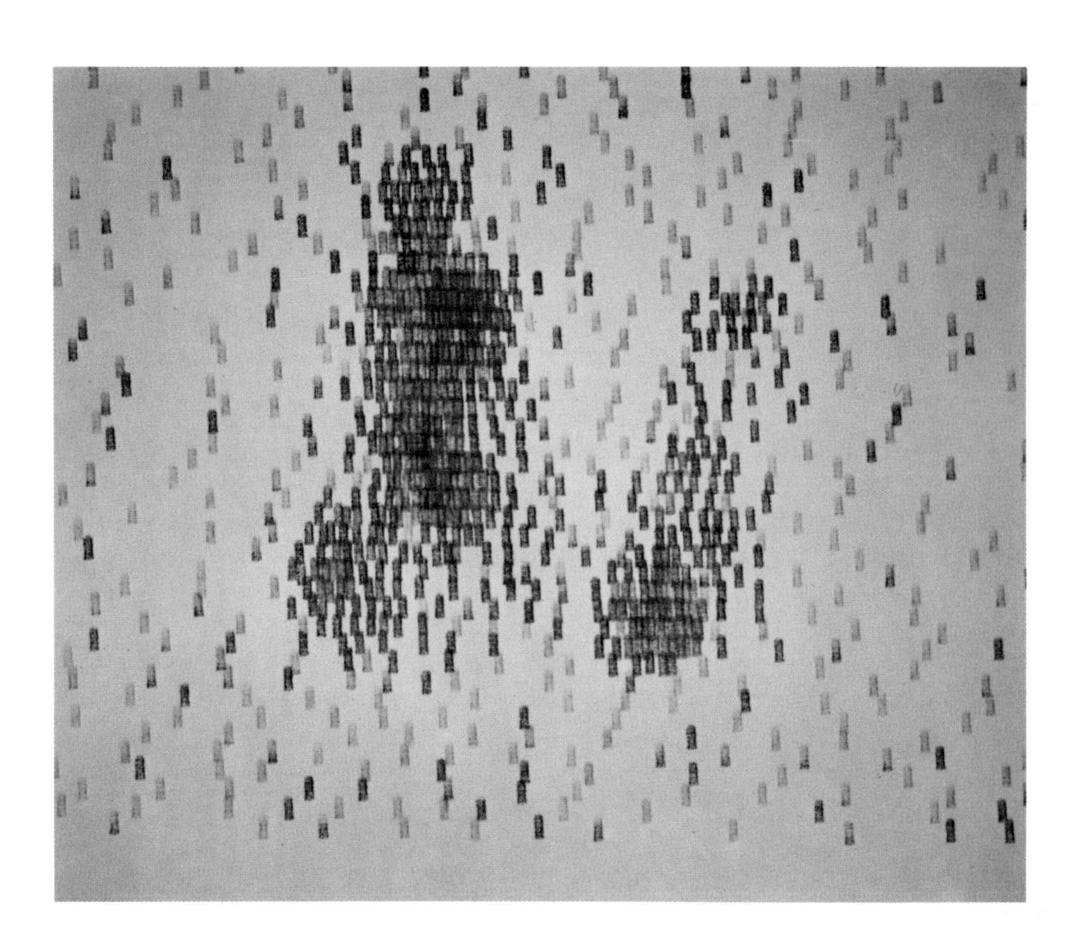

A thyroid scan can often detect an enlarged or diseased gland. Radioactive iodine given to a patient is taken up by his thyroid gland. A detector scanning the thyroid area registers the radioiodine uptake. The magnitude of uptake determines the size and activity of the gland.

Large doses of vitamin D, which enhances calcium absorption from the gut, are necessary to maintain normal blood calcium concentrations in patients with tetany. If excessive amounts of vitamin D are given, calcium levels in the blood may become too high and cause a harmful deposition of calcium in soft tissues, particularly in the kidneys. Therefore, proper use of vitamin D, or of any other vitamin for that matter, is important.

In persons with a certain genetic defect, a condition mimicking hypoparathyroidism, called pseudohypoparathyroidism, may be found. In these persons, the tissues fail to respond to normal amounts of parathyroid hormone. Also, softening of the bones, or osteomalacia, may result from a lack of absorption of vitamin D or calcium from the gastrointestinal tract because of an intestinal disease. Or, inadequate amounts of vitamin D during infancy or childhood may cause a similar bone-softening condition known as rickets.

The Adrenal Glands

The two adrenal, or suprarenal, glands exist with one located above each kidney. They are triangular in shape, and are about the size of a small Brazil nut. These glands derive their name from their positions adjacent to and above the kidneys, or renal organs. The adrenals are composed of an outer portion—the cortex—and an inner portion—the medulla. Although many people think these glands secrete only the hormone of the medulla, called either adrenaline or epinephrine, which helps to maintain blood sugar, increase the heart rate, and enhance nervous activity under conditions of stress, functionally the outer portion or cortex is the more important part.

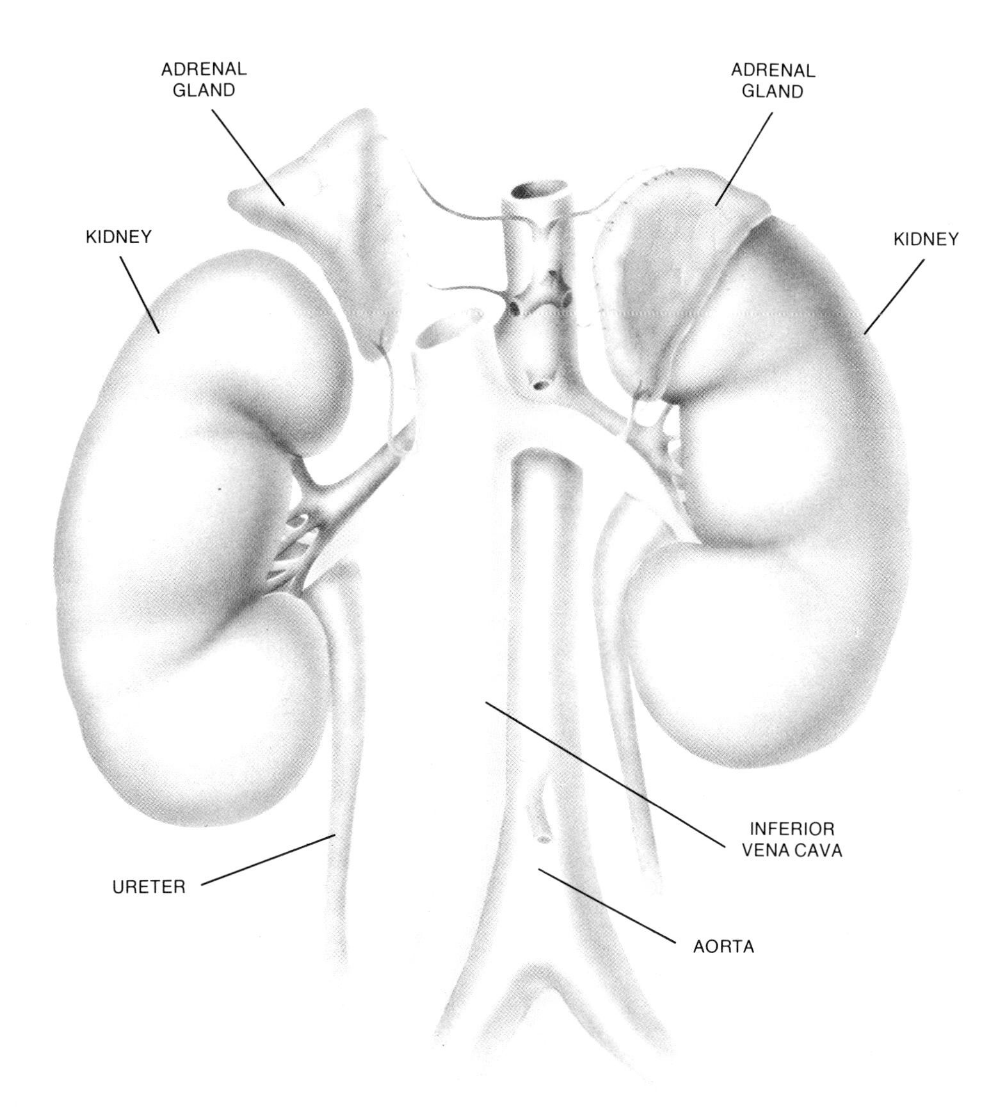

The adrenal cortex makes two main types of hormones, the more important of which is hydrocortisone, which is necessary for the maintenance of blood sugar, blood pressure, and, in conjunction with a very similar adrenal hormone, aldosterone, the proper salt content of the body (the electrolytes—sodium and potassium). These hormones are necessary if the kidneys are to retain proper amounts of sodium chloride and other salts. The adrenal cortex also produces hormones called androgens, which are chemically related to the male sex hormone testosterone and are necessary for maintaining the tissues of the body. They help to counteract the effects of hydrocortisone, which tends to break down body protein and turn it into sugar (glucose), the so-called catabolic effect. Reversal or resistance to the catabolic effect is called an anabolic effect.

Adrenal Disorders

The adrenal cortex may become overactive as the result of excessive growth, known as hyperplasia or hypertrophy. This effect is usually secondary to overactivity of the pituitary gland, which causes production of too much ACTH, the trophic hormone regulating the adrenals. Tumors of the adrenal gland may also cause this condition. In hyperplasia too much hydrocortisone, and sometimes too much of the androgens, is produced. Excess hydrocortisone results in the conversion of too much protein into glucose. This chain of events may cause diabetes mellitus and a bone-softening condition known as osteoporosis. A wasting of the muscle and other tissues follows and a thinning of the skin is particularly noticeable. Excessive amounts of hydrocortisone also cause accumulation of fat around the trunk and in the face, resulting in a round, "moon" face. The excess androgens cause excessive growth of body and facial hair, known as hirsutism. High blood pressure is also associated with the condition, caused by excessive retention of salt by the kidneys. This constellation of symptoms and signs is known as Cushing's syndrome.

Sometimes a tumor of the adrenal will develop that produces only aldosterone, which has primarily salt-regulating properties and little if any effect on protein or glucose metabolism. In this condition high blood pressure is seen, along with wasting of body potassium, without the other features characteristic of Cushing's syndrome. Treatment of these conditions may be accomplished by surgical removal of the adrenal glands or the tumor, or, in some cases, by administering a drug that causes destruction of adrenal tissue.

Underactivity of the adrenal glands usually results from destruction of the adrenal cortex by unknown causes. But destruction may result from tuberculosis, a fungus infection called histoplasmosis, or a cancer that has metastasized or spread, to the adrenals from another part of the body. When underactivity of the adrenal glands is present, a loss of body salts results in low blood pressure, and inadequate maintenance of the blood sugar gives rise to a condition known as hypoglycemia. Persons with these conditions have Addison's disease and must take hydrocortisone, or a similar drug, every day, increasing the dose when subjected to stresses. A deficiency in hydrocortisone is critical after more than a few hours; this hormone must be replenished if the body is to continue to function.

Tumors of the adrenal medulla cause excessive secretion of epinephrine, or the closely related hormone norepinephrine, which has marked blood-pressure-raising properties. Persons with these tumors, called pheochromocytomas, usually have very high blood pressure, often associated with episodes of sweating and palpitation resulting from the release of excessive epinephrine. The only effective treatment is removal of the tumor. Some cancerous tumors of this type may, if inoperable, be temporarily controlled by drug therapies. No clinical condition exists that is known to be associated with a deficiency of the adrenal medulla.

Hydrocortisone or cortisone-like drugs are used to treat many conditions, notably those associated with inflammation, such as arthritis or certain diseases of the skin. When these drugs are given in either therapeutic or excessive amounts over long periods, a condition mimicking Cushing's syndrome, with all its bad effects, results. Epinephrine dilates the bronchial tubes of the lungs and is used to treat asthma. This hormone can also be used as an emergency heart stimulant.

The Islets of Langerhans

The islets of Langerhans are very small glands, about a million in number, that are embedded within the substance of the pancreas. They make up about 1 percent of the total weight of the pancreas, the remainder of which is an exocrine gland that secretes enzymes into the digestive system. The islets of Langerhans, named for the German pathologist Paul Langerhans, who when he was a medical student in 1869 was the first to describe them, are composed of at least three varieties of cells and produce at least three types of hormone. The most prominent hormone, insulin, is secreted by the beta, or B, cells of the islets. Insulin is necessary for the control of sugar (carbohydrate), protein, and fat metabolism. Lack of this hormone causes diabetes mellitus. The alpha, or A, cells of the islets secrete a hormone called glucagon, which has effects opposite to those of insulin in that it raises blood sugar by causing a breakdown of protein and glycogen, a form of starch in which sugar is stored in the body. The delta, or D, cells of the islets produce the hormone gastrin, which regulates the secretion of acid by the stomach. All of these are peptide hormones.

Diabetes Mellitus

Diabetes mellitus is the most common endocrine disorder. An absolute deficiency of insulin is in-

A section of an islet of Langerhans is shown below. As discussed in the text, alpha cells in the islet produce insulin, and beta cells produce glucagon. Both substances are hormones.

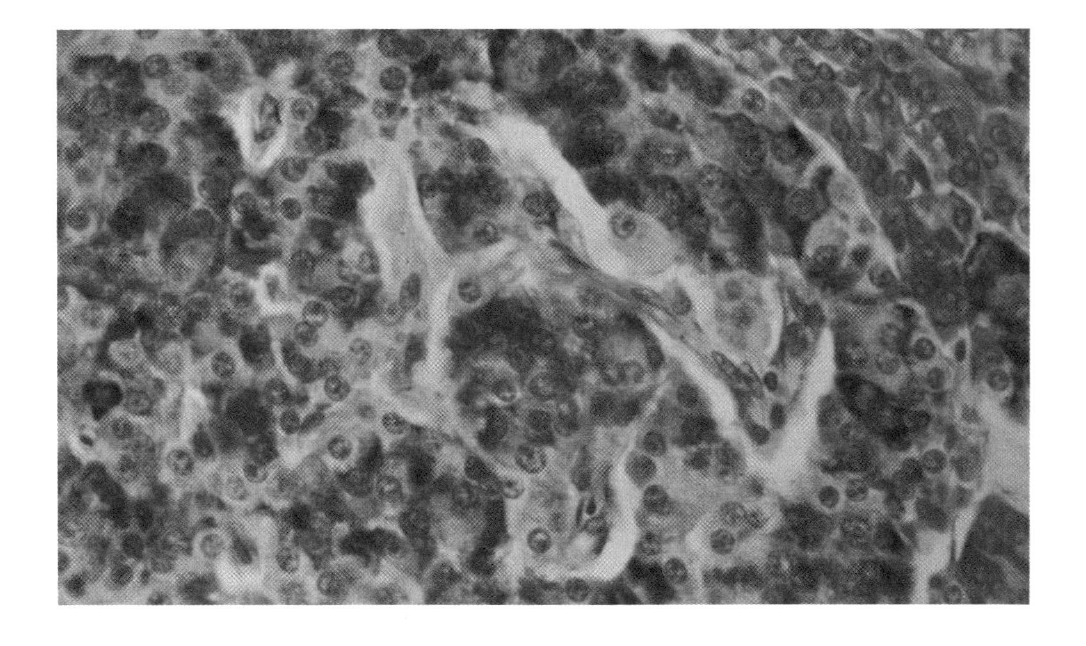

compatible with life for more than a few hours or days. However, the ability to prepare insulin extracted from beef and pork pancreas, established in 1921 by Frederick Banting and Charles Best at the University of Toronto, has proved to be truly life-saving for millions of diabetic persons.

Insulin has several functions, the most prominent of which is to enhance the transfer of glucose from outside to inside the cells. A cell lacking in glucose cannot function normally and begins breaking down its own protein. Unable to enter the body cells properly without the aid of insulin, sugar increases in the bloodstream. When its concentration gets above a certain level, the so-called renal threshold of the kidneys, the sugar spills over into the urine. Excessive urinary sugar promotes an abnormally high flow of urine, or diuresis, and the body loses excessive quantities of both sugar and water. These circumstances bring on the classic symptoms of diabetes: polyuria, or excessive urination; polydipsia, or excessive thirst or drinking; and polyphagia, or excessive eating. The latter two result from the body's attempts to make up for the excessive wastage of water and calories. Since insulin also suppresses the mobilization of fatty acids from fat tissue, a lack of insulin will also bring on an excessive mobilization of the body fat in the form of fatty acids. Once mobilized, the excessive amounts are converted in the liver to ketoacids. If the condition is not corrected by additional insulin injection, ketoacidosis results. If it becomes severe, the patient lapses into a coma and will die without proper and prompt treatment.

Milder degrees of diabetes, usually occurring in adult life, are associated with a partial deficiency of insulin. Persons with this deficiency have a moderate elevation of the blood sugar and may or may not spill sugar in their urine. While they are less subject to the onset of ketoacidosis, they are subject to other complications of diabetes that are primarily the result of degenerative changes in both large and small blood vessels. Even though their diabetes is apparently under control, these persons are subject for some reason to the development of hardening of the arteries at an accelerated rate and frequently experience heart attacks or strokes at a comparatively young age. Such diabetics are also subject to small vessel complications related to leakage of blood and protein from the capillaries. The leakage results in damage to the retina of the eye, which, when severe enough, causes blindness. Similar changes about the renal vessels may result in impairment of kidney function. The longer a person has diabetes, the more likely he is to incur such complications. For instance, more than 90 percent of persons who have diabetes for more than twenty years have one or more of these complications. Paradoxically, it seems that occasionally a patient with the mildest of diabetes may have the worst complications.

Diabetic women have a much greater risk than others of having stillborn babies and must be very carefully treated during their pregnancies. Since the diabetic has a great chance of developing cardiovascular disease, extra efforts should be made to avoid other risks, such as smoking, overweight, and excess cholesterol in the diet.

Diabetes is treated by a special diet, and the administration of insulin when necessary. Oral drugs that lower blood sugar by inducing pancreatic secretion of insulin, or by other unknown mechanism, have been used since the 1950s. But recent evidence has raised doubt that such oral drugs prevent the long-term complications of diabetes and has suggested that they may even have harmful effects.

Although diseases associated with excessive secretion of glucagon or glucagon-secreting tumors are extremely rare, they may cause a mild-to-moderate diabetes when they occur. Gastrin-secreting tumors cause a condition associated with excessive secretion of hydrochloric acid, resulting in severe ulcerations of the stomach and intestines and sometimes associated with a watery diarrhea, the Zollinger-Ellison syndrome. Insulin-secreting tumors are also rare but cause hypoglycemia when present. The only effective treatment of this condition is removal of the insulin-producing tumor. Persons who are chronically fatigued or nervous may think they have hypoglycemia but usually do not. Some of these persons may be depressed or anxious. Frequently, it is difficult to convince them that their symptoms are psychogenic.

The Sex, the Pineal, and the Thymus Glands

The pituitary gland, by means of the trophic hormones, controls the sex glands, the ovaries and the testes, which serve to reproduce the species. The ovaries produce estrogen and progesterone. Both are steroid hormones and are secreted in a cyclic manner, causing the buildup and later breakdown of the lining of the uterus that results in menstruation. Associated with this cyclic activity of the hormones is the production of the ovum, or egg, which is released from the ovary once a month about midway between periods of menstrual flow. These ovarian hormones, particularly estrogen, also cause the development of the secondary sex characteristics of breast enlargement, rounding and softness of the female figure, and hair in the pubic area. They also enhance development of the cells lining the vagina that produce a vaginal mucous secretion.

Estrogen and progesterone derivatives are components of birth-control pills and suppress ovulation by inhibiting release of FSH and LH from the pituitary, an example of the negative feedback mechanism. During pregnancy, the placenta, or afterbirth, also makes large amounts of estrogen and progesterone, as well as hormones known as placental lactogen and chorionic gonadotrophin. Lactogen stimulates lactation, or milk production, while gonadotrophin has a stimulating effect on the gonads (sex glands). The ovaries also produce small amounts of male hormones, (androgens). Sometimes these are produced in excessive amounts and interfere with ovulation and cyclicity of the estrogen and progesterone. The result may be irregular or absent menstrual periods, excessive growth of body hair, and lack of ovulation. Rarely, tumors of the ovaries can produce excessive amounts of estrogen or androgen and interfere with menstruation.

Only very small amounts of ovarian hormones are secreted in childhood. A biological clock, a poorly-understood mechanism probably residing in the brain, turns on the release of the pituitary trophic hormones so that the ovaries are stimulated about the time of puberty. Occasionally, tumors or other unknown mechanisms will cause this change to occur early, resulting in a precocious puberty.

The male sex glands produce testosterone, a steroid hormone similar in structure to estrogen. In addition to stimulating the formation of sperm by the testes, this hormone also causes an increase in the size of the penis, development of body and facial hair, and an enhancement of muscle tissue that accounts for the greater strength of men over women. Testosterone also brings on the enlargement of the larynx and thickening of the vocal cords that lowers the voice. Androgens, of which testosterone is the most potent, are also instrumental in producing the sex drive. Partial destruction of the testes may occur when they are infected by the mumps virus. This sometimes gives rise to sterility when both testes are involved but usually does not result in impotence. Impotence often has psychological causes.

The Pineal Gland

So named because of its resemblance to a tiny pine cone, the pineal gland is a small endocrine organ about the size of a peanut and is embedded deep between the two hemispheres of the brain. This gland was regarded by the 17th-century philosopher René Descartes as the seat of the soul.

The function of the pineal gland has only recently been discovered. To some extent it controls the sex organs, at least of the lower animals, responding to the presence or absence of light by producing melatonin, a hormone that suppresses the activity of the gonads. Exposure to light inhibits the production of melatonin, allowing more sexual activity in lower animals when exposed to light for long periods.

Whether this gland has a significant function in man is unknown, although it may be part of the biological clock system that turns on puberty. Tumors adjacent to the pineal in childhood have been associated with the development of precocious puberty, while tumors of the pineal have resulted in delayed puberty. The pineal calcifies after puberty and can be seen in the midline of the skull on X rays.

The Thymus Gland

The thymus gland is located in the upper part of the chest just behind the breastbone. It is about the size of a small pecan in adults but is about three or four times as large in children. The hormonal function of the thymus was unknown until recently, and researchers are continuing to learn more about it. It is now known to be of importance in the lymphatic system of the body, producing thymosin, a hormone that stimulates the development of this system.

Tumors of the thymus gland are occasionally associated with defective transmission of impulses from nerve to muscle, known as myasthenia gravis. Removal of these tumors sometimes results in improvement of the condition.

THE SEX GLANDS

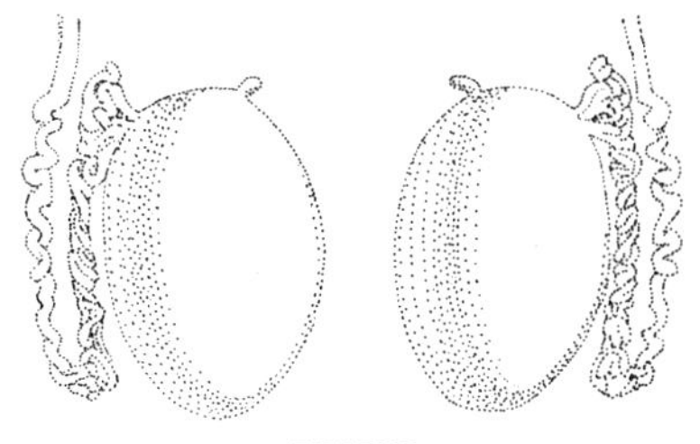

TESTES

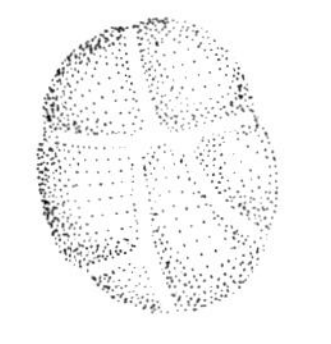

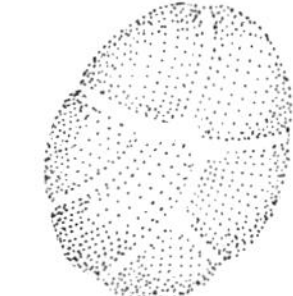

OVARIES

THE PINEAL GLAND

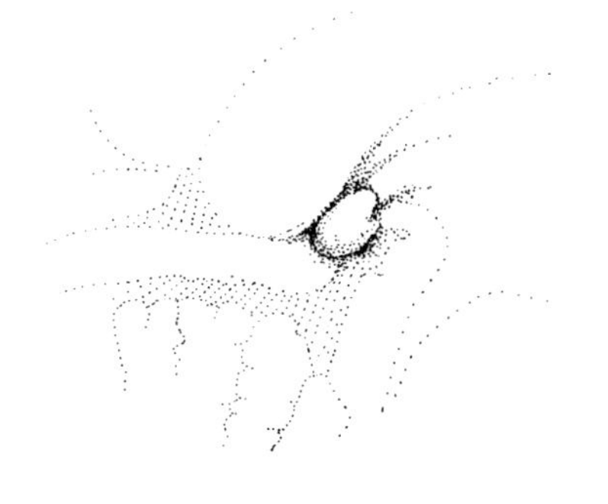

THE THYMUS GLAND

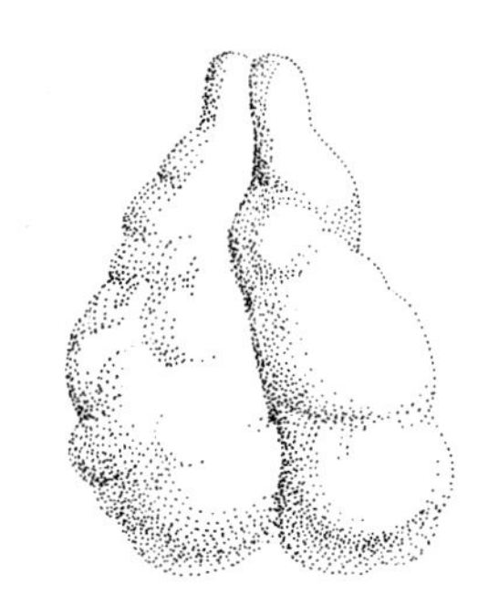

Non-Discrete Hormone Systems

Thus far this discussion of the endocrine glands has been confined to tissues or organs that are discrete and do not have other functions. However, some of the endocrine organs of the body cannot be so distinctly localized. Although the cells of these organs produce definite hormonal substances, they are scattered among other tissues. The gastrointestinal hormones are a good example of this arrangement.

Gastrointestinal Hormones

Interspersed in certain areas throughout the mucosa, or inner lining of the gastrointestinal tract, are cells that have definite endocrine functions, mostly concerned with controlling the gastrointestinal tract itself. One gastrointestinal hormone, gastrin, which has been mentioned in connection with the delta cells of the islets of Langerhans, is found primarily in the antrum of the stomach, near its juncture with the duodenum. As mentioned, the prime function of this hormone is to stimulate the secretion of acid by the stomach. However, gastrin also stimulates the secretion of pepsin, an enzyme of the stomach that breaks down protein, and the secretion of intrinsic factor, a protein produced by the stomach lining that aids in the absorption of vitamin B_{12}. It also has other effects that are less prominent.

Another gastrointestinal hormone, secretin, which can be extracted from the mucosa of the small intestine, stimulates the exocrine portion of the pancreas to secrete fluid and an alkaline substance, bicarbonate. It also increases the rate of bile secretion and its bicarbonate content can inhibit gastric motility, and stimulate secretion of intestinal juice. The most effective stimulus to the secretion of secretin is acidification of the duodenal mucosa, which occurs when the stomach empties its contents into the duodenum.

Pancreozymin-cholecystokinin is one hormone with two names. It was originally thought to be two hormones because of its widely differing actions on the gastrointestinal tract. The first name, pancreozymin, results from the fact that this hormone, too, can be extracted from the mucosa of the small intestine and causes the secretion of pancreatic enzymes; it can be separated from secretin. The hormone can also be extracted from the area of intestinal mucosa that contracts the gallbladder, giving rise to the latter name, cholecystokinin. In addition to these activities the hormone has been shown to stimulate the secretion of gastric acid.

Hormones of the Liver and Kidneys

Under this same classification of hormones that do not come from specific glandular tissues or organs, the liver may be considered to have endocrine gland actions, namely the production of somatomedin in response to its being stimulated by growth hormone. As mentioned in the discussion of the pituitary, somatomedin is the substance that acts on cartilaginous tissue to produce normal growth. Thus the liver may be considered to have an endocrine gland function.

Another organ that displays prominent endocrine function is the kidney. In response to a decrease in blood pressure or contraction of salt or plasma volume in the body, it produces an enzyme called renin. Renin is produced by the so-called juxtaglomerular apparatus, clusters of cells adjacent to the glomeruli. Renin acts on a substance produced by the liver to produce angiotensin II, a potent hormone that constricts the blood vessels in order to maintain blood pressure. Angiotensin II stimulates the adrenal gland to produce aldosterone, which has already been mentioned in the discussion of the adrenal glands. The kidneys also produce erythropoietin, which influences the formation of red blood cells.

The Prostaglandins

Another interesting group of hormonal substances that are not localized to an endocrine gland is the prostaglandins, so named because they were first discovered in the tissues of the prostate, an exocrine gland of the male reproductive system. Prostaglandins have their highest concentration in the seminal fluid produced by this gland but are found

During embryonic development, exocrine (ducted) glands (A) develop from chords of cells that form tubes through which secretions can flow. Endocrine (ductless) glands (B) lose their connections with surface-lining cells and come in close contact with blood vessels, into which endocrine secretions pass for circulation throughout the body.

A

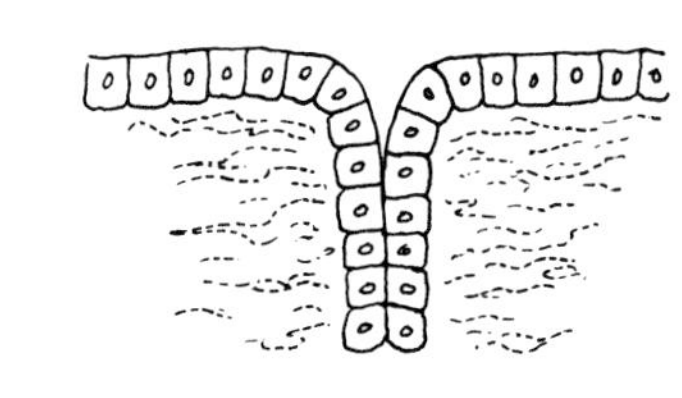

B

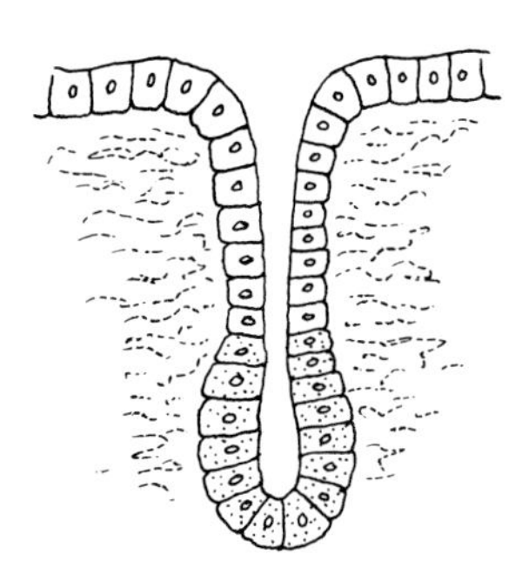

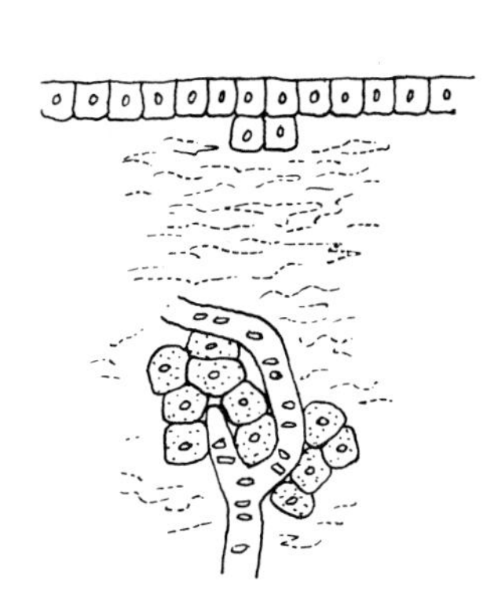

in many tissues of the body. These hormones have received much attention in the past few years because of their possible therapeutic use. The four main series of the prostaglandins are designated by the letters E, F, A, and B, corresponding to differences in their chemical structures.

The effects of the prostaglandins on the smooth muscle of the female reproductive tract are being extensively studied, particularly in relation to the possibility that prostaglandins may be used as birth control agents. There is evidence in several species that one of the prostaglandins causes luteolysis, the breakdown of the luteal follicle that develops on the ovary following ovulation. Certain prostaglandins have also been infused to induce therapeutic abortion.

Prostaglandins A and E are among the most potent substances known for dilating blood vessels and conceivably may play an important role in the control of blood pressure. These prostaglandins are also of possible importance as local mediators of blood flow. Other effects of prostaglandins include inhibiting the aggregation, or clumping, of the small formed elements of the blood known as platelets, stimulating the central nervous system, and inhibiting the breakdown of fat tissue. Prostaglandins also have an effect on the production of cyclic AMP.

In conclusion, prostaglandins have many different kinds of pharmacological activity, some of which may reflect physiological roles. Less orthodox hormonal roles have also been proposed. The absorption of seminal prostaglandins from the vagina after sexual intercourse provides an example in which substances produced by one person may act as circulating hormones in another. Although the evidence for this transfer of hormones is far from conclusive, the antifertility effects of prostaglandins are of great potential significance. The potential therapeutic significance of the prostaglandins is limited by the great variety of pharmacological effects that they possess, but the active research being carried on should clarify their functions over the next few years.

Scientists in medical research learn how the body uses biochemicals for growth and development. Hormones, prostaglandins, and other biochemicals were found through diligent research.

"Silent" Processes of Life

The endocrine glands are basic to the body's functioning. True, the beating of the heart, like the inspiration of air and the intake and elimination of food, is necessary to existence. But behind these obvious manifestations of eating, moving, and breathing lie the secret and silent endocrine processes that monitor growth, balance the ebb and flow of energy, preserve the sexual rhythms, maintain blood pressure, give protection from stress, and regulate the multitudinous inner activities that are the essence of animal life.

For further information on the subjects covered in this chapter, consult ***Encyclopædia Britannica:***

Articles in the Macropædia	*Topics in the Outline of Knowledge*
ENDOCRINE SYSTEM, DISEASES AND DISORDERS	424.G.1,2,3,4,5,6,7, and 8
ENDOCRINE SYSTEM, HUMAN	422.F
FLUID AND ELECTROLYTE DISORDERS	424.A.4
HORMONE	422.F.2
METABOLISM, DISEASES OF	424.A.2

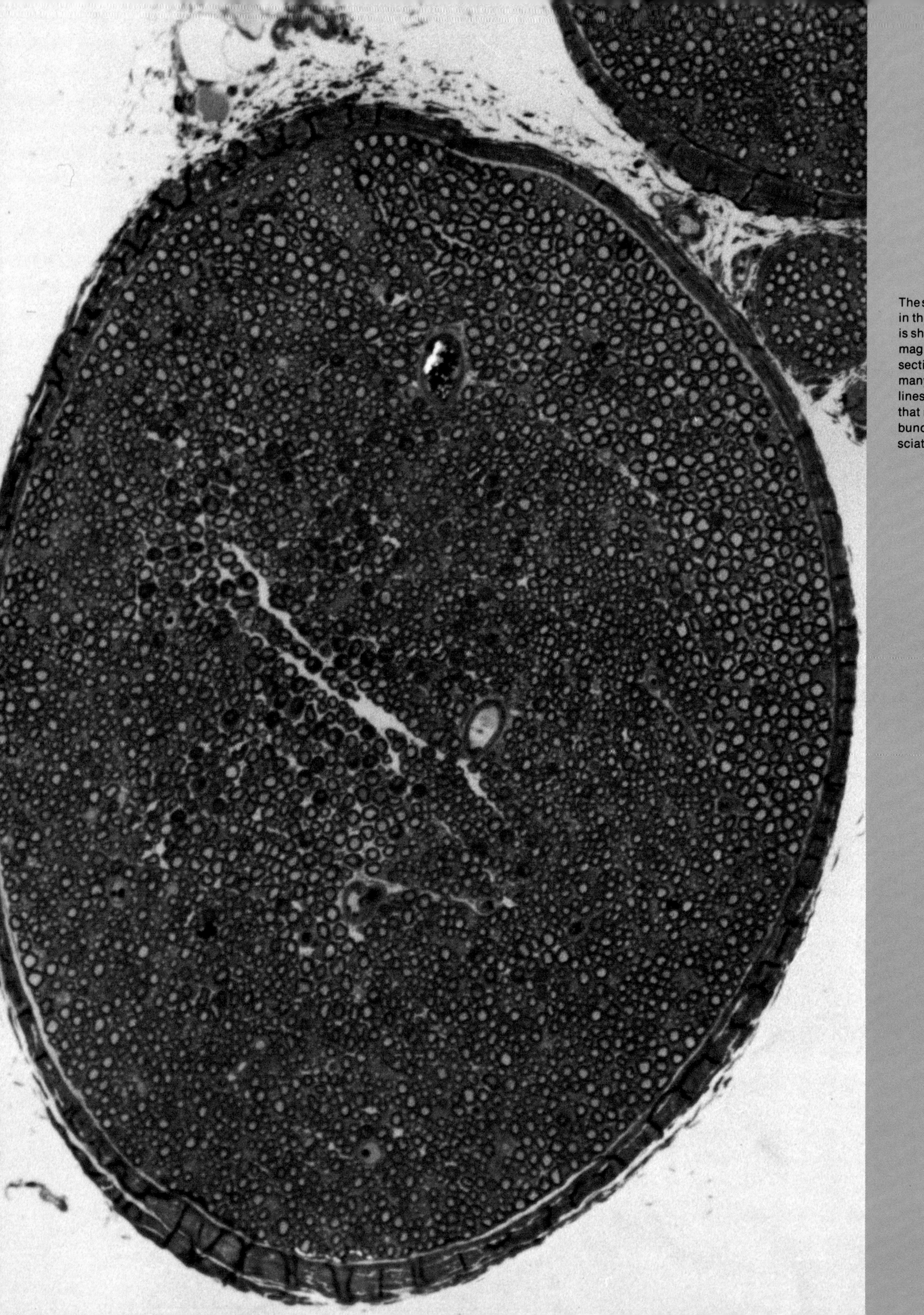

The sciatic nerve, largest in the mammalian body, is shown in this greatly magnified cross-sectional view. The many circles are outlines of the nerve fibers that make up a nerve bundle such as the sciatic nerve.

Kessler. *I like what Dr. Vick says in his first paragraph—that the body exists to support and nourish the brain.*

Van Doren. *Of course, he might have said the mind instead of the brain.*

Kessler. *There speaks the humanist and philosopher. The neurologist is more concerned with the brain. Actually, the mind—more the purview of the psychiatrist—is discussed in the next chapter, on the Integrative System. That's how fragmented medicine is.*

Van Doren. *I still wish he had said the mind.*

Kessler. *Philosophers are as stubborn as physicians.*

12

Nervous System

BY NICHOLAS A. VICK, M.D.

The nervous system has a unique role in the functioning of the body. Anyone who studies this system with any serious intent usually begins to believe that the remainder of the body exists solely to support and nourish the brain and spinal cord, together called the central nervous system, and the peripheral nerves that carry impulses to and from the central nervous system.

Over countless millennia, the brain has evolved into a complex organ composed of billions of cells with an exceedingly complicated pattern of organization. It is the part of the human body that endows man with his humanity. It is "through" the brain that we love, hate, aspire, and feel success or failure. Specific regions of the brain "produce" speech, see, hear, and have total control over movement and sensation.

In a very real sense, it is the brain that is not working properly in all emotional or psychiatric ills. However, it cannot be said that all behavioral and emotional difficulties are due to known structural or chemical derangements of the brain. These considerations simply serve to emphasize that, in man, "the mind" is a major activity of the brain, as flight is an activity of an airplane. All environmental stimuli, favorable or unfavorable, interpersonal or otherwise, are the province of the brain's function. All types of behavior, ranging from simple bodily movements to the most complex products of human reason or madness, are products of the brain.

Protected as it is by the bony casing of the skull, the brain is relatively inaccessible, compared, for example, with the blood, which is easily sampled and tested. This remoteness, to some extent, has slowed scientific understanding of the brain's function. More importantly, brain anatomy is highly complex. Each region of the brain has its own functions and its own structural and chemical peculiarities. In this, the brain stands in marked contrast to such organs as the liver, which is structurally and chemically similar throughout and where one portion does what another does. These regional differences in brain chemistry, anatomic complexity, and vulnerability to disease are extremely important when one begins to consider the various conditions that can cause malfunction. There are many diseases of the nervous system, and the symptoms are many and varied.

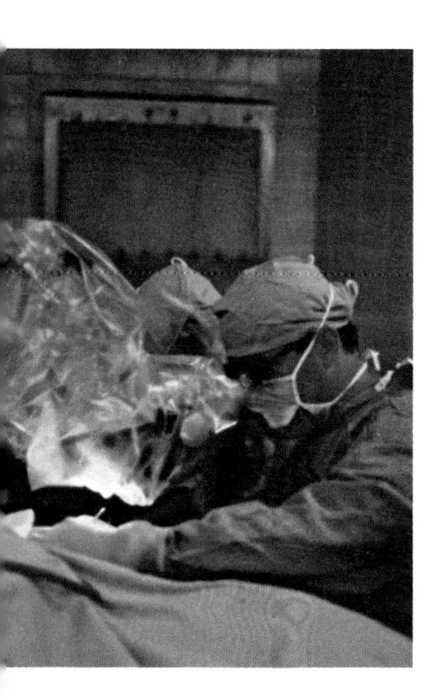

A neurosurgeon is shown using a special microscope for a better view of the delicate brain areas on which he operates.

The Practice of Neurology

In clinical practice, the nervous system is the domain of the neurologist and the neurosurgeon. (Psychiatry, dealing with mental disorders, is discussed elsewhere in this book.) These specialists, who work closely together, have different training and rather different outlooks, but their skills are complementary. The neurologist is concerned with medical, that is, nonsurgical, diseases of the nervous system. Common examples are mental retardation due to neurological defects, most varieties of stroke, epilepsy, multiple sclerosis, headaches, Parkinson's disease, and muscular dystrophy. The neurosurgeon, on the other hand, is trained to care for and operate on surgically correctable conditions, including, among many others, some types of intracerebral bleeding, hydrocephalus, or water on the brain, and tumors of the brain and spinal cord. Frequent consultation takes place between these two kinds of specialist, and their domains are subject to change. Thus, a disease that was once best treated surgically may be returned to the care of the neurologist as new medical treatment becomes available. Similarly, certain conditions once poorly treated medically or considered untreatable have become treatable in the operating room because of the surgeon's increased skill and knowledge.

In recent years subspecialization has produced a myriad of other neuroscientists, some physicians and some basic scientists, all delving into select and relatively restricted aspects of neural function. The ultimate value of this division and intensification of effort should be proved in future years as these scientists discover previously unknown causes and new treatments of neurological ailments.

One such disease that has been resolved satisfactorily is poliomyelitis, a virus disease of the spinal cord that produces varying degrees of temporary or permanent paralysis and is often fatal. This disease still cannot be treated adequately once it is established, but it can now be completely prevented by the use of vaccines.

Many diseases of the nervous system appear to be of genetic origin. They are due to inborn errors of metabolism, the complex chemical reactions in the body's cells that produce energy from the food we eat and the air we breathe. These reactions are regulated by protein substances known as enzymes. In certain conditions, a person may be born lacking an enzyme, and the resulting disruption of the body chemistry can profoundly impair brain function.

Certain kinds of mental retardation fall into this category. Some can now be treated with special diets. Others, even when the precise defect is known, can be dealt with only by prevention, beginning with premarital genetic counseling and the prevention of conception. For some of these diseases, persons carrying the potentially dangerous gene that determines the defect can be identified, and they can be warned of the possible danger if they have children. In some instances, tests can be used to predict whether an unborn child will be afflicted.

Disease of the nervous system is not rare. Strokes, of which there are several kinds, constitute one of the major health problems in the United States, along with cancer and heart disease. More than 700,000 cases of epilepsy exist in the United States, and the number of persons suffering from headaches is incalculable. Blindness and deafness, both ultimately due to failure of the brain to receive sensory information from the eye or ear, are also major problems, though they are not often due to brain disease. Head injuries, from automobile accidents or other kinds of trauma, are all too common and are one of the most frequent causes of death among young people. On the other hand, many diseases of the nervous system are relatively rare. Their rarity, and the fact that the cases are apt to be widely separated geographically, have made systematic research into their cause and treatment somewhat difficult.

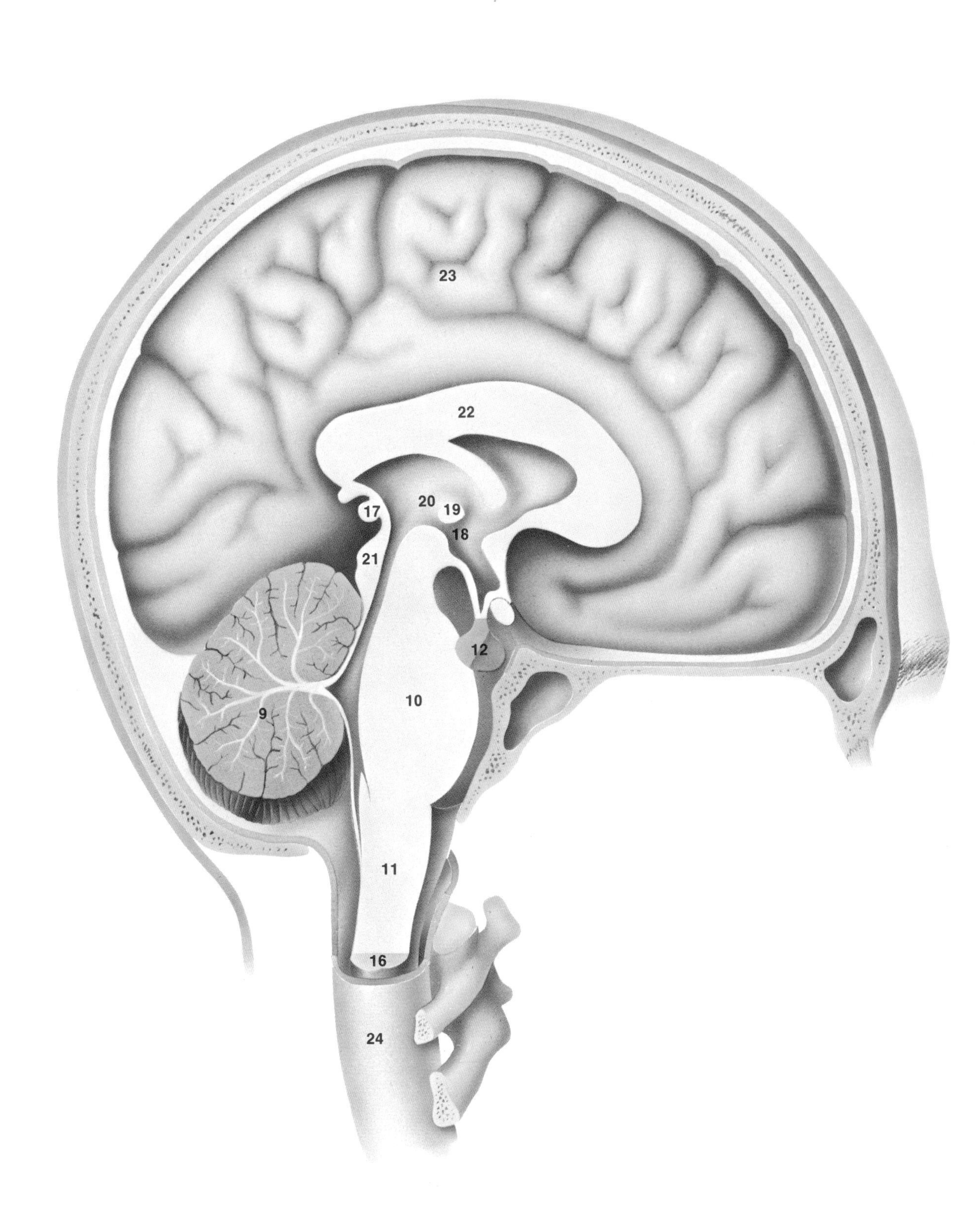
23
22
20
19
17
18
21
12
10
9
11
16
24

The Brain

1. Right cerebral hemisphere, 1
2. Gray matter, or cortex, 1
3. White matter, 1
4. Lateral ventricle, 1
5. Scalp, 2
6. Cranial bone, 2
7. Meninges, 2
8. Cerebrospinal fluid, 2
9. Cerebellum, 2, 3
10. Pons, 2, 3
11. Medulla oblongata, 2, 3
12. Pituitary gland, 2
13. Vertebra, 2
14. Internal carotid artery, 2
15. Cerebral peduncle, 3
16. Spinal cord, 3
17. Pineal gland, 3
18. Hypothalamus, 3
19. Thalamus, 3
20. Third ventricle, 3
21. Corpora quadrigemina, 3
22. Corpus collosum, 3
23. Left cerebral cortex, 3
24. Dura mater, 3

Function and Structure of the Central Nervous System

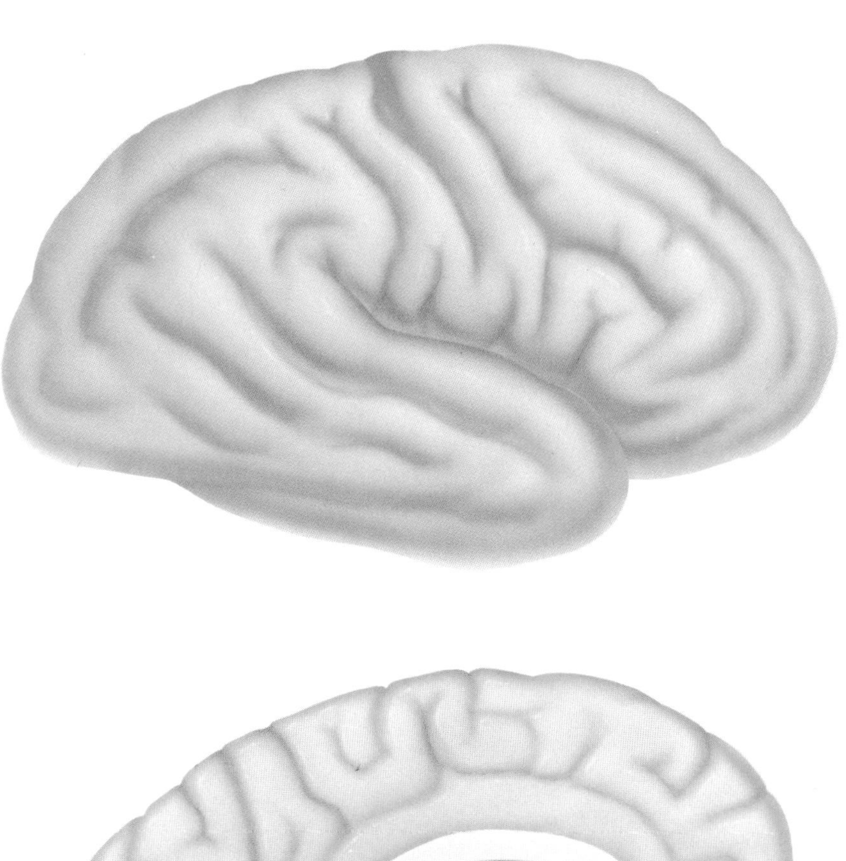

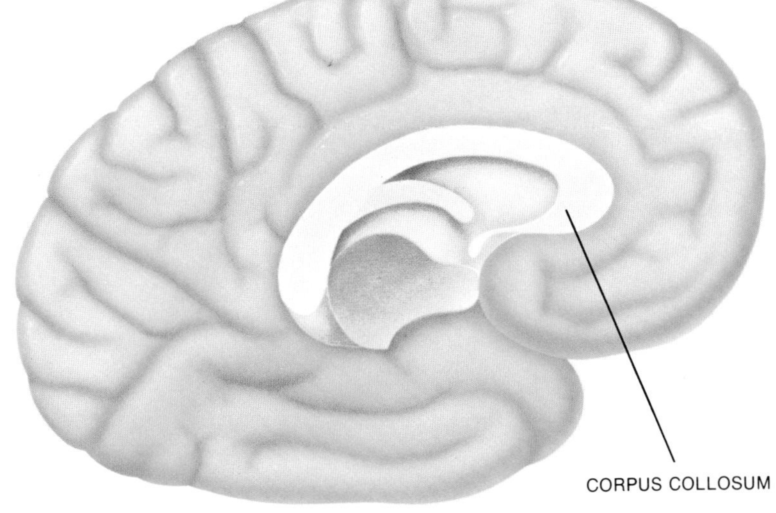

The cortex of both halves of the brain (the right half shown in the upper illustration) controls speech, writing, and the other uniquely human activities. The corpus collosum (shown in the lower illustration of a bisected brain) connects the two brain halves.

The anatomy, or structure, of the nervous system can be studied in the gross (life size) and under the microscope to magnifications of more than 200,000 times reality. In the last few decades, the development of high-magnification microscopes has allowed neurobiologists to observe the interior of brain cells. Certain important anatomical facts are relevant to any consideration of the function of the nervous system in health and disease.

The brain and spinal cord are well protected by the skull and the spinal column, and then further encased in three membranes, the pia mater, the arachnoid, and the dura mater. The space between the pia mater and the arachnoid contains spinal fluid.

Spinal Cord and Peripheral Nerves

The brain is the "telephone central" of the entire nervous system and is uniquely developed in man. Lower forms of life have proportionally small brains. Even monkeys, whose brains are most similar to those of humans, lack parts of the cerebral cortex present in man. These anatomical regions are generally believed to be those concerned with speech and other "higher" functions.

The functions of the spinal cord, and to an even greater extent the peripheral nerves, have undergone less change in the process of evolution. Some functions of these parts of the nervous system in man are quite "primitive" and are identical to those that occur in animals. The spinal cord is a direct extension of the brain, carrying the axons (the "telephone wires") of nerve cells from it. To a large extent, they control neurons in the spinal cord, which send their axons, in turn, to the muscles, skin, and other organs. A "nerve" is actually a kind of cable composed of vast numbers of these nerve cell processes, together with protective and supportive tissue.

Transmission of Nerve Impulses

Each nerve cell, or neuron, is an independent entity. It is connected to other neurons and to various organs through processes or projections called axons and dendrites. Axons carry impulses away from the cell body; dendrites carry impulses to the cell body. Some of these processes are two feet in length, reaching from the brain all the way to the end of the spinal cord in the lower back. The junctions or areas of contact between neurons are called synapses, and these are highly specialized structures. Most communication between neurons is chemically transmitted; the electrical impulse moving rapidly toward a synapse releases a "transmitter" chemical that "bridges" the minute gap in

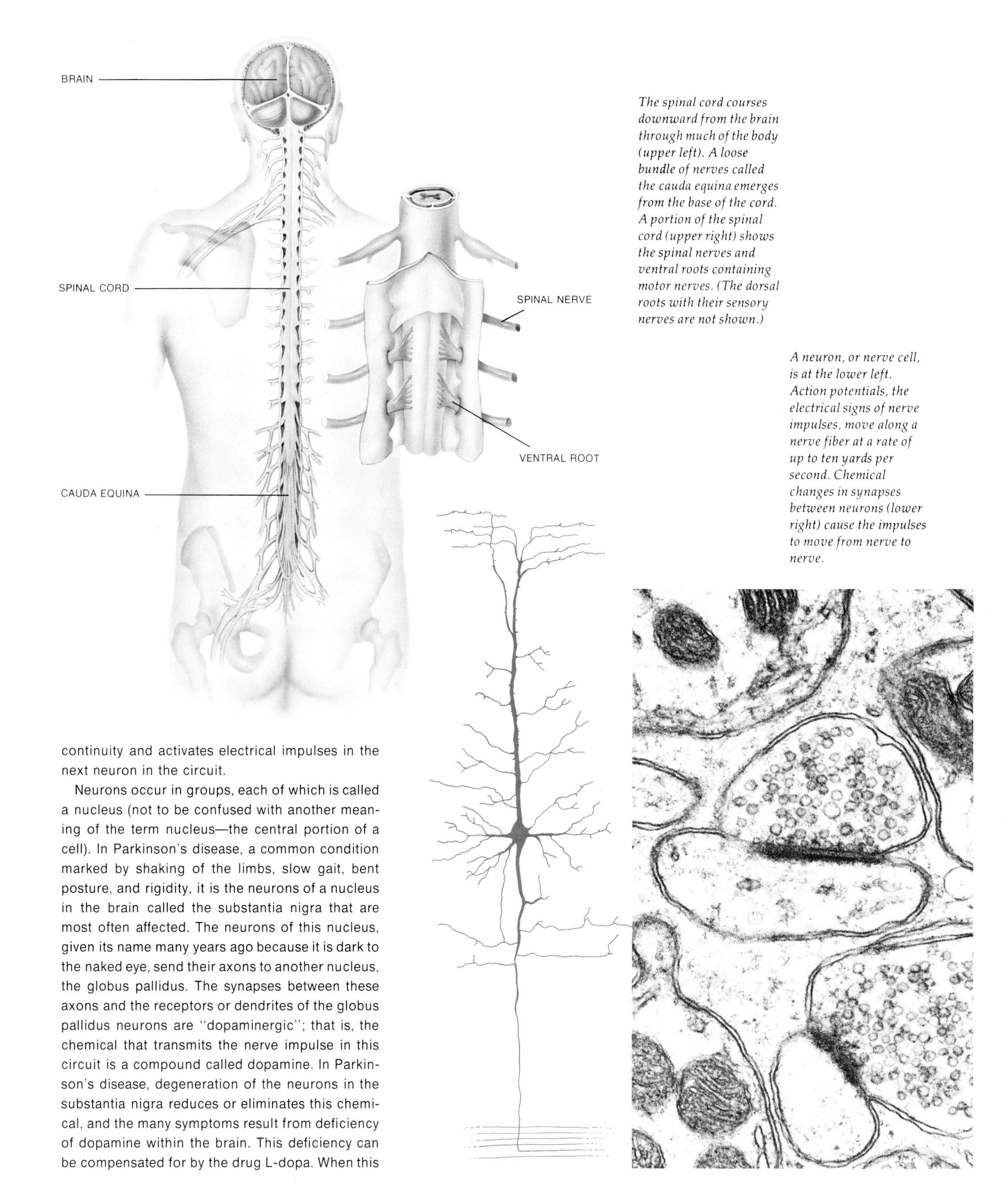

The spinal cord courses downward from the brain through much of the body (upper left). A loose bundle of nerves called the cauda equina emerges from the base of the cord. A portion of the spinal cord (upper right) shows the spinal nerves and ventral roots containing motor nerves. (The dorsal roots with their sensory nerves are not shown.)

A neuron, or nerve cell, is at the lower left. Action potentials, the electrical signs of nerve impulses, move along a nerve fiber at a rate of up to ten yards per second. Chemical changes in synapses between neurons (lower right) cause the impulses to move from nerve to nerve.

continuity and activates electrical impulses in the next neuron in the circuit.

Neurons occur in groups, each of which is called a nucleus (not to be confused with another meaning of the term nucleus—the central portion of a cell). In Parkinson's disease, a common condition marked by shaking of the limbs, slow gait, bent posture, and rigidity, it is the neurons of a nucleus in the brain called the substantia nigra that are most often affected. The neurons of this nucleus, given its name many years ago because it is dark to the naked eye, send their axons to another nucleus, the globus pallidus. The synapses between these axons and the receptors or dendrites of the globus pallidus neurons are "dopaminergic"; that is, the chemical that transmits the nerve impulse in this circuit is a compound called dopamine. In Parkinson's disease, degeneration of the neurons in the substantia nigra reduces or eliminates this chemical, and the many symptoms result from deficiency of dopamine within the brain. This deficiency can be compensated for by the drug L-dopa. When this

A Reflex Arc

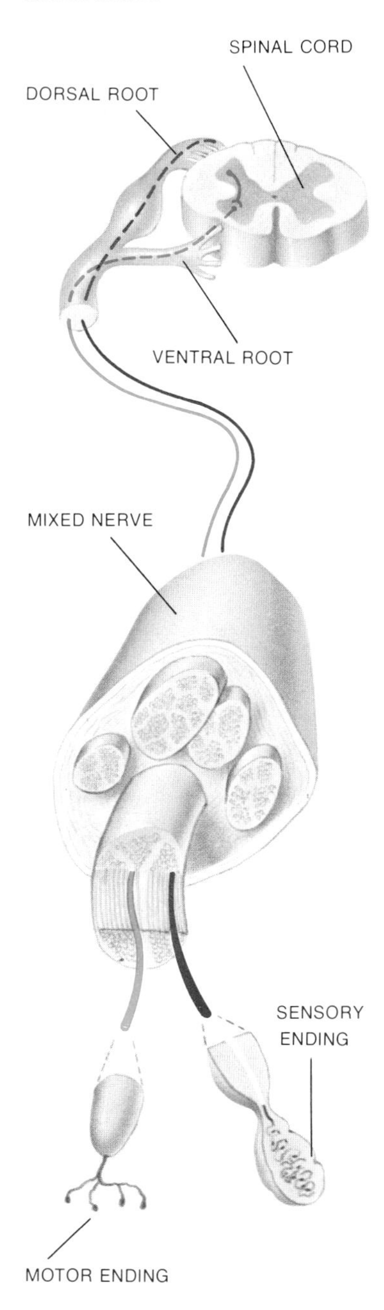

In a reflex action an impulse from a sensory nerve ending, caused by a rap near the knee for instance, travels through the sensory part of a mixed nerve to the spinal cord through a dorsal root. The impulse then passes to the motor part of the nerve, out of the cord through a ventral root, and to a motor ending where muscle contraction, such as a leg jerk, is stimulated.

drug is taken by mouth, it is converted to dopamine by the brain and can be extremely beneficial. Many other brain diseases are thought to be due to deficiencies of similar chemicals. It is not unlikely that within the next few years other drugs will prove to be of benefit in treating neurological conditions.

Reflexes

In the strict sense of the term, reflexes are neural actions that have relatively simple nerve circuits and that occur without the specific control of the higher centers of the nervous system. An example is the automatic response of a bare foot when it steps on a tack. The movement of withdrawal is immediate and requires no thought or conscious direction. The reflexes indicate the extreme speed with which the nervous system can operate. Once the foot has touched the tack, a whole series of events occurs within a fraction of a second. First, the sensory nerve endings in the skin are irritated mechanically by the tack. The mechanical irritation of the nerve fibers sets up an immediate electrical and chemical signal, which is carried by the sensory nerve fiber to the spinal cord. Then, a synapse involving the nerve cells concerned with muscle movement transmits an impulse that is sent back to the foot, activating the appropriate muscles and the movement of withdrawal. Many reflexes such as this are involuntary and are present even in the newborn child.

Other reflexes can be "learned" through association. An example is the puckering sensation in the mouth that often occurs at the sight of a lemon. Quite obviously, these reflexes are more complicated and are under the control of other parts of the nervous system.

Recently, much interest has arisen in the possibility of controlling behavior through conditioning or training of reflexes. This is sometimes called operant conditioning. Essentially, however, it is the same phenomenon that the Russian scientist Ivan Pavlov demonstrated years ago when he sounded a bell as dogs were presented with food and eventually conditioned them to salivate at the sound of the bell alone. There is some evidence that certain automatic, unconscious nervous activities, such as blood pressure variation, can be changed by such training.

Just how useful this mode of treatment will be remains to be seen. Many students of the nervous system doubt that it will be widely accepted. Control of behavior by such means raises larger philosophical questions, and a great deal of further investigation is clearly needed.

Diagnosis of Neurological Problems

In diagnosing a neurological disease, the physician depends to a large extent on the historical course of the disease—onset of symptoms, type of symptoms, previous treatment, and so on. As in other specialties, he supplements this with examinations and tests.

The neurological examination can be confusing to the patient. Most people understand a physician's basic purpose when he listens to the heartbeat. It is much harder to assess the detailed and complicated testing of strength, reflexes, sensation, balance, and other functions that the neurological examination entails.

The neurologist or neurosurgeon knows the anatomy of the nervous system in detail. By examining the patient systematically and assessing function, he arrives at what is known as a localizing diagnosis. That is, he is often able to tell exactly where the nervous system is malfunctioning. When this assessment is combined with the history of the symptoms, a diagnosis often can be made. Laboratory tests may confirm the diagnosis and make it more precise.

Spinal Fluid Examination

Spinal fluid—a crystal clear, watery liquid—is produced by specialized structures inside the ventricles, or cavities, of the brain. In man, about half a pint of spinal fluid circulates over the surface of the brain and spinal cord. It is formed from and is resorbed back into the blood very rapidly. Physicians can sample and test it by the technique of lumbar puncture, or spinal tap. Spinal fluid appears to have a protective function, cushioning the soft, jellylike brain and the spinal cord and thus reducing the risk of injury. It also has an important role in transporting various chemicals in and out of the brain.

Examination of the spinal fluid is often very important in diagnosing neurologic diseases. Knowledge of its chemistry, cell content, and pressure can provide a good indication of many of the things occurring within the brain. In a sense, examination of the spinal fluid is as important for the neurologist or neurosurgeon as blood counts are for the general practitioner or internist, and sometimes the findings are equally specific.

Spinal fluid pressure deserves specific mention. Pressure readings are done at the time of lumbar puncture, usually with the body horizontal. The spinal fluid is normally under a certain amount of pressure and, since the nervous system is confined within a fixed space, any increase in pressure above the normal is a strong indication of an abnormal space-occupying mass within the brain, such as a tumor or a cerebral hemorrhage. Increased pressure may also indicate an abnormal flow of spinal fluid, as in hydrocephalus. In young children, the bones of the skull are not tightly knit, and increased pressure often causes the head to enlarge.

Many diseases of the nervous system cause no increase in pressure, but the spinal fluid may be abnormal in other ways. In meningitis (infection of the meninges, or membranes, of the brain), white blood cells are often seen, and bacteria can be observed under the microscope and cultured in the laboratory. Frequently, excessive protein in the spinal fluid gives evidence of a disease process. In some diseases, certain types of protein increase as compared with others, and this may be of great diagnostic importance.

The spinal fluid examination is the one test that most laymen associate with neurological disease. Six other major tests used in neurologic diagnosis are blood tests, X rays of the skull and spine, electroencephalography (EEG), cerebral angiography, pneumoencephalography, and radioisotopic scanning. The first two are largely self-explanatory, but some explanation of the others is in order.

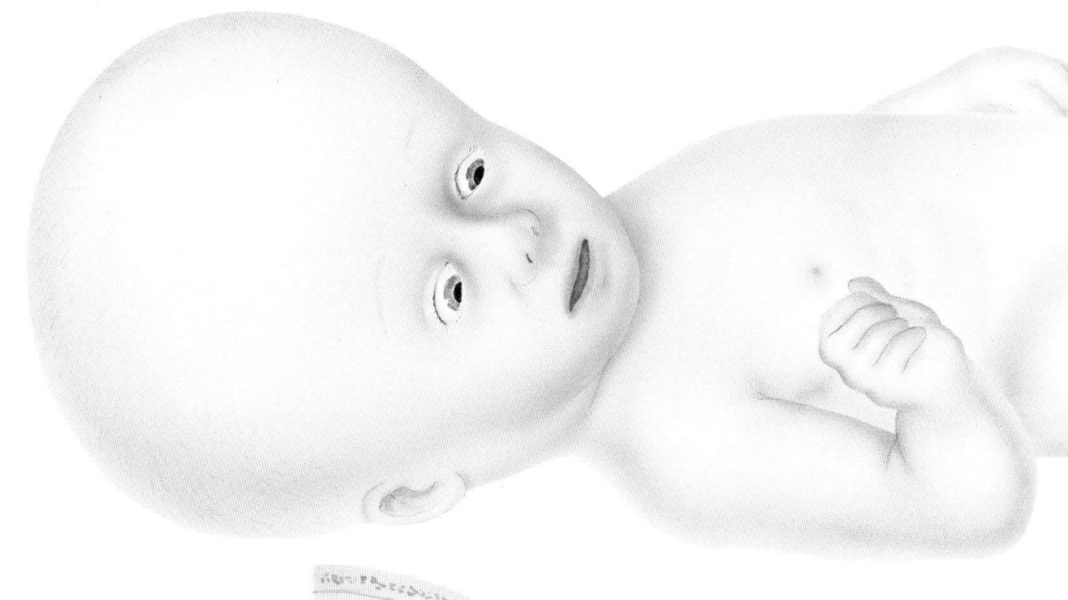

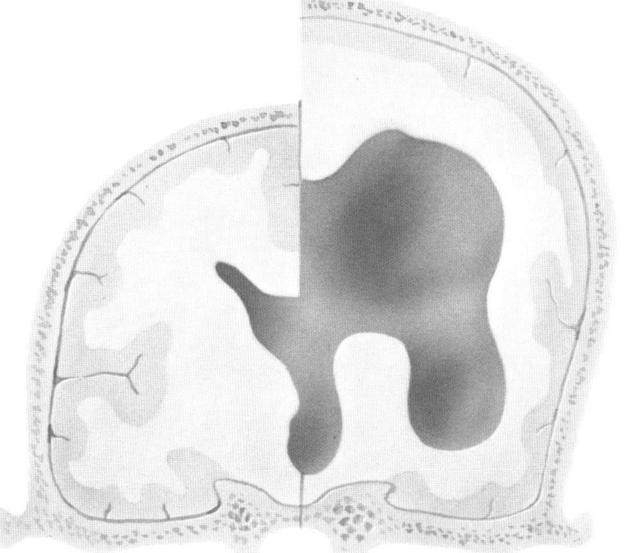

In hydrocephalus an accumulation of cerebrospinal fluid in the brain of a child (right half of the illustration at the left) causes enlargement of the child's loosely-knit skull bones.

Electroencephalography

Electroencephalography is usually one of the first procedures in neurological diagnosis. Though time-consuming, it is easily performed and causes the patient no more discomfort than an electrocardiogram.

The brain generates electrical activity as it functions. Each neuron develops a minute electrical potential and conducts this signal to other cells with which it is connected. This potential is generated by changes in electrically charged particles, or ions, especially of sodium and potassium, at the cell surface.

Huge populations of these neurons tend to discharge their potentials in certain rhythms, and neurologists can monitor these rhythms by picking them up and amplifying them electronically through wires applied to the scalp. The rhythms can be visualized by transmitting the amplified signals to a pen moving across paper. Such graphic records have permitted researchers to make an excellent determination of which impulses or rhythms are normal and which are abnormal. The exact origins of these rhythms are not entirely

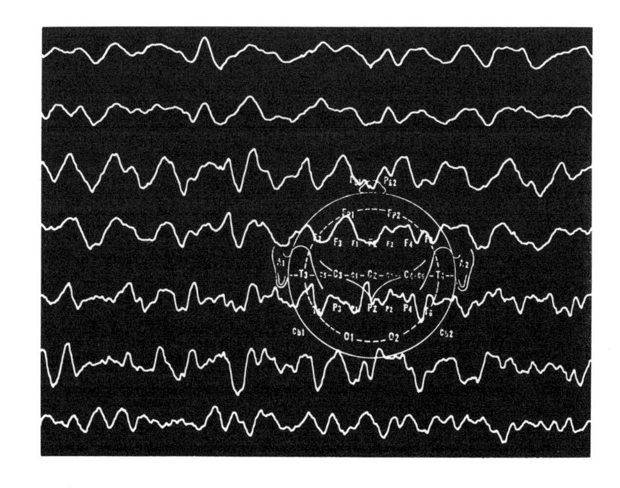

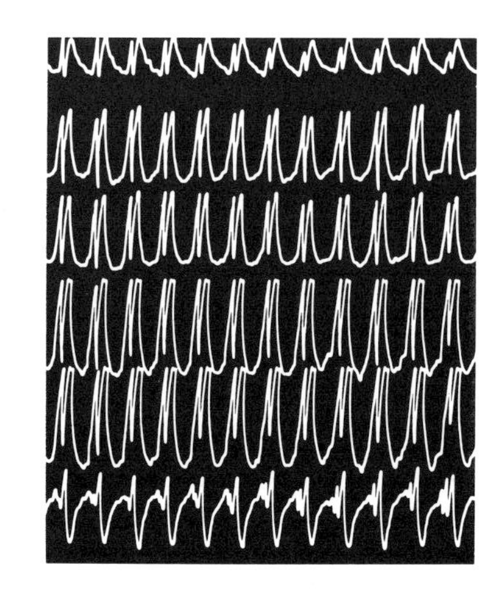

Electroencephalographic (EEG) tracings of normal electrical activity of the brain are shown (lower left). In epilepsy, however, portions of the EEG tracings are erratic (lower right).

clear, however. The so-called alpha rhythm, the most well known, is best developed over the occipital region—the part of the brain underlying the back of the head.

Generally, the electrical activity is symmetrical or equal over both sides of the brain. Disease states alter the brain rhythms. For example, a collection of blood in or over a region of the brain will usually depress or slow the electrical activity in that area. Coma, such as occurs with severe meningitis or drug overdose, will drastically slow activity in all brain regions. In epilepsy or other seizures, the region of abnormal brain irritability can be located by encephalography, and subvarieties of the disease can be diagnosed.

Cerebral Angiography

Cerebral angiography is an important technique used to outline with precision the course of the blood vessels of the brain and to determine whether or not they are obstructed. A dye material is injected directly into the carotid artery in the neck, the main artery supplying blood to the brain, or is introduced through a catheter inserted in the arm or groin and positioned, with the aid of a fluoroscope, into the carotid artery or another vessel leading to the brain.

An angiogram of the brain is made by injecting a radiopaque dye into the carotid artery of a patient and then X raying his skull as the dye circulates through the brain arteries.

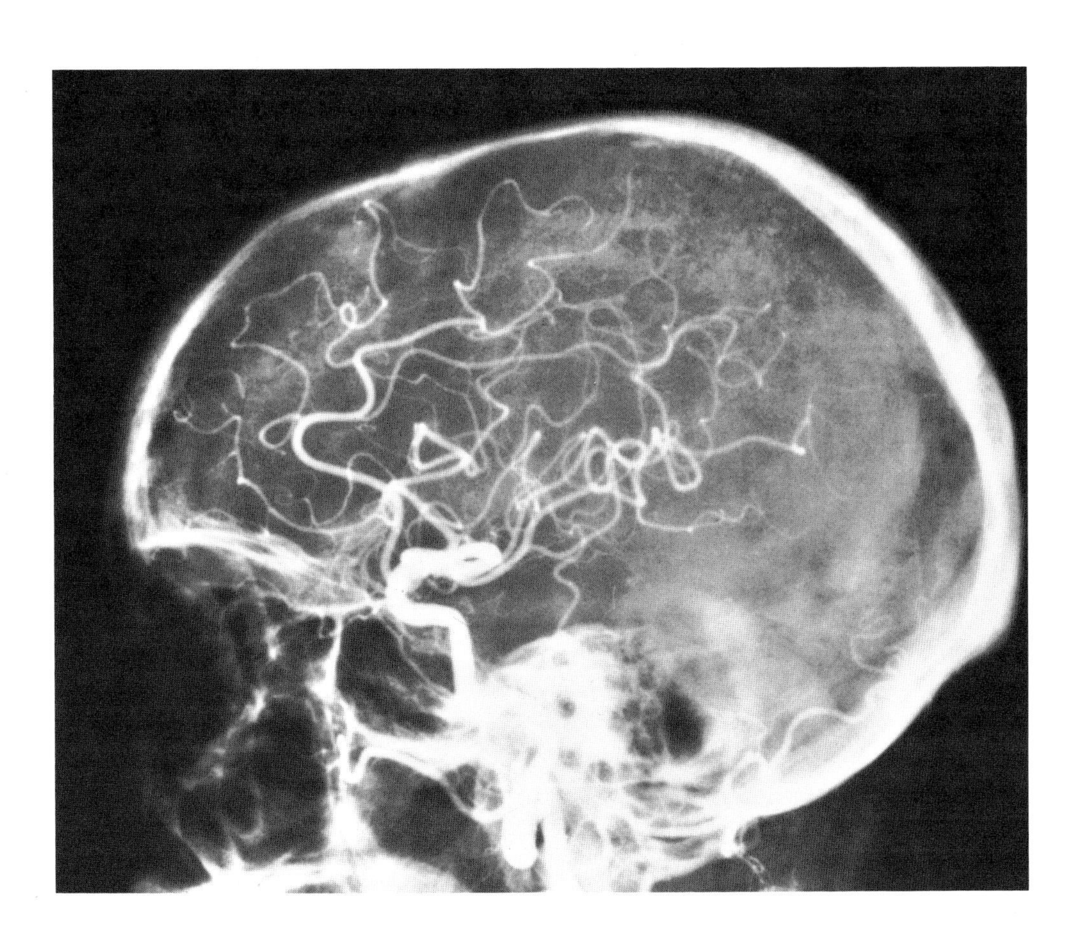

Rapid roentgenograms (X rays) are taken as this material circulates with the blood through the arteries of the brain. From these, various conditions can be diagnosed and pinpointed, including deviations from the normal vessel pattern, tumors (detected by their abnormal blood vessels and the displacement of normal ones), and defective arteries that could rupture and cause brain hemorrhage. Angiography is not used in all diseases of the nervous system but only when some structural deformity, such as a tumor or an abnormal blood vessel, is suspected. It is almost always used before brain surgery.

Pneumoencephalography and Brain Scan

Pneumoencephalography ("air study") is similar to angiography in principle and complementary because it allows visualization of the ventricles or cavities of the brain. Sterile air or other gas is placed into the spinal fluid spaces, usually by a technique similar to lumbar puncture. Air is less dense than the brain or spinal fluid and serves as a contrast material for X rays.

Again, this procedure is not needed for neurological diagnosis in all cases. More often a brain scan with a radioisotope is the screening procedure performed. A small and safe amount of radioactive material is injected into a vein in the arm. As the radioisotope circulates through the brain, special photographic techniques are used to detect any abnormalities in its distribution. Brain scanning is completely safe and painless, and it requires little from the patient other than to sit still and relax. It has come to be widely employed in recent years and often may be used in conjunction with electroencephalography in the initial diagnostic study before the major procedures, angiography and pneumoencephalography, are contemplated.

Tumors and Strokes

Of the conditions discovered by visual means, brain tumors are perhaps the most dreaded. Brain or spinal cord tumors destroy tissue locally but rarely spread elsewhere in the body. Some are quite benign and can be removed surgically. Others respond to radiotherapy (X rays, radiocobalt, etc.). Brain tumor symptoms may be similar to those seen in strokes, though generally their onset is not so abrupt, and they can occur in young people or children. Some malignant brain tumors are not caused by cancer of the brain itself; rather, they are due to the spread, or metastasis, of a tumor that originated in another part of the body. Currently, there is a great deal of interest in the possibility that brain tumors may be caused by viruses. Consider-

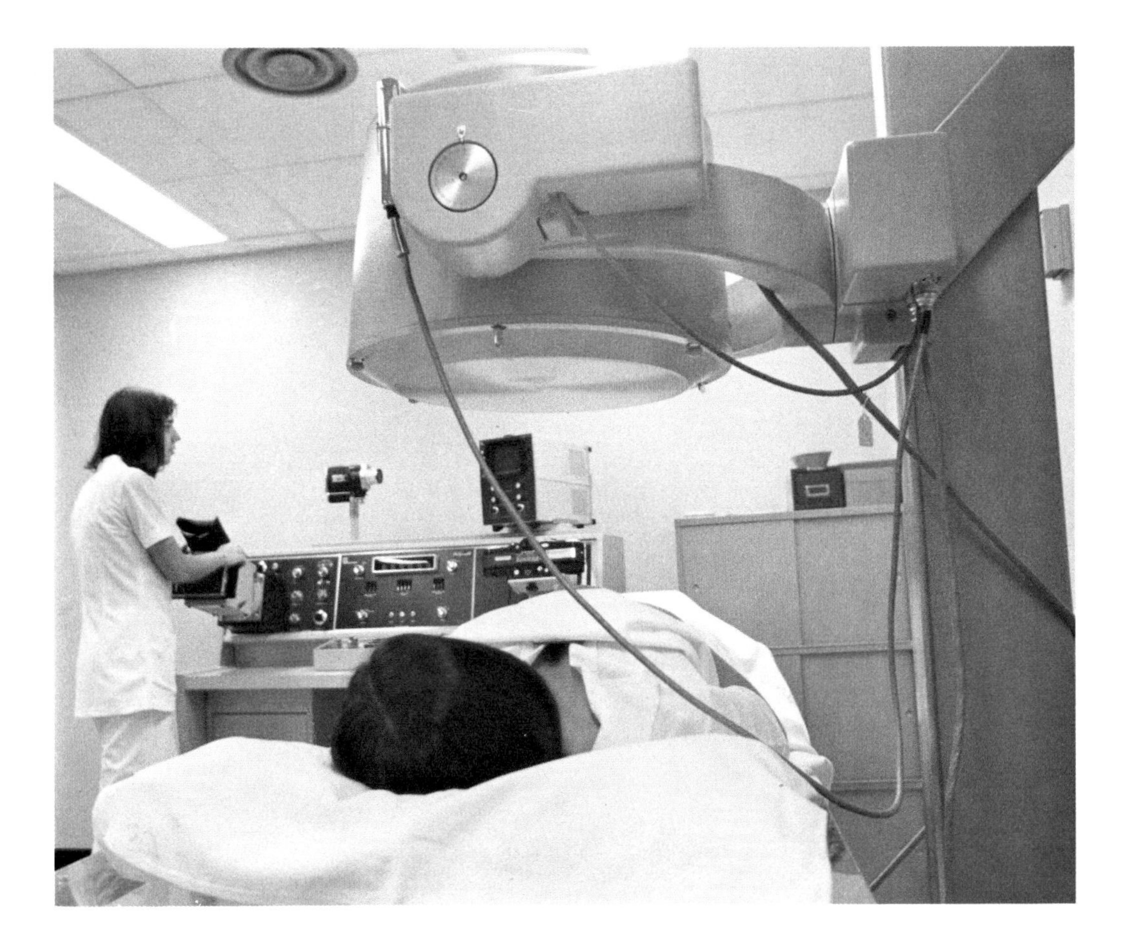

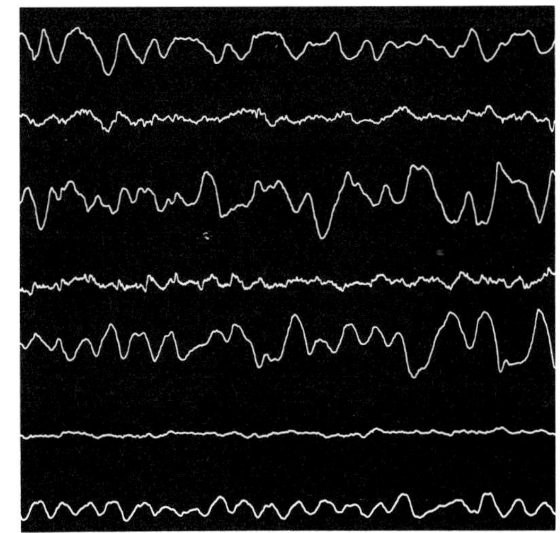

Radiotherapy with the radioactive isotope of cobalt or other means of radiation is sometimes used to treat brain cancer (left). An EEG tracing (right) can sometimes disclose the presence of a brain tumor.

able research is now being done on this hypothesis.

Strokes, of which there are various distinct subtypes, do not always require visual studies. Indeed, at times it is best not to test a stroke patient by any of these methods. The commonest type of stroke is cerebral infarction, the occlusion, or blocking, of a vessel leading to, or in, the brain by a thrombus, or blood clot. Destruction of brain tissue results because circulation of the blood is impeded and the brain cells cannot obtain the oxygen and nutrients they need.

Such an occurrence is identical to coronary thrombosis, except that in the latter case the affected blood vessel leads to the heart. The blood vessel disease responsible for the thrombosis is known as atherosclerosis, and some of the factors implicated in it are smoking, high blood pressure, and high blood fat. Treatment of cerebral infarction is unsatisfactory at present. Most students of the disease are convinced that only prevention of atherosclerosis will solve the problem.

Other causes of stroke, also difficult to treat, are rupture of a blood vessel in the brain (cerebral hemorrhage), usually caused by very high blood pressure and cerebral embolism, in which a clot travels through the bloodstream from another area of the body, such as a diseased heart valve.

Because different regions of the brain control different functions, the symptoms of blood vessel occlusion in the brain are more complicated and variable than in the heart. Perhaps the commonest symptom is hemiparesis, or paralysis of an arm and leg on the same side of the body, sometimes associated with loss of sensation in the weak limbs. If the left side of the brain is involved, speech may be impaired, a condition known as aphasia. Cerebral infarction in other areas can cause visual impairment, imbalance, severe dizziness, double vision, paralysis of all limbs, or even sudden death.

Since many strokes, particularly those caused by brain hemorrhage, are directly related to high blood pressure, physicians treat hypertension vigorously. Statistical studies show that reduction of blood pressure to normal reduces the incidence of stroke.

Non-hemorrhagic types of stroke have been treated, or possibly prevented, by anticoagulants, or blood thinners. Because the treatment differs with each type of stroke, the various types must be clearly differentiated.

Recovery from a stroke, as from a heart attack, depends largely on the amount of permanent damage that has occurred. Other factors include the type of stroke, the age of the patient, and his determination to regain function. The brain has little capacity to regenerate or renew its nerve cells, and a scar gradually forms where tissue was destroyed. However, alternative connections between nerve cells and sometimes even "supplemental" brain regions can partially take over the function of the damaged areas. The most rapid recovery usually occurs in the first few weeks or months after a stroke, although most patients can expect continued improvement for about a year. After that, further gains rarely take place.

Spinal Cord and Peripheral Nerve Disorders

The spinal cord and the peripheral nerves are subject to a number of diseases. In addition, diseases not usually considered to be neurological, such as diabetes and alcoholism, can have profound effects on the peripheral nerves. Damage to a nerve by trauma may result in paralysis or numbness in the part of the body serviced by the nerve.

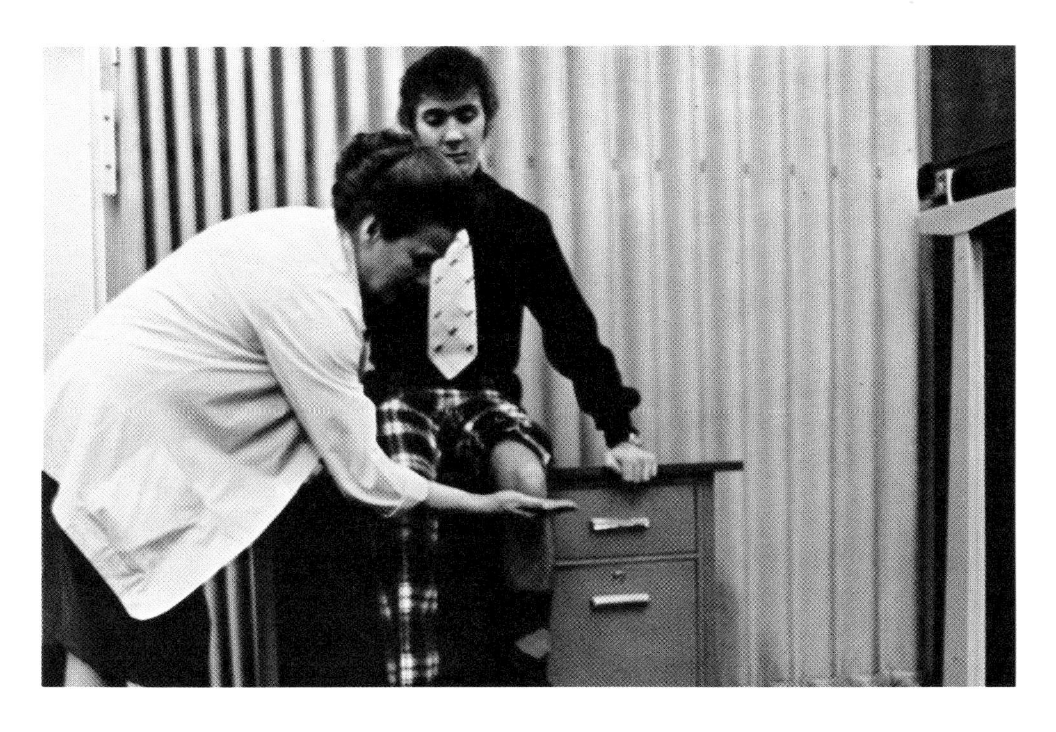

By testing a patient's reflexes a physician can detect the presence of nerve damage.

Many axons are wrapped in an insulating material, called myelin. Damage to this insulator can cause profound neurological symptoms even when the axon itself is relatively intact. Multiple sclerosis is a case in point, though in this disease it is the myelin of the spinal cord and not the peripheral nerves that is usually affected.

Examination with such simple devices as a reflex hammer or a pin can indicate whether there is nerve damage. Sometimes specialized diagnostic studies are done with instruments similar to the electroencephalograph. With these instruments, a neurologist can pinpoint the site of damage to a given nerve and assess the degree of dysfunction. Such techniques are complicated, however, and their use requires special training.

Headaches

As has been pointed out, different regions of the nervous system have specialized functions. Thus, for example, destruction of a given area tends to give rise to certain symptoms, regardless of what has caused the destruction. Often the sequence in which symptoms appear and the speed at which they develop allow a neurologist or neurosurgeon to know whether a given symptom is medically unimportant or indicates a serious condition. Headaches are a case in point.

Most people have experienced headache. Most understand intuitively that, if they have had headaches off and on for many years, a serious cause is unlikely. On the other hand, people who suddenly begin to have headaches should know that something may be seriously amiss. Headaches that recur over a long period are most frequently caused by migraine, tension, sinusitis, and eye disorders. A few are due to high blood pressure. These conditions are quite treatable. It is when headaches begin relatively abruptly that the physician must look for more serious conditions.

Increased pressure inside the head, a common cause of serious headache, can be due to many conditions, including infection, brain tumors, or certain types of strokes. Surprisingly, the brain itself is insensitive to pain. Nervous tissue has no "nerves." Pain usually results from stretching or distortion of the membranes covering the brain or from traction on cerebral blood vessels, which do have nerves. Tension headaches, common in anxious or depressed persons, are usually caused by excessive contraction of the scalp muscles that overlie the skull.

As a general rule, sudden headaches that indicate a serious condition are associated with other important symptoms, such as weakness, visual impairment, severe lethargy, disturbance of speech, or loss of sensation in the limbs. By combining his clinical observations with a history of a patient's symptoms and their development in time, the neurologist or neurosurgeon is able to arrive at an accurate diagnosis.

The Senses

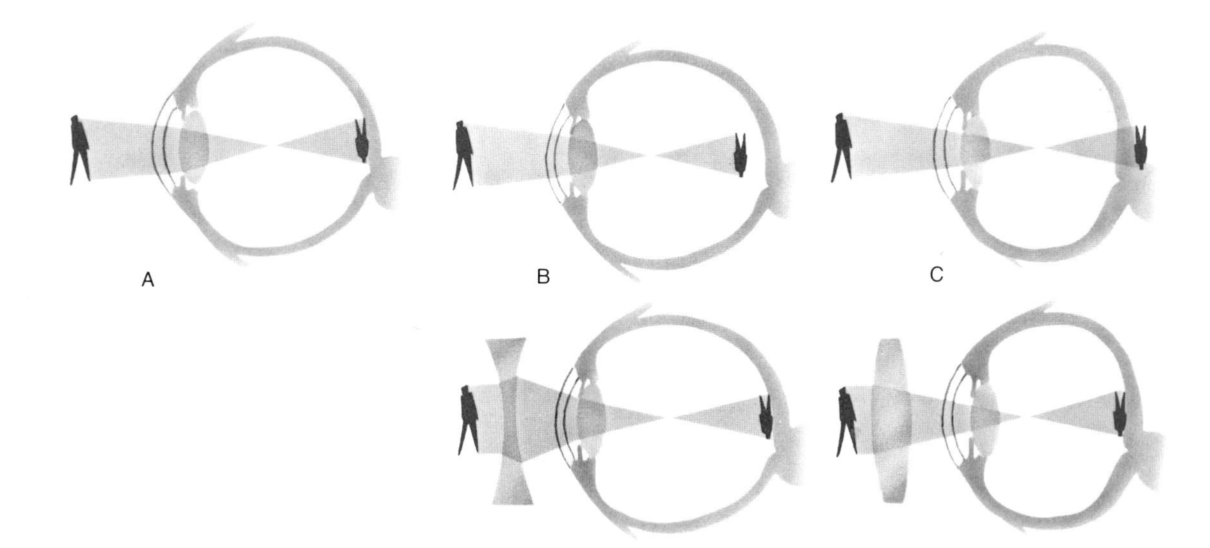

A. Normal vision.
B. Nearsightedness, or myopia, caused by a shortened eyeball and corrected with concave eyeglasses.
C. Farsightedness, or hyperopia, caused by an elongated eyeball and corrected with convex eyeglasses.

The foot and the tack constituted an example of sensation. Other sensations, such as perception of heat and cold, have a similar anatomical basis. Specialized sensory nerves respond to the stimulus and transmit impulses to the spinal cord and brain. These are generally called simple or ordinary sensations and are of obvious importance.

The body is also supplied with nerve endings that indicate position of the body, weight, and other subtle factors, and damage to these can cause profound impairment. If the body is to move properly, continual reflex adjustments must be made in muscle tone and power, and these, in turn, depend on information received from sensory nerves. Learning how to use and modify certain bodily movements and sensations to advantage is an important part of the training of an athlete.

Vision and Hearing

The eyes and ears are highly complex and cannot be dealt with here in any detail. From the point of view of the brain, however, these organs are simply receivers of certain types of information from the outside world. Light waves (in the case of the eye) or sound waves (in the case of the ear) are translated into electrical and chemical signals, and these are fed into the brain, where they are perceived. There are specific regions of the brain in which information from the eyes and ears is somehow "read" and conveyed to "consciousness." In a sense, the eye and the ear are like a camera lens and a microphone—highly important devices but useless without film or amplifier.

Disturbances of Vision

Diseases of the eyes or ears can cause blindness or deafness, but these disabilities can also occur when the eyes or ears are perfectly normal but when the brain regions that receive their signals are damaged. For example, a child with an untreated cross-eye, or strabismus, sees double because the eyes do not work together. To prevent this, the brain automatically suppresses or ignores the information from one eye. Without treatment, the eye ignoring the information becomes functionally blind; that is, even though the eye itself may be perfectly healthy, its input to the brain is never used.

Many experiments have been performed to assess the role of the brain in the phenomenon of sight. Patients with strokes or brain tumors involving regions of the brain concerned with vision may have serious visual disturbances. Sometimes such patients with perfectly normal eyes see nothing, do not realize that they do not see, or have extreme distortions of vision. They may see objects but not understand what the objects are or how to use the visual information they are receiving. They may have disturbances of color perception or may see only parts of objects without being able to "integrate" the visual information into useful patterns.

The commonest eye "disease" is a refractive error. Typically, this means nearsightedness (myopia) or farsightedness (hyperopia). These conditions result when the eyeball is either too long (myopia) or too short (hyperopia), and the light entering it is not focused properly on the retina, which lines the back of the eyeball. The retina is a very complex structure consisting of specialized nerve cells, capable of translating light waves into electrical and chemical activity. The processes of these nerve cells form the optic nerve, which leads into the visual region of the brain.

If they are not too severe, refractive errors can be corrected with eyeglasses. If the cornea, the clear covering over the pupil, is not perfect in shape, astigmatism, or distortion of vision, can result. This condition can also be corrected with glasses. The lens, which lies behind the pupil, has a role in focusing and magnifying. The lens may become cloudy with age or disease. This condition, known as a cataract, responds to surgery and the use of glasses. The iris, the colorful circle surrounding the pupil, is like a shutter on a camera because it permits the precise amount of light needed for proper vision to enter the eye.

The Hearing Mechanism

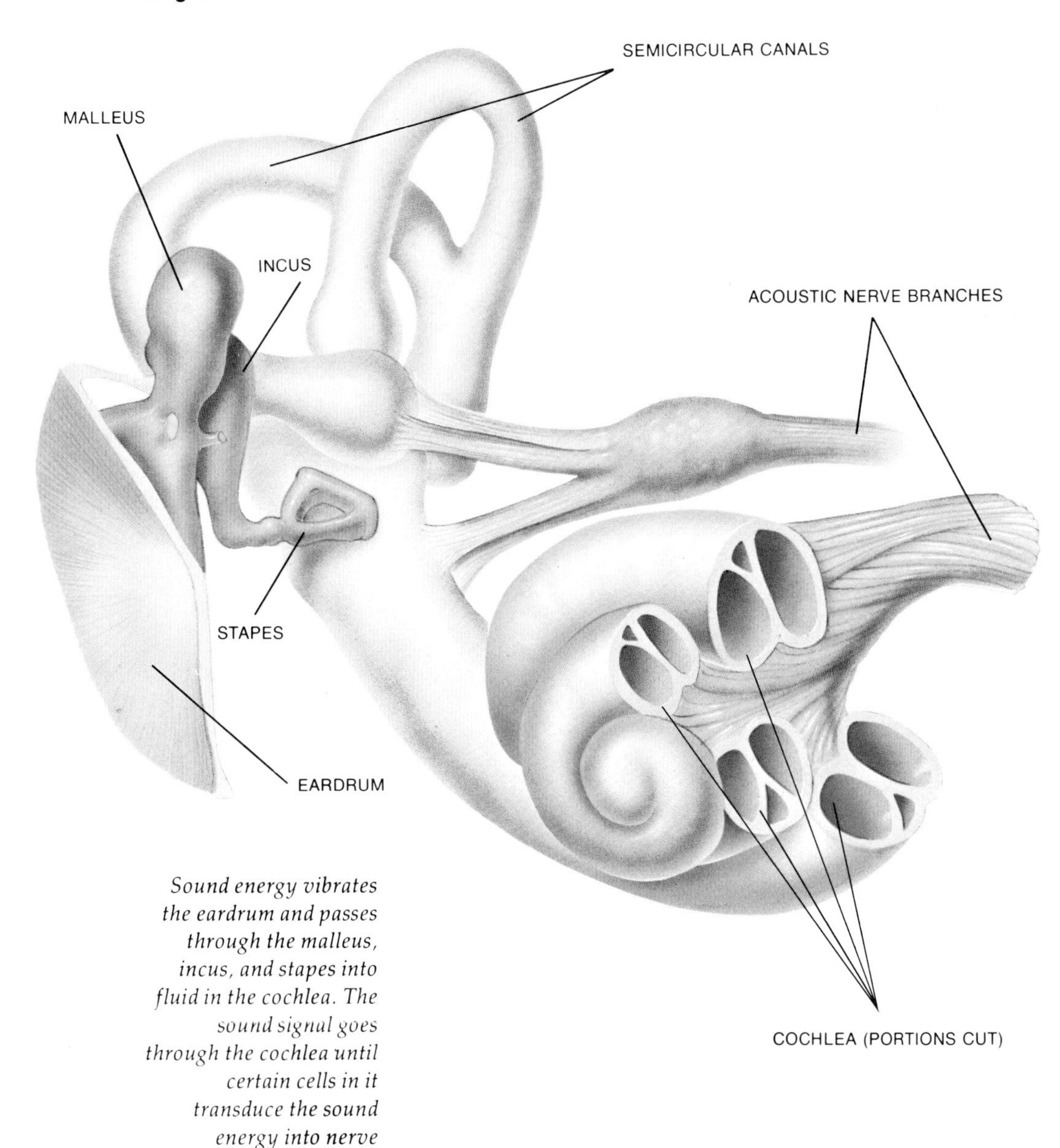

Sound energy vibrates the eardrum and passes through the malleus, incus, and stapes into fluid in the cochlea. The sound signal goes through the cochlea until certain cells in it transduce the sound energy into nerve impulses in the acoustic nerve to the brain. The semicircular canals are involved in body equilibrium.

Disturbances of Hearing

In a similar fashion, the ear is composed of a series of interrelated parts that transform the mechanical impulses set up by sound waves striking the eardrum, or tympanic membrane, into nerve activity. These nerve impulses are then conducted to specialized brain areas. Like vision, hearing can be impaired at any step, even by something as simple as excessive wax in the canal leading to the eardrum. The eardrum can be imperfect or perforated, the middle ear bones, called ossicles, can become arthritic and stiff, the nerve cells in the ear can be damaged by incessant, very loud sounds, or the nerve leading into the brain itself can be destroyed by tumors or infection.

Patients with hearing loss usually cannot tell which of these conditions they have, and the attention of an ear specialist—an otolaryngologist—is needed for proper diagnosis. Hearing loss is rarely caused by disease of the brain, though brain disease is often responsible for such closely related difficulties as dizziness or imbalance. The inner ear, in addition to its apparatus for conducting sound, contains closely related structures concerned with dizziness and balance. The symptoms of dizziness and imbalance arising from the inner ear must be differentiated from similar, if not identical, symptoms that occur when certain brain regions are impaired.

Other Senses

Other special senses include taste and smell. As in all sensory functions, the external stimulus is transformed into nerve impulses by nerve cells or processes, located, in this case, in the nose and mouth. As with sight and hearing, these impulses are transmitted to specialized brain regions that control these functions. The mechanism of touch is roughly similar.

The Autonomic Nervous System

We have been discussing that part of the nervous system concerned with conscious movement and sensation, but there is a second, distinct system that controls the myriad activities taking place within the body day and night without any direction or conscious awareness. This is called the autonomic nervous system because it is autonomous, independent of the conscious mind. The original Greek source of the word means "self-governing." It could just as well be called the automatic nervous system, because it acts automatically whether a person wills it or not.

This is the system that widens the pupils of the eyes in darkness and constricts them in the presence of light, that maintains a flow of saliva to moisten the membranes of the mouth even during sleep, and that opens the pores for sweating when it is hot and closes them when the weather turns chilly. It slows the heartbeat during sleep. It maintains the wavelike motion that propels food through the gastrointestinal tract. With the help of hormones, it regulates the constriction and dilation of the small blood vessels.

Anatomy of the Autonomic Nerves

The autonomic nervous system is a true nervous system. Like the nerves that control movement and sensation, it originates in the central nervous system, sending out nerve fibers to their sites of action and transmitting its messages through chemical action at the nerve endings. As might be expected of a system that operates mainly outside the conscious mind, however, the autonomic nerves originate in the primitive parts of the brain and in the spinal cord.

The autonomic system is divided into two distinct parts. One is called the sympathetic division; the other, the parasympathetic division. The parasympathetic nerves have two points of origin. Most of them originate in the parts of the brain known as midbrain, pons, and medulla oblongata. The others originate at the other end of the central nervous system, in the sacrum or lowest section of the spinal cord. The sympathetic nerves originate in the spinal cord between the first thoracic and the second or third lumbar segments.

The nerve fibers of these two divisions do not go directly from the brain or spinal cord to the organs they control. Rather, the fibers of origin connect with a second fiber or group of fibers at a junction called a ganglion. Logically enough, the former are called preganglionic fibers and the latter are called postganglionic. It is the postganglionic fibers that stimulate or inhibit their related organs.

In the sympathetic division, the ganglia are located in interconnected chains that run alongside the spine. Thus the sympathetic preganglionic fibers are short and the postganglionic fibers are long. In the parasympathetic division, the ganglia are usually very close to or on the innervated organs, so the preganglionic fibers may be very long and the postganglionic fibers short.

Activation of Autonomic Nerves

For the most part, the sympathetic and parasympathetic divisions work in opposition to one another. That is, where one division dilates a particular organ, the other constricts it, or where one excites, the other inhibits. There is no simple pattern to this opposition. For example, the sympathetic nerves dilate the salivary glands and constrict the iris of the eye, and, of course, the parasympathetic nerves have the opposite effects.

In general, however, the sympathetic division is usually concerned with activities that expend energy while the parasympathetic division is attuned to conserve energy. In anger or fright, the sympathetic nerves, acting as a unit, increase the heart rate, raise blood pressure, draw blood from the skin and viscera into the skeletal muscles, dilate the pupils, widen the air passages in the lungs, and in general prepare the body for action. Many of the parasympathetic nerves, on the other hand, conserve energy by such actions as slowing the heart, lowering the blood pressure, and stimulating the movements and secretions that aid digestion.

Each division of the autonomic system regulates the organs it governs by releasing a transmitter hormone. The sympathetic nerves characteristically release norepinephrine. This and a very similar

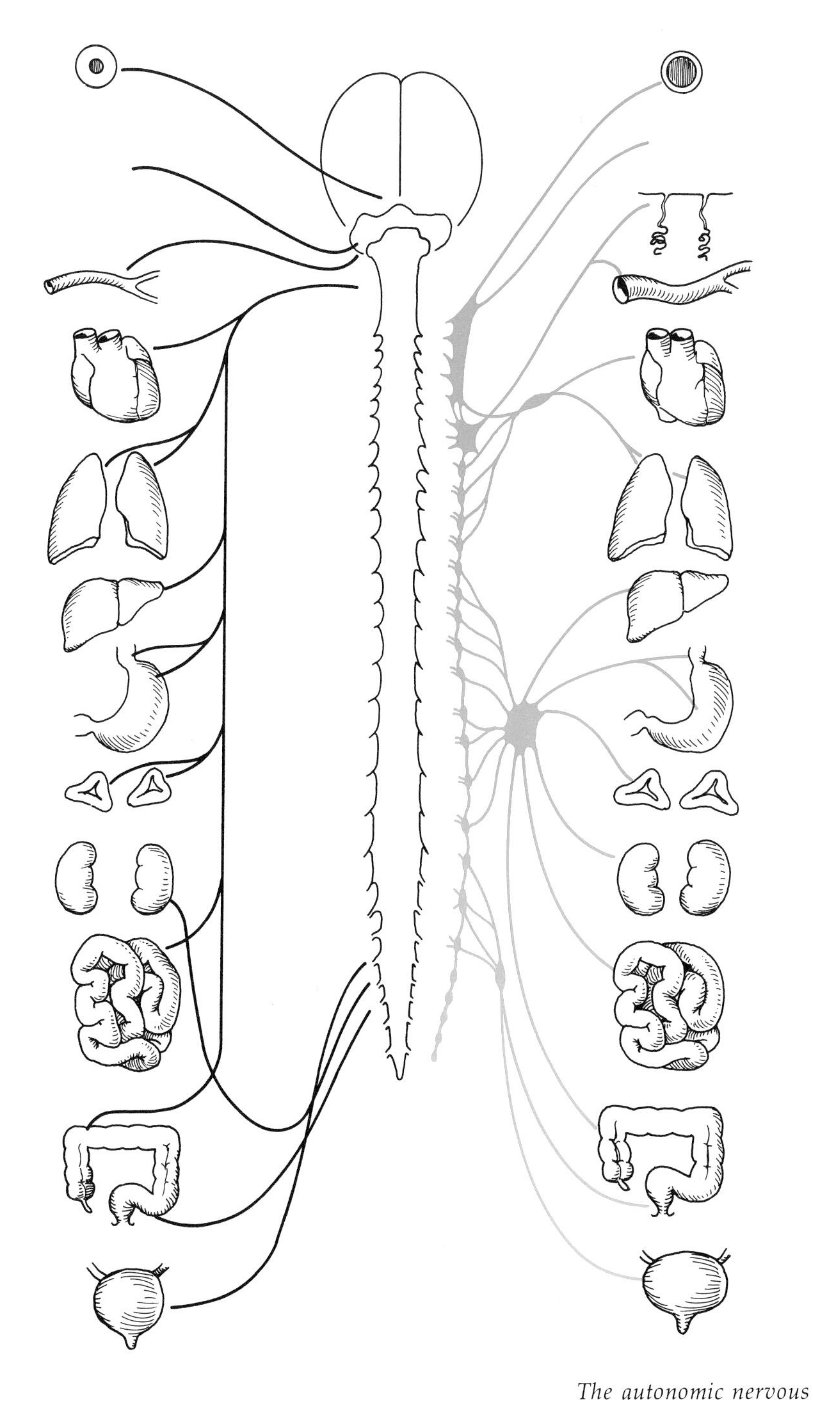

The autonomic nervous system consists of sympathetic fibers (shown in the right half of the illustration) and parasympathetic fibers (in the left half). As they innervate muscles of certain tissues and organs, sympathetic and parasympathetic nerves act as antagonists. For example, a sympathetic nerve causes dilation of the pupil of the eye; a parasympathetic nerve triggers its contraction.

chemical, epinephrine, are the principal hormones of the adrenal gland, and another name for epinephrine is adrenaline. The adrenal gland and the sympathetic system have a common embryonic origin and are closely allied functionally. The parasympathetic nerves, on the other hand, release a transmitter chemical called acetylcholine.

The differences in activity initiated by the autonomic nerves lie not only in the nerves themselves and in their transmitter chemicals, but in a complex reaction that includes the neural stimulus, the receptor site of the organ stimulated, and the programmed response of that organ to the stimulus.

While it is true that the autonomic system is quite distinct, this does not mean that it is entirely independent of the rest of the nervous system. A high degree of integration and regulation of all nerve function takes place in the higher brain centers. Probably there is no conscious activity that is not reflected in an autonomic response of some sort, and there are few significant autonomic actions that do not impinge on conscious activities and feelings. A prime example of this crossover is the high degree of awareness everyone has of autonomic visceral effects—including the "butterflies" in the stomach nearly everyone experiences as he steps onto a stage before an audience.

The Autonomic System in Disease

The effects of the autonomic system on the body are so extensive that the system is involved, at least in a secondary capacity, in many disorders. However, autonomic activity is an important determinant in only a few ailments. Among these are peptic ulcer, spasm of the bile duct, and constipation.

Scientists have been able to synthesize many drugs that either mimic or block the activity of the autonomic transmitters, norepinephrine and acetylcholine, so that physicians can modify the effects of autonomic impulses. These drugs have a number of uses in treating and diagnosing illness. Epinephrine, for example, is a powerful cardiac stimulant and is frequently used to resuscitate patients in acute heart failure. Atropine and its many synthetic cousins block the action of acetylcholine, and physicians employ them, among other uses, to control excessive activity of the gastrointestinal tract and to dilate the pupils for eye examinations.

Unfortunately, the actions of drugs that affect the autonomic system are widespread, as are the effects of the system itself. The difficulty or impossibility of limiting the actions of such drugs to one area of the body or one particular disorder limits their use in treating disease.

The Marvel of the Nervous System

The nervous system is the material basis for the sum total of human activity, from the most simple of bodily functions to the most subtle of thoughts. The complexity of the nervous system and the scope of its role is often appreciated only when it fails to work properly.

In an effort to develop models of neural function, biologists have provided calculations that dramatically emphasize the amazing capacities of the nervous system. For example, the number of transistors needed to equal all the nerve cells in a human brain would require a space some 10 billion times greater than that of the brain. And the transistors' total energy requirements would be 10 billion times greater than neural energy needs. Only in functional speed can computers compare with the nervous system.

Though we have learned to analyze and even duplicate some of the molecular aspects of the nervous system, the secret of its organization and the integration of its components still elude us completely. They remain one of the great mysteries of life.

For further information on the subjects covered in this chapter, consult **Encyclopædia Britannica:**

Articles in the Macropædia	*Topics in the Outline of Knowledge*
CEREBROSPINAL FLUID	422.K.2.c
EAR AND HEARING, HUMAN	422.K.6; 433.C.3
EAR DISEASES AND HEARING DISORDERS	424.M.8
EYE AND VISION, HUMAN	422.K.5; 433.C.2
EYE DISEASES AND VISUAL DISORDERS	424.M.7
HEADACHE	424.M.10
NERVOUS SYSTEM, HUMAN	422.K.1
NERVOUS SYSTEM DISEASES	424.M.1,2,3,4,5, and 6
SENSORY RECEPTION, HUMAN	422.K.7; 433.C.1

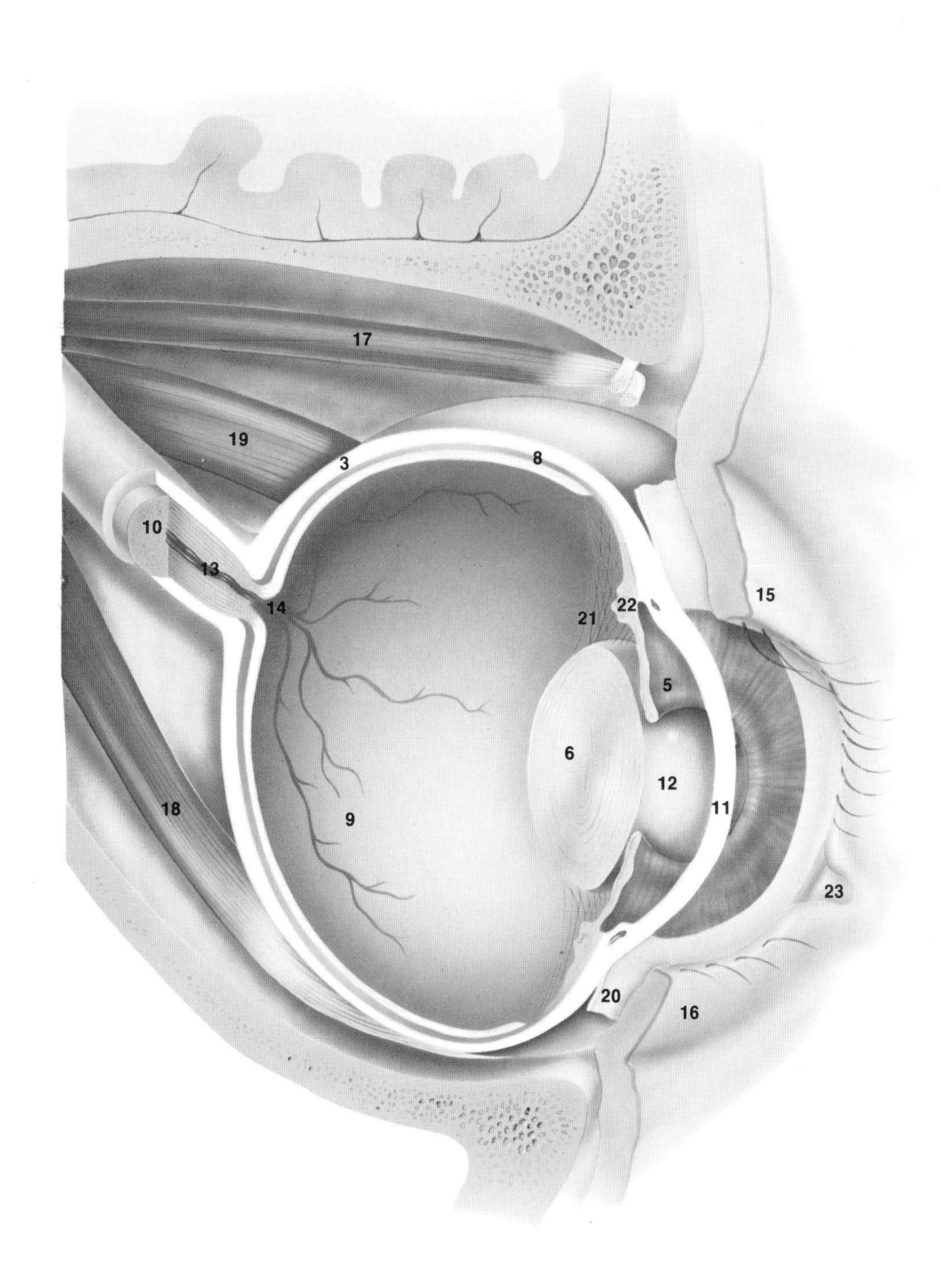
17
19
3
8
10
13
14
15
22
21
5
6
12
11
18
9
23
20
16

The Eye

1. Rectus superior, 1
2. Rectus lateralis, 1
3. Sclera, 1
4. Lens seen through pupil, 1
5. Iris, 1
6. Lens, 2
7. Obliquus inferior, 2
8. Choroid, 2
9. Retina, 2, 3
10. Optic nerve, 3
11. Cornea, 3
12. Pupil, 3
13. Retinal artery, 3
14. Blind spot, 3
15. Upper eyelid, 3
16. Lower eyelid, 3
17. Obliquus superior, 3
18. Rectus inferior, 3
19. Rectus medialis, 3
20. Conjunctiva, 3
21. Radiating fibers, 3
22. Ciliary process, 3
23. Tear duct, 3

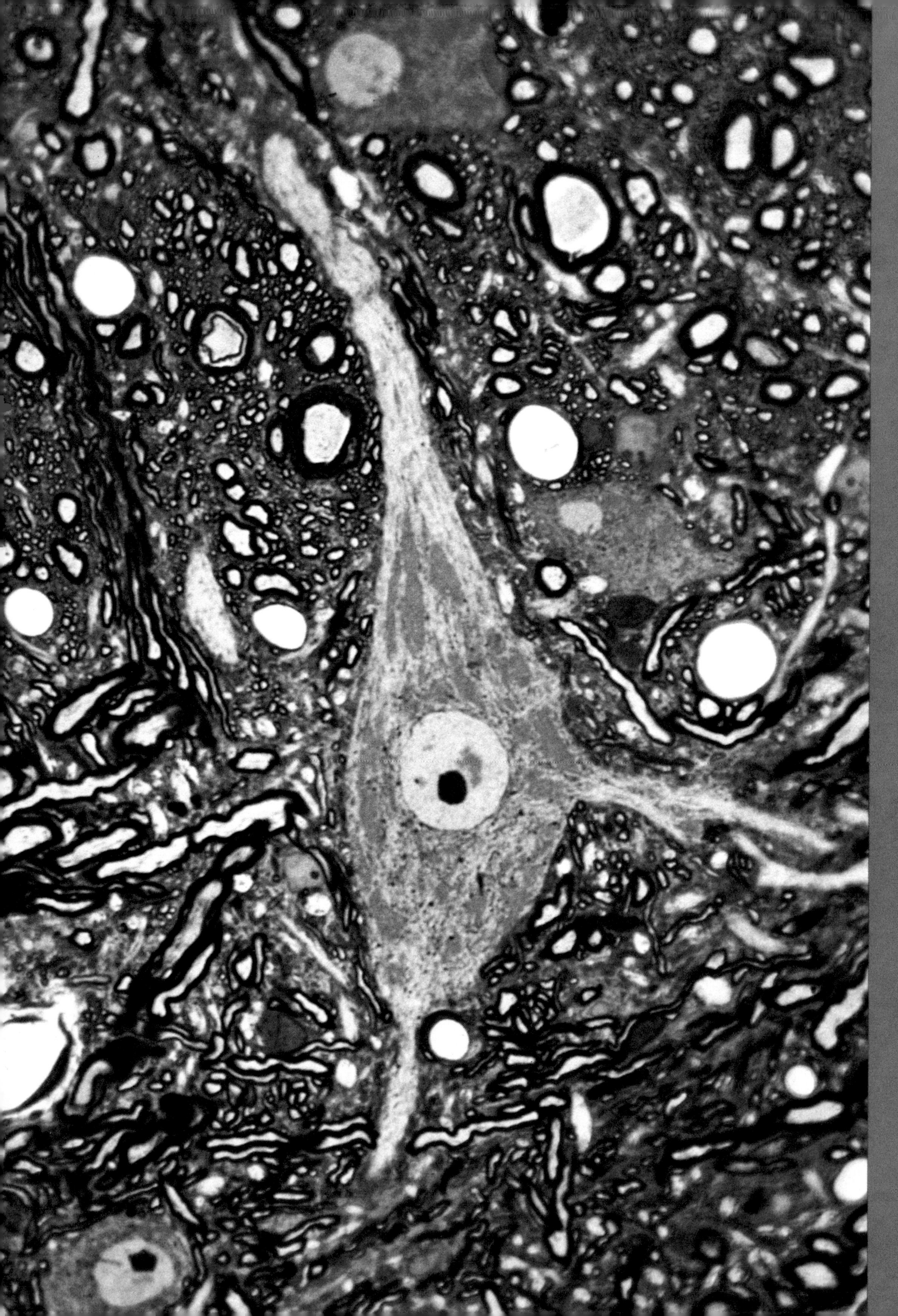

Many nerve cells, like the one shown greatly magnified at the left, are involved in the integration – the unification – of all the body's activities. Each person's responses to environmental stimuli are individualized by the paths taken by nerve impulses in the nervous system. Courtesy Nicholas A. Vick, M.D.

Kessler. *In the first chapter we talked about the "you" that hates to visit the doctor.*

Van Doren. *Here, at the end, we've finally come to that "you," or "I." The "I" that possesses "my" body. The "I" that understands how "my" body works.*

Kessler. *The "you" that has read the book and asked questions and sought answers.*

Van Doren. *The "I" that desires to be normal and be mentally healthy.*

Kessler. *The "you" and "I" that together have helped make a book about health. It's a good book, I think.*

Van Doren. *Thank you, doctor!*

13

The Integrative System: The Psyche

BY ROY R. GRINKER, SR., M.D.

Almost everyone wants to feel that he is "normal." Normality is a source of pride and a valued attribute, but the definition of what is normal may be unclear. In fact, a person's ethnic background, the society in which he lives, and the cultural rules of that society determine whether he is "normal" or not. For example, slight anxiety and depression may seem reason enough to seek psychiatric help in an "upper class" society that is generally well structured and where the patterns of "normal" behavior are well understood and accepted. However, in a "lower class" society that is more loosely structured and more inclined to reject many social mores, a greater tolerance for eccentric behavior or thought may exist. More basically, everyone is more or less neurotic because of the inner conflicts between his animal heritage and the humaneness toward which he strives.

What Is Mental Health?

Definitions of mental health are many and vague. The simplest and perhaps the most accurate indicates that a person who loves well, plays well, works well, and expects well—is optimistic—is "normal" and enjoys a state of mental health.

Another definition describes "normal" as healthy self-identity—that feeling of being at-one with oneself and of functioning adequately in society. Obviously, many variations of health, personality, character, behavior, and creativity combine to make humanity a highly variable commodity. In the process of day-to-day living each "normal" person strives to adapt to his talents, that is, know what he is, and to discover his most suitable niche in life.

Mental function depends in the first instance on the adequate behavior of body systems, such as the circulation, the nervous system, and the hormones, all of which are discussed in this book. The mind or mental processes, with which we are concerned here, involve thinking, feeling, and total behavior—the integration or working of the whole. We should think of the mind as the conductor of an orchestra, not as a soloist performing by himself.

Psychosomatic Influences

The mind and body function in a reciprocal relationship. Each influences the other in its total function. Even the word for this relationship, *psychosomatics,* indicates unity. However, for convenience, the effects of mind and body on each other will be discussed separately.

Body Over Mind. A large number of diseases and disabilities affect mentation or thinking. Among the most serious—and obvious—is brain damage through malformation, trauma, infection, deficient blood supply, or senile deterioration. The earlier in a person's life the damage occurs, the more serious and lasting are the effects on thinking and behavior. Unfortunately, brain tissue does not have the capacity to regenerate. However, modern methods of teaching can partially compensate for some defects by helping the afflicted person to learn to use undamaged portions of his brain.

Congenital or acquired deafness has a profound effect on thought processes since it cuts the person off from normal communication. Manual, or sign, language or lipreading must then be used to establish logical thought. Hormonal imbalances affect the brain and influence feeling and thinking.

Many diseases and injuries that do not directly involve brain tissue still have a profound effect on mental function. Some conditions that take a heavy toll are disfiguring burns or lacerations, open heart surgery, amputation of a limb or breast, hemophilia, diabetes mellitus, and infections, especially those that are chronic, such as tuberculosis, leprosy, and viral infection of the liver. These and many other conditions wound the person's self-image, often to the point of making him feel inferior, damaged, depressed, and useless.

Early and appropriate treatment of all such disorders is vital, if damage to the mind is to be prevented.

Mind Over Body. Looking at the reverse side of the coin, emotional disturbances may alter bodily functions.

In the past, man had to cope with episodes of stress. Today, in our highly mobile, competitive, and congested society, a far greater variety of unexpected problems constantly subjects him to complicated, stressful situations to which he has difficulty adapting. Running away from the so-called urban rat race to bucolic communes or to the "drug scene" is not a solution. Escape is difficult for modern man; he has no place to hide.

Stress stimuli, however, must be meaningful to stir a person to bodily preparation, action, or overaction. His ego filters the stimuli and determines what they mean to his personal integrity, often erroneously. His initial feelings may change after he has made a thoughtful reevaluation.

In the 1940s and 1950s, some investigators mistakenly believed that each of several psychosomatic diseases was triggered by a specific stress stimulus. Now it is known that persons react in a stereotyped manner regardless of the nature of the stress. It is the response that is specific, not the stimulus. In general, the hormonal responses of the body are similar whether the emotion involved is anxiety, anger, or depression. Each person recognizes his individual signs and knows, for example, that if his heart beats fast, or if he begins to cry, something is disturbing him.

Age Factor in Response to Stress

Stress responses vary with age. Each phase of the life cycle—childhood, adolescence, young adulthood, adulthood, and old age—has its own general predispositions, type of response, and coping devices.

The adolescent, for example, is concerned principally with moving from childhood to adulthood. His major problems relate to the maturation of his sex drives. The young adult is confronted with career choices. The middle-aged cope with the discrepancies between their goals and desires and their actual accomplishments. The elderly worry about chronic disability, dying, and death.

Exposure to meaningful stress stimuli in the form of anxiety, grief, loss of loved ones, feelings of hopelessness and helplessness is, of course, the common lot of all persons. All such stimuli result in bodily or somatic reactions, but the response of each individual depends on his innate, inherited makeup and conditioning in early life.

In an absolute sense, no person is altogether normal. Everyone exhibits minor mental disturbances that bother others, sometimes to the extent that, under certain circumstances, such behavior may be considered evidence of mental illness.

Neuroses

In each total personality, some character defects occur, minor psychopathic trends appear, and patterns of function recur in appropriate situations. In other words, a recognizable life-style, a "repetition compulsion," binds each person. By these behavioral responses we identify ourselves to ourselves and to other people.

In responding to events, each person experiences minor and transient episodes of anger, anxiety, and depression or elation. Such behavior is neither evidence of illness nor of the need for help. These reactions are typically human and normal—or what might be called normally neurotic.

In contrast to such normal neurotic reactions, neuroses result from more intense and prolonged states of conflict between the push of personal drives and the patterns acceptable to the particular society and culture of which one is a part. These conflicts create anxiety. Sometimes this anxiety reaches such an unbearable level that a person sets up defenses against it.

The classification of neuroses is based on these defenses. A fearful, or phobic, person may want to avoid particular stress situations such as heights, a closed space, or the sight of a rat. The source of his anxiety may be an early traumatic experience, or the phobic object may be a symbol of some forgotten or repressed feeling. Such conditions are known as anxiety neuroses.

Another form of neurosis is exhibited by obsessive-compulsive neurotic persons. These persons display stereotyped rituals or repetitive thinking through which they attempt to erase an earlier memory in some magical way. Repetitive handwashing may be an effort to erase thoughts or acts that the conscience considers dirty. Shakespeare used this handwashing ritual to great effect in his play *Macbeth.*

Perversions may represent early sexual phases that were not abandoned or brought under control as the person reached sexual maturity.

Delinquency may represent a heroic act to avoid unacceptable feelings, the denial of depression, and/or an acting out of a sense of futility.

Psychoses

Whereas neuroses are endurable and may be borne without help—though with varying degrees of discomfort—throughout life, psychoses cannot. These conditions exhibit a break with reality or disorder of identity that drastically interferes with a person's ability to function in his society.

Many types of psychoses relate to incidental or accidental experiences, such as infections, trauma, drugs, toxic substances, nutritional deficiencies, arteriosclerosis, and senility, to name a few. Others have no physical cause that can be identified with certainty, at least in our present state of knowledge.

Mental Illness

When conflicts in a person's life produce unbearable anxieties, serious mental problems can ensue.

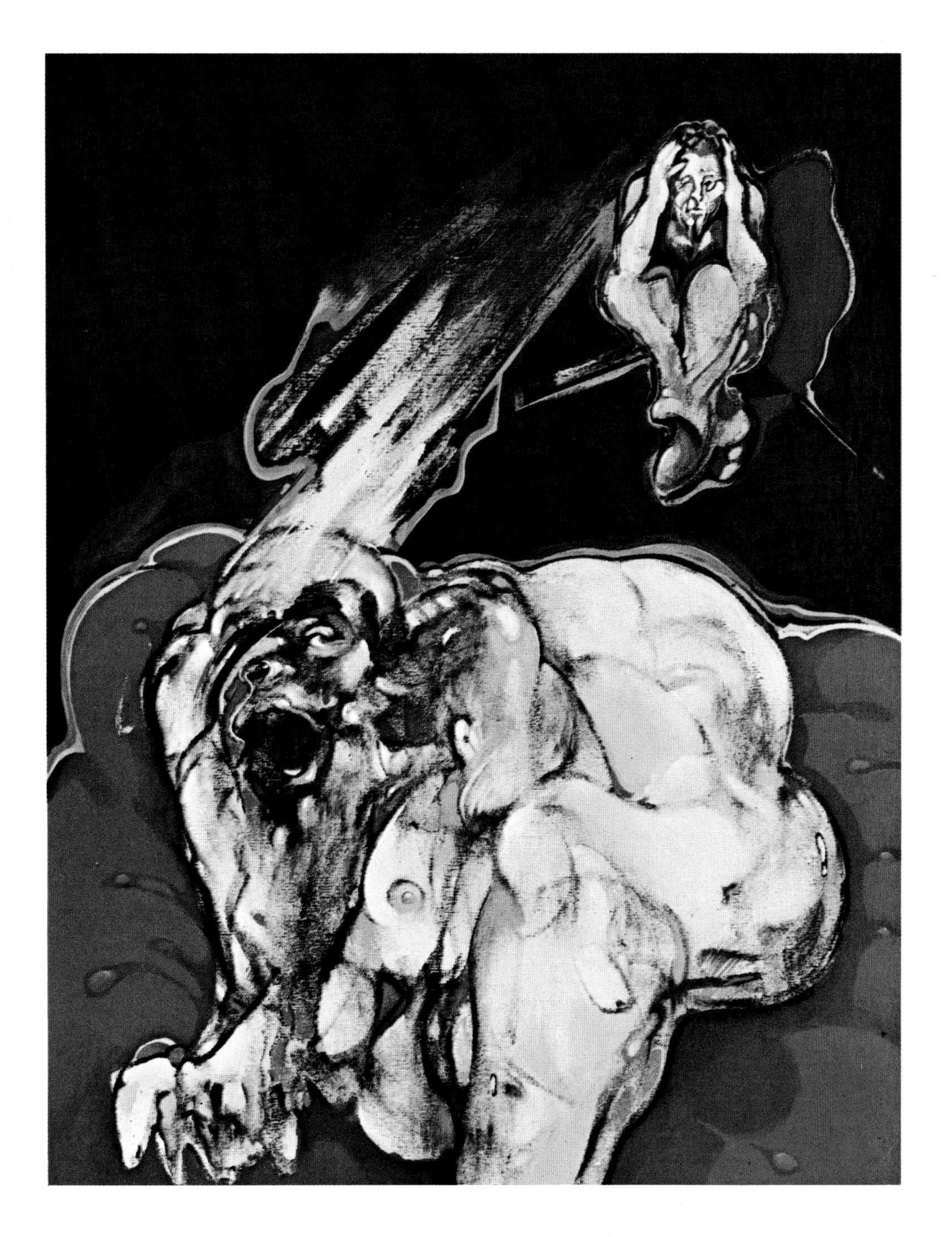

Some of the more serious forms of mental illness include symptoms of hallucinations, paranoid delusions, or withdrawal.

To some extent, the rapid changes in our society have been reflected in the symptomatology of psychiatric patients. The incidence of conversion hysteria, the process by which consciously unacceptable emotions become transformed into physical manifestions—writer's cramp or hysterical blindness, for example—has decreased with the increasing sophistication of the population, especially with respect to sexual matters. Changes in technology have affected the content of paranoid, or persecutory, delusions and hallucinations.

Society has also altered its view of what constitutes acceptable behavior. Forms of behavior, such as nudity, which at one time would have been considered highly eccentric if not pathological, are now tolerated, at least within certain subcultures. Similarly, there are subcultures where aggressive behavior and rebellion against conformity are tolerated or even, indeed, valued. In such cases, the dramatic acting out of excited psychotics may be indistinguishable from the "normal" behavior of the group, at least to the outsider. Certainly many persons with serious mental illness are able to live quiet and sometimes creative lives in communities that have a high tolerance for eccentricity.

Depressions

Ancient writings describe depressions, and these descriptions differ little from those in the current literature. Job's depression is vividly portrayed in the Bible.

Depressions, or the "blues," are disturbances of mood. Most persons experience them in the wake of frustration and disappointment, but such episodes are usually mild and transient. However, severe depressions account for 15 percent of state hospital admissions and 50 percent of private psychiatric hospitalizations.

Depression affects twice as many women as men. Women often experience hopelessness, sadness, and boredom, based on dissatisfaction with their role of mother and housekeeper. Frustrated ambitions and the cessation or diminution of sexual and procreative functions are prominent factors in producing depressions among males.

Depressive symptoms consist of varying degrees of sadness, dejection, and apprehension. Doubts, fears, and self-accusations become overwhelming. Depressed persons are often hypochondriacal and may have a mixture of symptoms. Sleeplessness mounts, and depression is most evident in the morning. It improves as a day of supposed uselessness comes to an end. Appetite is diminished, and there is often considerable loss of weight. Some depressed persons are agitated, pacing back and forth, asking the same questions over and over as if wanting assurance but not listening to the answers.

Unquestionably, depressions run in families. Depressed family members may serve as a source of identification and become a model for the developing child. Increasingly, it has been recognized that overt, typical depressive syndromes can and do occur in young children. In the elderly, removal from employment, the loss of old neighbors, or the death of relatives may give rise to depression, appearing as a reaction characterized by "giving up." On the other hand, depression may be the first sign of an undiscovered physical illness or the initial indication of an impending psychiatric break, as in schizophrenia.

Manic-Depressive Syndrome. Depressions are often associated with mania, a feeling of excitement and euphoria. This is sometimes called "circular insanity" or the manic-depressive syndrome. The cycles vary; a "high" may be followed immediately by a "low," or there may be a normal period in between. These attacks may occur seasonally and for no consciously known cause. On the other hand, a series of minor precipitating causes may mount over a period of time until depression suddenly manifests itself.

Depression and mania tend to run in families, suggesting a strong biogenetic factor related to the cyclical nature of the illness. Hormones secreted by the adrenal glands have been implicated, and to some extent this implication influences the choice of drug treatment. Manic-depressives also show abnormally low, diabeticlike blood-sugar levels. There appears to be a correspondence between variation in mood, blood-sugar level, and the person's response to psychotropic, or mind-influencing, drugs. These findings suggest a hormonal variability requiring shifts in the way the patient is treated over his lifetime.

Mania

Mania is characterized by agitation, unrestrained behavior, and volubility. Often manic persons are destructive toward others as well as toward inanimate objects. In mania the span of attention is brief and shifts with the slightest noise or movement. Such patients move about constantly. They are often jovial, but generally they become boring.

Attacks of mania are self-limiting. When the attack is aborted by medication, the patient appears to miss his manic behavior and high state of mind and to resent slowing down.

Schizophrenia

Schizophrenia is sometimes called the "cancer" of psychiatry. It has been known since antiquity and occurs all over the world.

Victims of schizophrenia have a normal lifespan, and they are often a burden to their family, friends, and society. Frequently, the illness appears to improve, a condition called remission, but many psychiatrists believe that the condition has no "cure."

The dramatic and disruptive manifestations of schizophrenia, once considered characteristic of the disease, have been observed less frequently in recent years. The reasons for this are speculative. To a considerable extent, it can probably be attributed to the widespread use of psychotropic drugs, which suppress the extreme and bizarre symptoms. Another explanation may lie in the growing practice of early discharge from mental hospitals and the maintenance of patients within the community. Unlike long-term inmates of institutions, nonhospitalized patients tend to retain hope and to maintain some orientation toward an active, rational life.

The cause of schizophrenia remains one of the most hotly debated topics in the field of psychiatry. Researchers have implicated a number of factors, ranging from a chemical imbalance in the brain to an unsatisfactory family life, but no one theory has been proved beyond reasonable doubt. Family studies, especially those made with identical twins, appear to point to a genetic causation. One widely held explanation is that many persons carry an inherited or genetic predisposition to the illness, but in most cases it requires the presence of some life stress to produce signs of schizophrenic behavior. Such precipitating factors include growth adjustments, injuries, the strains of puberty, loss of dependence on parents or others, school competition, breakup of the family by death or divorce, and drug abuse, to name only a few. Children with such a predisposition are often observed to be shy, withdrawn, prone to daydreaming, unhappy, or depressed. Occasionally children develop a full-blown schizophrenia, usually characterized by withdrawal, or autism, and failure to develop normal communication and motor skills.

Schizophrenia manifests itself in a wide variety of symptoms. These include detachment from reality, disturbances of perception, hallucinations (which are false perceptions having no basis in "reality," such as hearing "voices"), distorted thinking, anxiety, loneliness, and depression, sometimes leading to an attempt at suicide. Different schizophrenic patients may exhibit different sets of symptoms, and one patient may display various symptoms over the course of his illness.

Characteristic of all such persons, however, is disorder of thought. In mild cases, this may be slight or sporadic and difficult to detect in ordinary conversation, simply because one can fill in the communication gaps. Yet, if one listens carefully to such persons, the thought "slippage" can be discerned. This disorder increases under stress, but it can also be observed when stress is not present; this seems to indicate that stress intensifies the condition but does not cause it. Thought disorders may be apparent in one or several psychological acts, such as perceiving, speaking, recalling, or writing. Most schizophrenic patients exhibit thought disorder in the course of communicating with others. The presence of thought disorder may reflect a more general impairment or dysfunction in the capacity to organize and entertain ideas and experiences or to maintain the organization of ideas and percepts that had once been established.

Types of Schizophrenia. Numerous attempts have been made to classify various types of schizophrenia. Some are based on symptomatology, some on the course of the illness, and some on other factors. Only three subcategories will be mentioned here.

Acute schizophrenic psychosis is characterized by sharp, suddenly appearing psychotic episodes that occur in persons who have previously exhibited good judgment. These persons soon enter into a remission and subsequently show only slight evidence of disorganization. They are able to return to society and resume their former way of life.

Chronic schizophrenic psychosis is characterized by a long history of inner turmoil and social incompetence with clear psychotic symptoms. The illness develops gradually, and the person may never have an acute episode.

Paranoid schizophrenic psychosis is characterized by frank delusions, fixed, false opinions that cannot be altered by logic or explanation and about which the person has no insight. These delusions are persecutory or grandiose in content. The delusion is usually clearly focused, and the remainder of the person's behavior may not be bizarre. Such persons are very alert, cautious, and suspicious. If the delusion is such that the person may harm others—if, for example, he believes he is threatened by someone—confinement may be necessary.

Schizophrenics generally require professional help. In many cases, ambulatory or outpatient management may be sufficient. Others may be so disorganized that hospitalization is required, especially in the acute phase.

In most modern hospitals, the treatment includes drugs, such as phenothiazines, often in conjunction with psychotherapy, group therapy, and/or family therapy. Brain surgery and shock treatment are now rarely used. Recently, therapeutic claims have been made for therapy using massive doses of vitamins, but the value of megavitamin therapy has not been proved.

Before the patient is discharged, careful plans must be made regarding living conditions and realistic career goals. At best the psychosis may be terminated, though many psychiatrists believe that the schizophrenic tendency remains.

Borderline Syndrome

The borderline syndrome combines features of normality, neurosis, and psychosis. Its manifestations may include compulsive, angry emotions, depressions with suicidal tendencies, alcoholism, compulsions and phobias, and temporary confusion, but the principal characteristic of persons in this category is that they have difficulty in attaining and maintaining positive relationships with other people.

The borderline person is socially awkward. Psychological tests and long-term follow-up records of such patients reveal that they are not truly schizophrenic, although severe defects in behavior are present. This syndrome or illness results from some unknown developmental variant. Modification of such behavior is difficult, and the aim of therapy is to help the person adjust to his defects.

Suicide

Suicide is always a potential danger for any person suffering from a psychiatric disorder. Depressive patients constitute the highest risk, and the greatest danger occurs during the onset of a depression or when it is clearing. In the depths of a serious depression, patients simply do not have the energy to contemplate and commit suicide. Suicide prevention is one of the most potent reasons for the hospitalization of mentally ill persons. Even then, self-inflicted death is always a calculated risk.

Exact statistics on the number of deaths from suicide are difficult to obtain. In some cases all the facts may not be fully reported because self-destruction still carries a certain stigma. Sometimes it is almost impossible to determine whether a death was accidental or suicidal. For example, did the driver of a speeding automobile have an accident, or did he wreck the car on purpose? According to the best determinations, suicide is the eighth most common cause of death among the population as a whole and the second most common among young people. Suicide tends to recur within families, and one person will often make repeated suicidal attempts.

The psychiatrist attempts to differentiate between suicidal gestures, used to gain added attention and sympathy, and genuine attempts. The gesture may accidentally succeed, and the genuine attempt may fail. Therefore, the prediction of suicide is extremely difficult.

Suicidal persons usually are rational. Their depression is generally focused on real life problems that are beyond their capacity to cope. Inability to sleep, or insomnia, frequent complaints of physical symptoms, or hypochondriasis, and alcoholism suggest hidden suicidal tendencies, especially when they are associated with an outlook of hopelessness. If possible, these tendencies should be brought into the open through direct questioning by the physician, and appropriate treatment with adequate medication should be instituted.

Many large cities have organized groups that are available around the clock to talk to potential suicides by telephone, allowing them to speak freely and trying to persuade them to postpone or drop their suicidal plans. Some success has accompanied these efforts. An expressed determination or promise not to commit suicide within a specified time gives the therapist a reasonable opportunity to treat the primary cause of depression. Open questioning of suicidal persons is not harmful. However, persons who have suicidal tendencies but are not willing to discuss suicidal intent are usually determined to kill themselves. In such cases, prevention is impossible.

Peer Group Effects

A peer group is composed of persons within the same age classification—children, adolescents, young adults, adults, or aged persons. Each group has its own problems with their own causes, course, and treatment.

Many popularly held ideas about peer groups are little more than myth. For example, children do not "grow out of" emotional difficulties. When unusual behavior is observed by family and friends, the affected child should be professionally diagnosed and appropriate treatment outlined. Elderly people with emotional problems need not be written off as irrevocably senile with brain deterioration. They too can be helped by medical treatment, as well as by psychotherapy and family therapy.

Adolescent and young adult groups have received the most public attention. Some are a source of considerable upheaval. Although it has been sanctioned, institutionalized, and even encouraged in Western society, parent-adolescent conflict is not inevitable. The adolescent must disengage himself from parental domination, but this development does not necessarily require the total renunciation of parental values or extreme turbulence. Adolescents need to develop detachment and emotional relationships with others outside the family unit, but a "generation gap" can be avoided if the young person is treated with respect and understanding.

The adolescent or young adult who becomes delinquent or destructive may require hospitalization. Ideally, the hospital should offer a protective, sympathetic, and controlling environment where the patient can learn new psychological and social skills and develop a sense of responsibility. Various behavior therapies in which the patient earns privileges in return for modifying his behavior have proved especially effective with this age group. Such persons respond best to treatment if they have brought psychological assets and strengths gained from early life experiences to the therapeutic relationship.

In recent years considerable attention has been centered on adolescents and young adults who abandon conventional society to enter communes or other alternative social organizations. For some this represents a needed pause before they are ready to enter the larger society. Most "make it" on their own. Some who do not have brought basic difficulties with them from their earlier life and will eventually require professional psychiatric care.

Failure to cope consistently with stressful situations in life can result in mania, depression, or other forms of mental illness.

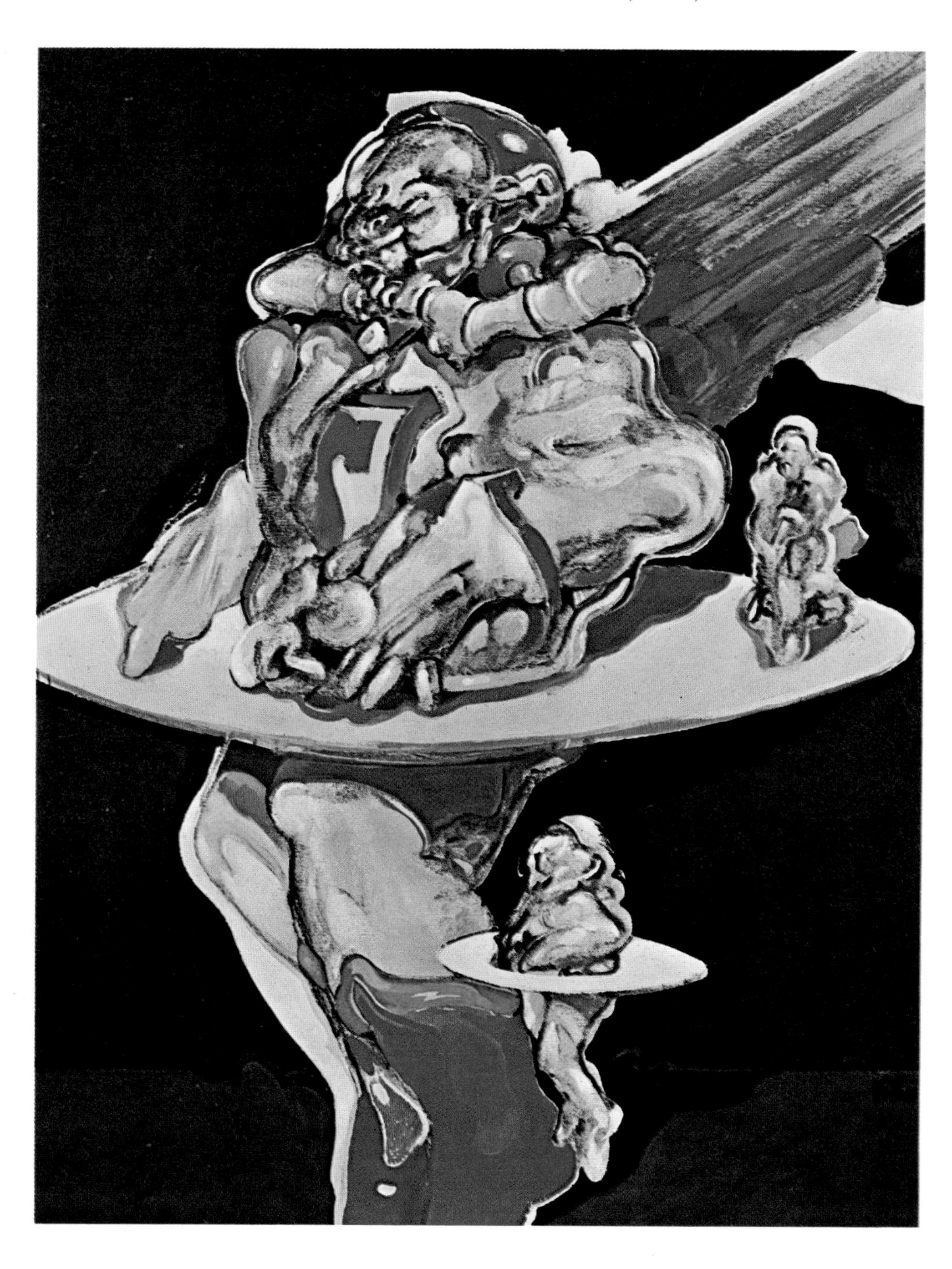

Psychiatric Treatment

Preventive psychiatric treatment has three phases: (1) primary, which includes genetic counseling, prenatal care, adequate postnatal nutrition, and teaching the mother methods of "good mothering"; (2) secondary, involving early recognition of deviance in childhood and providing the troubled child and family with early treatment that may prevent the development of serious disabilities; and (3) tertiary, which is the appropriate treatment of persons diagnosed as mentally ill in an effort to prevent further deterioration.

Unfortunately, the major problem of prevention appears to be that mental health educators have little or nothing specific and practical to offer the public. Many so-called principles of mental hygiene are vague slogans rather than strategies of behavior that can be put into practice.

We know that environmental and situational stresses play an important part in determining generalized mental distress, but the hard truth is that no known method exists for preventing the major psychoses. No known rules exist for influencing the emotional development of children. Although basic research on these matters is increasing in quantity and quality, it is unlikely that the situation will change significantly in the next decade. Progress will come through a gradual accumulation of knowledge rather than from a dramatic experimental breakthrough.

Active Treatment

Active psychiatric treatment deals with persons who show signs and symptoms of mental illness. It may take many forms, and there are many schools of thought as to what constitutes the best procedure. All workers in the field agree, however, in condemning the permanent warehousing of psychotic persons in "snake pits." Fortunately, this is largely a thing of the past.

Psychotherapy—treatment by psychological means—is a chief psychiatric technique. The patient is provided with the opportunity to express his problems and is given support, direction, and suggestion, usually on a one-to-one basis. This method, of which psychoanalysis is one form, is time-consuming and frequently expensive.

Psychotherapy is old. Hippocrates believed that baths, rest, a beautiful environment, and permissive treatment would relieve many persons. The witch doctors of Africa use suggestion, reinforced by impressive surroundings, strange instruments, nonunderstandable words, and high fees, and often achieve good results. Some therapists today occasionally use the ancient technique of hypnosis.

In group therapy patients interact with each other, usually under the guidance of a trained therapist. The participants tend to gain added support from one another, and group experiences enhance the individual's social abilities. Clinical improvement among patients in group therapy is sometimes superior to that achieved through individual psychotherapy, and hospitalization is often shortened.

Role of Psychotropic Drugs

The extensive use of psychotropic drugs in the treatment of mental illness is a relatively new development and has revolutionized the practice of psychiatry, particularly in mental hospitals. Specific drugs are now available to counteract almost all the major forms of mental illness. One large group of such drugs, called tranquilizers, are used to quiet patients exhibiting intense anxiety or aggressive behavior. Another group of agents, the energizing drugs, are used to counteract depressive states. All of these pharmacological agents have led to increased interest in long-term hospital patients, who previously had been considered untreatable. In many cases drug therapy has made it possible for them to return to society and sometimes to an active life.

Drugs do not "cure" mental illness, but they may alleviate the symptoms, assist the patient in getting through a crisis, and make him accessible to psychotherapeutic techniques that cannot be used on a person in a highly disturbed state. Despite their undoubted benefits, however, all of these drugs must be used with extreme care; maintenance doses must be carefully calculated, and the therapist must be alert for possible undesirable side effects. Some psychotic depressions that react poorly to medication may require electroconvulsive treatment, although use of an electric shock carries some dangers and is used much less frequently than it was a decade or two ago. Another once-popular form of treatment, psychosurgery, involving operations on parts of the brain believed to be responsible for undesirable behaviors, has fallen into disrepute among many psychiatrists.

Role of Mental Hospitals

One of the chief aims of modern psychiatry is early discharge from hospitalization or, preferably, avoidance of hospitalization entirely. Many persons with mild evidences of inner discomfort or asocial behavior can be treated adequately on an outpatient basis. This is the thrust of the movement for establishing community mental health centers and clinics.

In certain cases, however, hospitalization may be essential. The mentally ill person may harm himself or others, or his illness may be in a crisis state so that he cannot care for himself. Many forms of treatment can only be conducted in a hospital. Often brief hospitalization may be sufficient. For example, after crisis treatment of about three weeks for an acute schizophrenic breakdown, a person can often be managed successfully as an outpatient.

As a result of improved procedures and the new drugs, the census of patients in state mental hospitals has declined dramatically, and the average length of stay has been shortened considerably. Some states have begun to phase out their large mental hospitals in favor of short-term inpatient facilities and outpatient clinics. However, though the patient census has declined, admissions have risen; in many instances these are readmissions, and the result is a kind of "revolving door policy" in which the patient receives little more than immediate crisis care. It is possible that in some cases this policy has been carried too far. Another new thrust in state hospital systems has been to emphasize voluntary admission rather than involuntary committal and to pay greater attention to the civil rights of mental patients.

Antagonisms Toward Conventional Treatment

An area of some tension between society and the psychiatric profession relates to the treatment of patients on a low socioeconomic level who have an acute mental crisis or break. Many psychiatrists believe that such persons are not suitable for psychotherapy, but this is not true. However, many such persons do not consider psychiatrists as physicians, are unwilling to submit to psychotherapy, and demand medication at once so they can leave the hospital and return to their work or homes. Once they are discharged, it is often extremely difficult to persuade them to continue treatment on an outpatient basis.

A somewhat different view of psychiatric treatment is represented by the sociotherapists, who emphasize group or family therapy in and out of the hospital. They repudiate the concept of "mental illness" in favor of "disturbances resulting from problem living," and they denigrate diagnostic terms and specific therapies. The so-called encounter groups are a popular outgrowth of this movement. This type of therapy is shorter and less expensive than the conventional kind, but its effectiveness has not been proved. Unscrupulous and untrained persons have entered the field, and patients must be cautious about joining such enterprises unless they are under strict clinical supervision.

Lingering insecurities and minor personality defects are the common lot of man.

Psychiatric Research

In the latter 1950s psychiatrists began to realize that their specialty was an integral part of the vast field of behavioral sciences. No longer could psychiatry be separated from the larger areas of biological, psychological, social, and economic behavior. At the same time, clinical psychiatrists began to participate in social action, and research psychiatry absorbed theories involving the whole behavioral field. This development came belatedly. Psychoanalysis had long dominated the field of psychiatry and led to resistance against a more general theory of mental health and illness.

Obviously, no scientist can cover all the behavioral sciences, but the psychiatrist can better develop his own field if he understands its relation to others. The social sciences in general stress the importance of human symbolic functions, and these form the essence of humanity, individually, in groups, and in society as a whole. In this view, blunting of development, disturbances in the ability to adapt to society, and a failure to react conservatively and constructively to disturbing stimuli constitute disease.

Consideration of Individual Uniqueness

Individuals respond to life and its stresses in different ways. Some can cope with problems that completely overwhelm others. Science has few explanations to offer for this. Family training, environment, and genetic predisposition are all advanced as theories, but none is entirely satisfactory. The fact is that each person is unique, and his responses are peculiar to him and to him alone. Many people respond in generally the same way to many stresses and stimuli, but no two persons respond in exactly the same way. Each person's response is as distinctive as his fingerprints. Some people find it extremely difficult to adapt to even the most minor stresses of life. This does not mean they are inferior. This discussion of the integrative, or psychiatric, system has attempted to explain how and why some people behave as they do and how, if necessary, they can be helped.

For further information on the subjects covered in this chapter, consult ***Encyclopædia Britannica:***

Articles in the Macropædia	*Topics in the Outline of Knowledge*
HYPNOSIS	434.D.3
MENTAL HEALTH AND HYGIENE	438.D.5
PERSONALITY, MEASUREMENT OF	438.B and C
PERSONALITY, THEORIES OF	438.A
PSYCHIATRIC TREATMENT, CONCEPTS OF	438.D.4
PSYCHONEUROSES	438.D.2
PSYCHOSES	438.D.1

CREDITS

Illustrations: Charles S. Wellek – 21, 22, 23, 24, 25, 27, 36, 39, 40, 41, 42, 43, 45, 62, 63, 64 (top), 67, 74, 75, 80, 81, 84, 85, 86, 87, 90, 96, 122, 123, 124, 125, 126, 128, 129, 138, 139, 148, 154, 155, 156, 160, 170, 171, 172, 173, 177, 178, 179. Arv Tessing – 44, 64 (bottom), 66, 100, 136, 137, 140, 146, 158. James Curran – 65, 69, 77, 98, 99, 102, 103, 119, 150. Tak Murakami – 68, 82, 89, 91, 131, 163. George McVicker – 78, 101, 147, 164. Sandy Greenlaw – 118. John Murphy – 142. Ron Villani – 135, 185, 186, 191.

Photographs: Jerry Straus – 9, 10, 11, 14, 15, 16, 17. S. Bluefarb, M.D. – 26, 30, 31, 32. Charles Lieberman – 28, 29, 101, 108 (top), 117, 141 (top), 149, 176. Francis H. Straus II, M.D. – 68, 83, 89, 92, 161. John J. Fennessy, M.B. – 70, 79. Sharon Thomsen, M.D. – 72. Gerald Rogers, M.D. – 79. F. Manarchy – 91, 129, 130, 175 (left). Charles McAvey – 104. H. Kwaan, M.D. – 108 (bottom), 109, 110, 111, 112, 113, 114, 115. L.J.D. Zaneveld, K.G. Gould, W.J. Humphreys, W.L. Williams – 134, courtesy *J. of Reproductive Med.* Camera M.D., Inc. – 147, 159. Charles S. Wellek – 168. Ruthmary Deuel, M.D., U. of C. – 173, 175 (right). E. Duda, M.D. – 174. Chicago Blood Donor Service – 117. Midwest Population Control Center – 141 (bottom).